BIOCHEMISTRY OF ANTIBODIES

BIOCHEMISTRY OF ANTIBODIES

Roal'd S. Nezlin

Institute of Molecular Biology
Academy of Sciences of the USSR, Moscow

Translated from Russian by
Michel C. Vale

Translation edited by

Fred Karush

Department of Microbiology
School of Medicine
University of Pennsylvania
Philadelphia, Pennsylvania

PLENUM PRESS • NEW YORK–LONDON • 1970

Library of Congress Catalog Card Number 69-12534
SBN 306-30430-9

The original Russian text, first published by Nauka Press in Moscow in 1966, has been corrected by the author for this edition. The present translation is published under an agreement with Mezhdunarodnaya Kniga, the Soviet book export agency.

Роальд Соломонович Невлин

Биохимия антител

BIOKHIMIYA ANTITEL

A Division of Plenum Publishing Corporation
227 West 17th Street, New York, N.Y. 10011

United Kingdom edition published by Plenum Press, London
A Division of Plenum Publishing Corporation, Ltd.
Donington House, 30 Norfolk Street, London W.C. 2, England

Printed in the United States of America

Preface

When the history of immunology in the twentieth century is written, the decade of the 1960's will, in all probability, stand out as the period of greatest advance in the development of molecular immunology. It is appropriate and useful, therefore, that a scholarly and integrated presentation of this progress should be made available in English. The translation of Dr. Nezlin's "Biochemistry of Antibodies" from Russian admirably fulfills this need in the form of a scientific monograph directed to medical and biological scientists.

The appearance of this monograph also serves to emphasize the conceptual unification of diverse immunological phenomena which has emerged from progress in molecular immunology. This unity is a consequence of the key role played by the antibody molecule (either in solution or cell-bound) in every biological process properly described as immunological. Indeed, immunology as an independent natural science can be described as the study of the structure, interactions, and biosynthesis of the antibody molecule. Because of this central role of the antibody molecule the "Biochemistry of Antibodies," through its exposition of molecular immunology, provides the essential foundation for the fruitful exploration of immunobiological phenomena. There is little doubt that the knowledge presented in this monograph will be instrumental in the elaboration of the functional activities of the varieties of cell types involved in the immune response and in such potential medical applications of immunology as the prevention of the rejection of organ transplants, the immunotherapy of malignancy, and the control of fertility.

The translation of Dr. Nezlin's book has been an international enterprise in which my participation, however minor, has been a privilege and a source of personal satisfaction.

Fred Karush
Philadelphia, Pennsylvania
1970

Foreword

About 75 years have passed since it was discovered that after addition of bacterial toxins animal serum acquires the capacity to neutralize the effect of these toxins. It was found that serum derives its antitoxic properties from newly synthesized specific protein components – antibodies. Not only toxins, but also almost all other proteins, as well as certain other substances, including synthetic preparations, are able to stimulate synthesis of specific antibodies. According to the current definition, antibodies are proteins belonging to the class of immunoglobulins, which are able to conjugate with the antigens responsible for their formation.

Antibody synthesis is one of the most important protective mechanisms against infections. Antibodies also account for reactions of hypersensitivity, which play a leading role in the development of pathological processes. Hence, it is not surprising that for a long time antibodies, as one of the chief participants in immunological reactions, have been of interest not only to microbiologists and immunologists, but also to physicians, who found in these proteins an extremely powerful and specific therapeutic agent.

In recent years, scientists in various fields of molecular biology, such as chemists, studying the structure and properties of proteins, biochemists, studying protein biosynthesis, and finally, geneticists, have devoted an ever increasing amount of research to antibodies. This interest is accounted for first of all by the fact that antibodies serve as a convenient model for the study of many general biological problems. Antibodies are highly specific proteins which form complexes only with the antigens causing their formation. They owe this property to a very small area of the molecular surface – the binding site. Hence, an elucidation of the structure of the binding site would be a definite contribution to the solution of the more general problem, i.e., the relationship of the specific properties of the protein to its structure. Let us take yet another example. Antibody formation begins only after addition of an antigen. It is clear that a clarification of the question of the site of action of the antigen would represent a definite step toward an explanation of the regulator mechanisms of protein biosynthesis. Of considerable importance is the fact that quite simple methods for the isolation and accurate quantitative determination of pure antibodies have been developed. Finally, the elusive and puzzling aspects of the problem will always be a serious psychological enticement for investigators.

A large quantity of data concerning the properties, structure, and biosynthesis of antibodies has been accumulated from the assiduous efforts of many research laboratories. The information available requires systematization and discussion, and it is to this end that the present monograph is devoted. In the first two chapters two important methodological issues are discussed: quantitative determination and isolation of antibodies in pure form. Special attention is given to the most efficient methods – those employing immunoadsorbents. The properties of antibodies are treated in the third chapter. Recent striking achievements in research on the structure of antibodies are the topic of the fourth chapter. The fifth chapter presents a summary and discussion of a large amount of data collected in research on the biosynthesis of antibodies, the sixth chapter explores the genetic aspects of antibody formation, in particular, the rapidly advancing research on allotypes, and finally, the last chapter is devoted to a discussion of contemporary theories on antibody synthesis. The author expresses his appreciation to Doctor of Biological Sciences, N. V. Kholchev, Professor V. I. Tovarnitskiy, and Professor L. A. Tumerman for their valuable comments, and to L. M. Kul'pin for his assistance in planning the book.

Contents

CHAPTER I. Quantitative Methods for Determination of Antibodies .. 1

Heidelberger's Quantitative Precipitation Method ... 3
Quantitative Determination of Antibodies with the Aid of Immunoadsorbents 9
Equilibrium Dialysis for the Quantitative Measurement of the Hapten–Antibody Reaction 15
Fluorescence Methods for Antibody Determination... 18
Literature Cited 21

CHAPTER II. Isolation of Pure Antibodies.... 27

Nonspecific Methods........................ 27
Specific Methods 31
Isolation of the Antigen–Antibody Complex..... 31
Dissociation of the Antigen–Antibody Complex .. 32
Temperature Change 32
Change in Ionic Strength 32
Changes in pH of Medium 33
Other Methods for Dissociation of a Specific Complex.................. 34
Separation of Antibody from Antigen after Dissociation of a Specific Complex 36
Separation of Antibody from Soluble Antigen 36
Separation of Antibody from Insoluble Antigens......................... 40
Isolation of Antibodies with the Aid of Fixed Antigens (Synthetic Immunoadsorbents) . 42
Antigens Adsorbed on an Insoluble Support..... 42

Antigens Linked to an Insoluble Support via a Chemical Bond 43
Cellulose-Based Immunoadsorbents 45
Immunoadsorbents Based on Synthetic Polymers 51
Literature Cited 54

CHAPTER III. Properties of Antibodies 65

The Chief Types of Immunoglobulins 65
Other Serum Immunoglobulins 69
Microglobulins in Plasma and Urine 69
Size and Shape of Molecules 71
Molecular Weight 75
Electrophoretic Properties 77
Optical Properties 82
Ultraviolet Absorption 82
Optical Rotatory Dispersion 82
Fluorescence Polarization 84
Chemical Properties 85
Amino Acid Composition 85
Terminal Amino Acid Residues 88
Carbohydrate Component of Immunoglobulins . . . 91
Valence of Antibodies 91
Antigen Properties 93
Determinants Common to All Immunoglobulins . . 93
Determinants Characteristic of the Principal Types of Immunoglobulins 93
Determinants of Subgroups of the Principal Types of Immunoglobulins 94
Allotypical Determinants 94
Determinants Exposed by Recombination of Heavy and Light Chains 94
Internal Determinants 94
Individual Determinants of Immunoglobulins and Antibodies 95
Heterogeneity of Antibodies 96
Literature Cited 111

CHAPTER IV. The Structure of Antibodies . . 127

Fragments Obtained by Proteolytic Cleavage of Immunoglobulins 127
Papain-Cleaved Fragments 128
Fragments Obtained by Hydrolysis with Pepsin and Other Proteases. 134
Classification of Proteolytic Fragments. 139
Subunits and Peptide Chains Obtained by Chemical Treatment of Molecules of Immunoglobulins and Antibodies . . 139
Splitting of Disulfide Bonds 139
Separation of Peptide Chains 143
Physicochemical Properties of Peptide Chains 146
Molecular Weight. 146
Absorption at 280 mμ 148
Solubility 149
Chemical Composition of Peptide Chains of Immunoglobulins 149
Antigenic Properties of Peptide Chains 152
Peptide Chains of Pathological Immunoglobulins 155
Role of Peptide Chains of Antibodies in the Antigen Reaction. Molecular Antibody Hybrids 156
γG-Globulin Subunits Containing Both Types of Chains 166
Cyanogen Bromide Fragments 168
Peptide Chains and Subunits of γA- and γM-Globulins 169
Classification of Peptide Chains 170
Structural Model of γG-Globulin Molecules . . 171
Interchain Disulfide Bonds. 174
The Antibody Binding Site 175
Chemical Structure 175
Size of Antibody Binding Sites 179
Primary Structure of Antibodies and Immunoglobulins 184

Primary Structure of Certain Sections of Immunoglobulins. Amino Acid Sequence in Polypeptide Chains of Immunoglobulins 184
Peptide Maps of Antibodies 187
Dependence of the Secondary Structure of Antibodies on the Primary Structure .. 192
Literature Cited 194

CHAPTER V. Antibody Biosynthesis 209
The Fate of Antigen 209
In Which Organs and Cells Are Antibodies Synthesized? 216
Antibody-Synthesizing Cells.............. 220
RNA Synthesis in Lymphoid Cells during Antibody Synthesis 228
Attempts to Induce Antibody Synthesis by RNA Fractions Isolated from Lymphoid Tissue of Immune Rabbits 235
Dynamics of Antibody Synthesis 240
Latent Period 240
Antibody Production..................... 245
Kinetics of Antibody Formation 245
Sequence of Synthesis of 19 S and 7 S Antibodies 249
In Vitro Antibody Synthesis 255
Methods for Culturing Antibody-Synthesizing Tissues and Cells 255
Primary and Secondary Responses in vitro ... 260
Course of Antibody Synthesis in vitro 262
Quantity of Antibodies Synthesized in vitro 264
Microglobulins Synthesized in a Tissue Culture 266
Synthesis of the Carbohydrate Components of Immunoglobulins 271
Participation of Ribosomes in the Biosynthesis of Antibodies. Cell Free Systems Polyribosomes 272
Inhibitors of Antibody Synthesis 280
Literature Cited 286

CHAPTER VI. Genetic Aspects of Antibody Synthesis 305
Effect of Hereditary Factors on Synthesis of γ-Globulins and Antibodies 306
Hypo- and Agammaglobulinemia 306
Inheritable Variations in Antibody Synthesis . . . 308
Immunoglobulins in Newborn Animals 311
Genetically Controllable Antigenic Determinants of Immunoglobulins (Allotypes) 312
Rabbit Allotypes 313
Genotypic Factors of Human Immunoglobulins. . 315
Allotypes of Mice. 320
Genes Controlling Antibody Synthesis 321
Regulation of Gene Expression 322
The Unipotency of Lymphoid Cells 322
The Balance of Synthesis of Heavy and Light Peptide Chains 324
Suppression of Synthesis of Immunoglobulin Allotypes 325
Synthesis of DNA by Lymphoid Cells and Its Stimulation 326
Stimulation of DNA Synthesis 330
Literature Cited 334

CHAPTER VII. Theory of Antibody Synthesis 347
Template Theory of Antibody Synthesis 348
Theory of the Pre-existence of Information for Antibody Specificity 351
Hypothesis of the Somatic Origins of Variability 353
Theory of Repression–Derepression. 356
Evolutionary Aspects of Antibody Synthesis 362
Literature Cited 366

Index 371

Chapter I

Quantitative Methods for Determination of Antibodies

In the study of the properties, structure, and biosynthesis of antibodies, it is of considerable importance to have a method for quantitative determination of these serum proteins. As with methods for determination of any other substance, this method must also fulfill certain requirements. It must be specific, accurate, yield reproducible results, possess high sensitivity and be sufficiently easy to employ.

In serum, antibodies are mixed with other proteins. Hence, they can be determined by any visible reaction occurring only after conjugation of the antibodies with a homologous antigen (agglutination, lysis, precipitation, etc.), or after extraction of the antibodies from serum with the aid of an antigen [42a, 61a, 71]. The antibody–antigen reaction is specific and highly sensitive; hence, these particular requirements are fulfilled. However, the requirements of accuracy and reproducibility of results present more difficulty: in most cases, only a very rough estimation of the content of antibodies in a given sample can be obtained. In many serological tests, the highest serum or antigen dilution still capable of manifesting an effect is determined. Most often, two successive dilutions are used, and consequently the error can be as high as 50-100% [2].

Recently, a method of agglutination of erythrocytes with antigens adsorbed on them has come into wide use. This method, which was first proposed by Boyden and further developed by Stavitsky et al. [29, 30, 34a, 37, 69], is distinguished by its universality, its extremely high sensitivity, and its simplicity. In many cases, 0.001-0.002 mg antibody is sufficient to obtain a good result using tannic acid-conjugated erythrocytes. However, this method is not strictly quantitative and reproducibility is not always good. There is always the possibility of nonspecific agglutination, and hemolytic reagents cannot be used. For certain antigens, in particular, ovalbumin, the hemagglutination method is not adequately sensitive. It should also be borne in mind that the adsorbed antigen may be eluted and thereby distort the true values. This shortcoming can be eliminated if the antigen is coupled with erythrocytes via a stable bond [39]; for example, by using bis-diazotized benzidine [67, 69], 1,3-difluoro-4,6-dinitrobenzene [56], or toluene-2,4-diisocyanate [46].

Many investigators have used particles of different adsorbents to which the antigens are coupled for the agglutination reaction. Such adsorbents are carbon, kaolin, collodion, bentonite, latex, etc. (cf. Chapter II). However, these methods have many of the same shortcomings as the passive hemagglutination method [44].

Olovnikov's method [26] of agglutination of benzidine-protein complexes of antigens shows some promise as a method for antibody determination. It does not have the disadvantages of passive agglutination, in particular, its low reproducibility, yet is still as sensitive. To determine antigens, a suspension of insoluble particles obtained by polycondensation of immune serum proteins with diaminodiphenylamine can be used. This method is particularly suitable for the detection of diphtheria toxin [27].

Methods employing precipitation in a solution or a gel have been widely used for antibody detection [17, 38, 52]. Gel-diffusion methods have been used particularly often for detection of the most varied antigens and antibodies [1, 28, 33, 62, 63]. There exist many versions of this method (e.g., Ouchterlony, Oudin [1, 28, 33, 34b, 62, 63]) which is extremely sensitive, especially in its micromodifications [3, 15], and gives a quite complete qualitative characterization of the systems being studied. Methods combining gel-diffusion with electrophoresis, i.e., in immunoelectrophoresis [4], and radioautographic immunoelectrophoresis, which is especially sensitive [77], offer an even wider range of possibilities. The possibility of titration of antibody and antigen in an electric field has been demonstrated [55]. However, of all the precipitation methods, only Heidelberger's [49, 53] is able to give a truly quantitative estimation of the antibody content in serum.

In microbiological and immunological investigations, complement fixation is widely used [66]. In certain modifications of this sensitive test, the results can be expressed in weight units of protein [20, 21, 31]. However, difficulties are encountered in the use of this method in biochemical experiments because of the considerable complexity of the system, and the lack of standardization in its components.

With certain biological methods, very small quantities of antibody can be detected. These methods include: neutralization of bacterial and virus antigens, immune lysis of bacteria and erythrocytes [70], anaphylactic reactions [64], elimination of a labeled antigen from the blood of animals [65], and hemagglutination [19]. For example, for immune lysis only 10^{-6} mg antibody nitrogen is required, and 0.0012-0.0033 μg antibody nitrogen per ml serum can be determined from the elimination of antigen from the blood. If a hapten is coupled with a bacteriophage, neutralization will be observed on addition of an antihapten serum as well as an antiphage serum. On the basis of this principle, extremely small quantities of antihapten antibodies (on the order of 10^{-2}-10^{-5} μg/ml) can be determined [47a, 58a]. However, it is difficult to obtain strictly quantitative and consistent results with these tests.

Methods whereby the antibody content expressed in weight units of protein in a given volume of serum or other liquid can be determined directly are the most suited to the purposes of the biochemist. At the present time, there are two such methods available: a quantitative precipitation method, and a method employing immunoadsorbents. Both of these methods will be discussed later in more detail. In addition, certain advantages are offered by equilibrium dialysis, a method presently popular in immunochemistry, whereby the interaction of low molecular weight substances (haptens) with antibodies can be accurately described, and by recently proposed fluorescence methods for determination of antibodies in solutions. These latter may possibly be able to provide answers to several questions in the biochemistry of antibodies.

HEIDELBERGER'S QUANTITATIVE PRECIPITATION METHOD

The first accurate method for quantitative antibody determination was developed by Heidelberger [49, 53]. The method is based on the determination of the amount of protein in a precipitate formed after addition of a very slight excess of antigen to serum; the quantity of antigen in the precipitate is then subtracted. This analytical test has come into wide use, and its development marked a very important stage in the advance of immunochemistry. A detailed description of the method is given in "Experimental Immunochemistry," by Kabat and Mayer [52].

As increasing amounts of antigen are added to the immune serum, the quantity of the precipitate increases to a maximum, and then decreases again as a result of formation of soluble antigen–antibody complexes. All antibodies are precipitated by a very small excess of antigen. Hence, the exact quantity of antigen needed for this excess is determined. If the antibody concentration is high, a known amount of antigen can be added by small portions to a definite serum volume, the precipitate being filtered off after each addition. The sum of the portions (added until antibody is no longer precipitated) will be approximately the value sought. Gradually increasing quantities of antigen may be added to a series of test tubes containing equal volumes of serum. The amount of antigen and antibody remaining in the supernatant is then determined in those test tubes containing the largest quantity of precipitate. Of course, before the experiment is set up some rather complicated preliminary operations must be performed. However, these can be avoided without appreciable loss of accuracy if the antigen–antibody complex used has been well studied previously. Thus, serum or ovalbumins are often used as the antigen, and rabbits are used as the experimental animals. When a very slight excess of albumin is added to immune serum, the precipitate will contain approximately 10% antigen, i.e., about 2.5 molecules of antibody with a molecular weight of 160,000 for each molecule of ovalbumin with a molecular weight of 42,000, or somewhat more than 3 antibody molecules per molecule of serum albumin with a molecular weight of 67,000. Hence, if the ratios are correctly chosen, all the antigen will be precipitated in the test tube containing the maximum amount of precipitate; its content will be 10% of the protein in the precipitate.

The procedure for determination involves, briefly, the following [11, 52]: Antigen solutions with increasing concentrations (10-200 μg/ml) are added in 0.1-0.5 ml portions with a pipette to

TABLE 1. Amount of Precipitate Obtained from Addition of Ovalbumin (OA) to Immune Rabbit Serum [49]

Added OA, mg nitrogen	Precipitated OA, mg nitrogen	Total quantity of precipitate, mg nitrogen	Quantity of antibody in precipitate, calculated from the difference, mg nitrogen	Ratio of antibody to OA nitrogen in precipitate	Supernatant sample
0.009	All	0·156	0.147	16.3	Antibody excess
0.015	"	0·236	0.221	14.7	" "
0.025	"	0·374	0.349	14.0	" "
0.040	"	0·526	0.486	12.2	" "
0.050	"	0·632	0.582	11.6	" "
0.065	"	0·740	0.675	10.4	Antibody excess, traces of OA
0.074	"	0·794	0.720	9.7	No antibody, no OA
0.082	"	0·830	0.748	9.1	No antibody, < 0.001 mg OA nitrogen
0.000	0.087	0·826	0.739	8.5	OA excess
0.098	0.089	0·820	0.731	8.2	" "
0.124	0.087	0·730	0.643	7.4	" "
0.135	(0.072)	0·610	(0.538)	(7.5)	" "
0.195	(0.048)	0·414	(0.366)	(7.6)	" "
0.307	(0.004)	0·106			" "
0.490		0·042			" "

Note: The figures in parentheses may not be completely accurate.

several centrifuge tubes. Then a constant volume of serum is added to each tube. After mixing, the tubes are placed in a temperature-regulated incubator at 37° for 60 minutes, and then in a refrigerator at +4° for one day. If the antibody concentration in the serum is low, the samples should be kept in the cold for several days [52]. The samples are then centrifuged at 2000-3000 rpms in the cold, and the supernatant is drawn off with a capillary tube, or decanted if the precipitate is well-formed and in large quantity. The precipitate is washed twice with a cold solution of 0.85% NaCl and twice with water. After dissolving the precipitate in 1*N* NaOH, protein is determined by one of the usual methods, for example, the Lowry method. The amount of antigen added (which for serum albumin and egg albumin is 10% of the protein in the precipitate) is subtracted from the amount of protein in the tube containing the largest quantity of precipitate. The difference thus obtained is equal to the antibody protein content in the given serum volume.

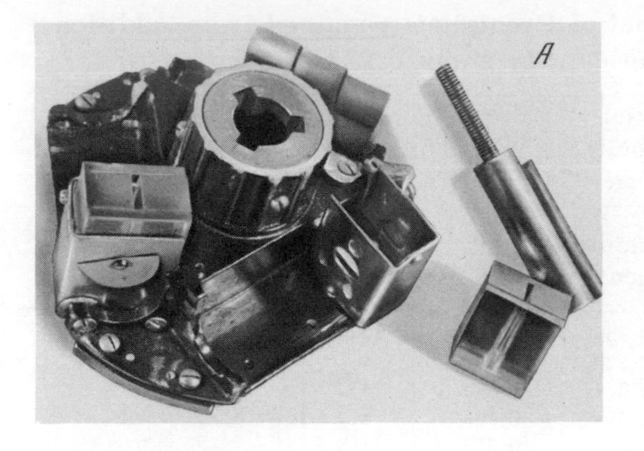

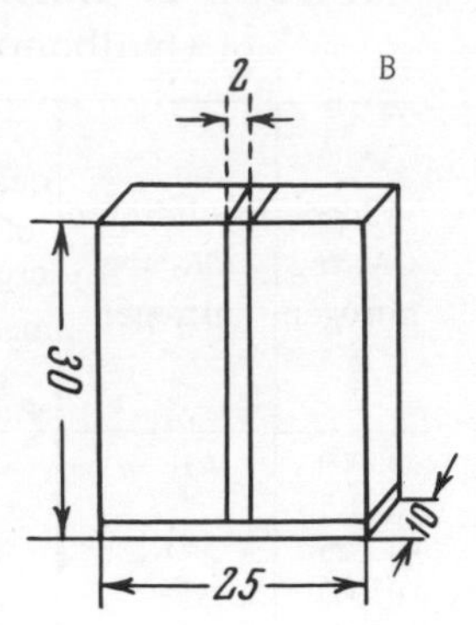

Fig. 1. Use of the photocolorimeter FEK-M for microdeterminations [23]. A) General appearance of the cuvette holder modified for microcuvettes; on the right, microcuvette and shaft securing the cuvette holder; B) sketch of microcuvette, dimensions in mm.

Table 1 presents results of one of Heidelberger's experiments on quantitative precipitation. Figure 7 (p. 51) presents curves obtained from this type of analysis, illustrating the dependence of the amount of precipitate on the amount of antigen added to immune serum or to an antibody solution.

The sensitivity of the method is dependent on the sensitivity of the method used to determine protein [58]. When Lowry's method is used, the lower limit will be about 10-20 μg protein. However, micromethods developed for protein assay give reliable results for contents which are several orders lower. These methods employ capillary or slit cuvettes [57]. The widely used Soviet photoelectric colorimeter FEK-M [23] can be used for this purpose. Figure 1 shows the cuvette holder of the instrument in a modification for slit-type cuvettes, and the cuvettes themselves. Figure 2 shows calibration curves for macro- and microdeterminations of protein. It is evident from the figure that the curves closely coincide. Reproducibility is good. Only one-eightieth of the lowest amount of protein detectable by the macromethod (on the order of 1-2 and even 10^{-1} μg in a sample) is required in the micromethod.

A method for determination of millimicrogram quantities of protein in precipitates with the aid of Bromsulfalein (Sulfobromophthalein sodium) has been developed [43]. If 10 μl each of serum and antibody solution are taken, the volume of the final solution for photometry will be 1 ml. For samples of 1 and 0.1 μl, microcuvettes are used.

The chief advantage of the Heidelberger method is that it gives an accurate estimation of the amount of antibody expressed in protein weight units in a given volume of serum. However, there are several points to be taken into account in using this method. First, it must be kept in mind that the precipitate is soluble (primarily owing to the solubility of the antibody). Sometimes the amount of precipitate lost from washing can be very high (30-300 μg/ml) [36]. Consequently, when the quantity of precipitate is small, a large error is possible. In addition, it should be observed that the precipitate contains a small quantity of other serum proteins, in particular, complement (usually not more than 2-4%), as well as antigen and antibody [7, 61]. If a large quantity of precipitate is obtained, a small error is admissible; however, the error increases the smaller the quantity of precipitate. Dilution of the serum with a physiological solution also increases the error since the complement concentration also decreases [61]. There is also the danger of distorted results from contamination of the precipitate with radioactive impurities when labeled antibodies are determined by coprecipitation, whereby, for example, tissue extracts, a known quantity of antigen, and an unlabeled precipitin are added to the sample liquid and the activity of the precipitate is counted. In such cases, even a small amount of impurity is sufficient to give the impression that new antibodies have been formed.

Another complication arises from the fact that knowledge of the precise antigen—antibody ratio in the precipitates is necessary since this ratio changes with increase in the molecular weight of the antigen. It is especially important to keep this in mind if antibodies are studied during immunization, since antibodies with a molecular weight of ~1 million are usually formed during the first days after immunization, after which the molecular weight of the antibodies formed diminishes to 160,000 (cf. Chapter V). Moreover, the capacity to react with antigen can vary. All these factors together are responsible for differences in the antigen—antibody ratio in the precipitate, and for differences in the shape of the precipitation curve.

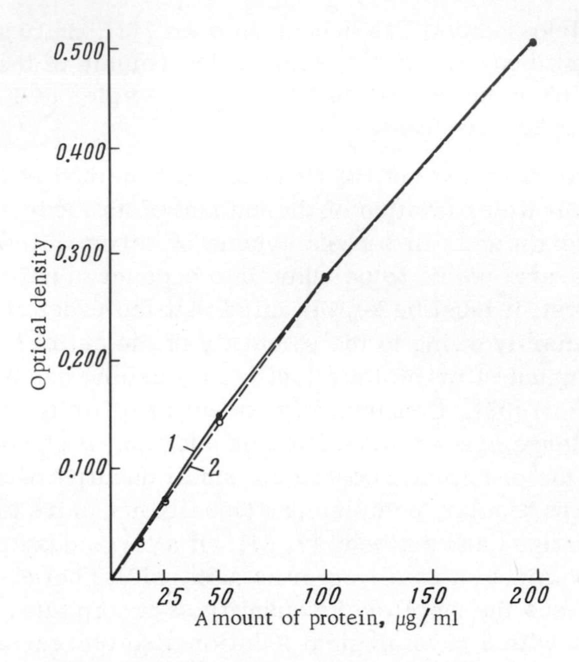

Fig. 2. Dependence of optical density on the amount of protein analyzed, in macro- and micromodifications of the Lowry method. 1) Macromethod (total volume of mixture 4 ml); 2) micromethod (total volume 79 μl).

Finally, difficulties are encountered in attempts to determine the quantity of so-called nonprecipitating antibodies, i.e., those which are not precipitated by addition of antigen. Coprecipitation may also be employed in this case; i.e., an antigen and a precipitin, which yield a known quantity of precipitate, are added to this type of antibody and the quantity of nonprecipitating antibody is estimated from the increase in the amount of precipitate. However, since a strictly quantitative determination by this method is difficult, the use of immunoadsorbents (see below) is to be preferred in such cases.

Antibodies may be quantitatively determined by precipitation on paper as well as in solution [5, 32], but this method has the same shortcomings as the Heidelberger method.

QUANTITATIVE DETERMINATION OF ANTIBODIES WITH THE AID OF IMMUNOADSORBENTS

One of the most convenient methods in current use for quantitative detection of antibody employs immunoadsorbents. The essential features of this method are as follows: first, the antibody is adsorbed from a given volume of serum or other sample liquid by means of an antigen firmly coupled to an insoluble support (immunoadsorbent); second, after washing the immunoadsorbent, the antibodies are eluted, and their quantity determined from the protein content in the eluate.

The most advanced immunoadsorbent method employs antigens linked to cellulose by a N-(m-nitrobenzyloxymethyl)-pyridine residue. This method was proposed and studied in detail in recent years by Gurvich [7] in the USSR. To fix the antigens, paper discs [6], cellulose powder [8], and very small cellulose particles in suspension (immunoadsorbent suspension) [12] have been used. The cellulose suspension is most suitable, since on this type of support the immunoadsorbent capacity is greatest (see Chapter II for a more detailed treatment).

The procedure for antibody determination with the aid of such immunoadsorbents is as follows: An accurately measured volume of precentrifuged sample serum (0.1-0.3 ml) is added to a twice-washed (with 0.85% NaCl) precipitate of the immunoadsorbent suspension (10 mg dry weight). After stirring 10-15 minutes practically all antibody is bound to the cellulose-linked antigen [22]. Then the unbound proteins are washed away with either weak buffer solutions (pH 7.0) or with 0.85% NaCl alkalized to pH 7.0. For washing, the adsorbent may be transferred to special vessels (small paper-bottomed drums) which are than placed in a chamber for descending chromatography on a sheet of paper one end of which is immersed in the washing solution [12]. The adsorbent may be washed in the same test tubes in which it was saturated with antibodies. After suspending the adsorbent in the washing solution (5 ml) the suspension is centrifuged for 3-5 minutes at 3000-4000 rpm. Eight such washings are sufficient. The bound antibodies are eluted at acid or alkali pH values. At alkali pH, 1 ml of 1*N* NaOH is added to the adsorbent, whereby all antibodies

TABLE 2. Quantitative Determination of Antibody from the Increase in Protein in an Immunoadsorbent Suspension

Amount of Immunoadsorbent	Amount added, ml			Protein split off, μg	Nonspecific adsorption, μg	Antibody content, μg
	0.85% NaCl	normal rabbit serum	immune rabbit serum			
Control: nonspecific serum (elution with 1*N* NaOH [7])						
Fixed albumin, 6.6 mg	1	–	–	104	–	–
Ditto	–	1	–	128	24	–
"	–	–	1	808	24	680
Control: nonspecific adsorbent (elution with 1*N* HCl)						
Fixed casein, 3 mg	–	–	1	36	36	–
Fixed albumin, 3 mg	–	–	1	1950	36	1914

and some antigen are split off. The amount of the latter is estimated in a control experiment. A 1*N* HCl solution elutes almost all antibodies without decomposing the molecules of the fixed antigen. Hence, the background in the determination is very low. Our control experiments, carried out with radioactive antibodies, showed that the effectiveness of elution with 0.1*N* HCl was approximately 90%. After sedimentation of the adsorbent by centrifugation, the protein content in the eluate is determined by one of the common methods, for example, Lowry's method.

Normal serum can be added to the adsorbent as a control for determination of nonspecific adsorption. Nonspecific adsorption can also be controlled on another fixed antigen. This procedure is convenient if the sample is very small. The control adsorbent is first added to the sample, and after it is precipitated, the adsorbent used in the experiment is added to the supernatant.

Table 2 gives results obtained from a determination of antibody in serum with the aid of immunoadsorbents with different controls.

The method discussed has many advantages. Above all, it is

strictly quantitative: the final results are expressed in weight units of protein. Secondly, since the antigen itself is insoluble, the antigen—antibody complex does not dissolve in the excess antigen. Reproducibility is very good; the standard deviation does not exceed 4% [12]. This to a great extent can be attributed to the standardization of the immunoadsorbent preparations. The same adsorbent can be used in prolonged experiments since its suspension keeps for a long time in the cold, especially in the presence of antiseptics, for example merthiolate (1:10,000). In addition, the method is simple and convenient to use, and does not require any special equipment other than that usually found in a biochemical laboratory. And, what is very important, the determination is rapid (several samples can be analyzed in 2-3 hr).

Sensitivity is dependent primarily on the method used to determine protein in the eluate. If the usual Lowry method is used, the lower limit is 10 μg antibody protein. The sensitivity can be increased to 1 μg protein or less by using a microcuvette (see above). Minute quantities of antibody can be determined with adsorbents if the antibodies are radioactively labeled. This is of especial importance for the study of antibody biosynthesis, where up to 0.005 μg antibody can be detected [13] by measuring the radioactivity of the eluate if end-window counters are used. The sensitivity can be increased by one order by using scintillation counters. For example, in one of our experiments we used eluates of ^{14}C labeled antibodies which had been synthesized in a tissue culture. In one case, the eluate was applied to a metal planchet and counted with an end-window counter. In another instance, the same quantity of eluate was applied to a disc of Whatman 3MM paper and after careful drying in a stream of hot air the activity was counted on the French-made liquid scintillation counter "Carbotrimeter." The counting efficiency was 17-fold higher on the scintillation counter.

Nonspecific adsorption of protein on fine cellulose particles is comparatively slight: one gram of cellulose adsorbent retains no more than 1 mg protein [12]. Consequently, the nonspecific background is relatively low, especially when acid eluants, which do not detach the antigen from cellulose, are used; sensitivity is thereby also increased.

Perhaps one of the most important advantages of this method

TABLE 3. Determination of the Quantity of Protein Adsorbed by a Fixed Antigen of *Rickettsia prowazekii* from the Serum of a Patient with Exanthematous Fever (Agglutination Titer 1:1600) [24]

Antigen fixed on paper	Amount of antiserum applied, ml	Total protein content on disc, μg	Nonspecific adsorption, μg	Amount of adsorbed antibody, μg
Suspension of	–	30	–	–
Rickettsia prowazeki	43.3	69	–	14
	500	94	–	29
Suspension of	–	21	–	–
Q-fever	43.3	41	20	–
	500	56	35	–

is that not only precipitating, but also nonprecipitating antibodies can be determined. An example of determination of antibody (in μg) in serum before and after heating for 20 minutes at 75° [8] is given below:

Determined by the Heidelberger method	Determined from the protein increase on an immunoadsorbent
Before heating	
1.63	1.55
2.3	2.40
After heating	
0.12	1.53
0.10	1.20

As is seen, the adsorbent was capable of extracting antibodies which had not been precipitated by the Heidelberger method.

Immunoadsorbents have been successfully used for determination of antibodies to different antigens. Data were given above on the determination of antibodies to various serum proteins. Good results were obtained in a determination of the antitoxin to *Clostridium oedematiens* and antidiphtheria antibodies [9, 10].

The quantitative determination of antibodies to viruses and

Rickettsia with the aid of immunoadsorbents has also been shown to be possible [24]. This is of special interest, because the methods used in virology for determination of antibodies are not strictly quantitative [72]. In these experiments, the viruses were fixed on paper discs via a N-(m-nitrobenzyloxymethyl)-pyridine chloride residue. It was found that the fixed viruses of influenza A, poliomyelitis, ornithosis, and *Rickettsia prowazekii* were capable of extracting antibody from the corresponding serums. An experiment with *Rickettsia* antigens is given as an example (Table 3).

The development of a quantitative method for determination of antibodies specific to viruses involves certain difficulties since the percentage of virus protein in the antigens usually used is relatively small; in consequence, the adsorbent capacity is low. As a result, the fixed antigen is rapidly saturated, and it is impossible to ascertain whether all the antibodies are extracted from the given volume of serum. However, by using purified antigens, it is possible under certain conditions to maintain a proportionality between the amount of serum applied and the increase in protein on the adsorbent [24]. Figure 3 gives results of an experiment in which a concentrated suspension of purified influenza A virus was used. As is seen from the figure, a direct proportionality between the amount of serum applied and the amount of antibodies detected is initially maintained. From this it may be assumed that within the given range, all antibodies are extracted from the serum by the adsorbent. In consideration of the foregoing, it can be calculated that, for example, 50 μl serum contains 25 μg antibody, and consequently 1 ml contains 500 μg (the serum titer, determined by hemagglutination inhibition, was 1:1028).

Besides the immunoadsorbent described above, other types of fixed antigens have also been used in quantitative antibody determination. Gostev and Shagunova [3a, 34] used antigens fixed by drying on paper, and antigens coupled via a diazo bond to polystyrene were used by Pressman [76] for determination of antibodies labeled with radioactive iodine. This method, of course, was extremely sensitive. These investigators also point out the considerable nonspecific adsorption on the adsorbent.

Recently, Hirata and Campbell [50] proposed a method for determination of antibodies to bovine serum albumin by adsorbing them on an antigen denatured by heating at pH 3.7. The antibody content was found from the difference between the amount of antigen

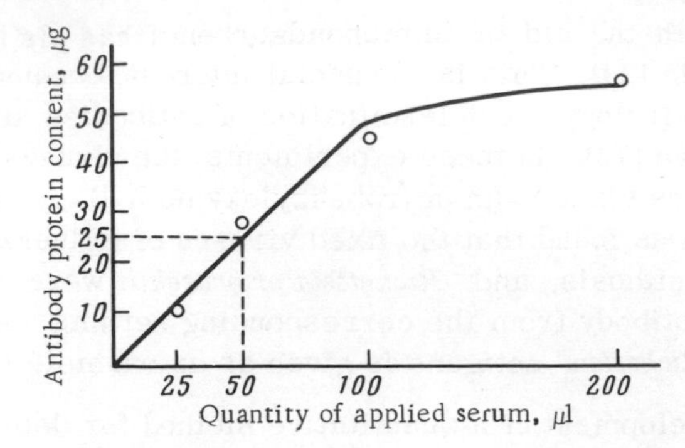

Fig. 3. Dependence of the yield of adsorbed antibodies detected on discs with a fixed amount of influenza virus on the quantity of anti-influenza serum applied to the disc [24].

in the sample and the amount of protein in the precipitate. The adsorbent used had a high capacity (0.8 mg antibody protein per mg adsorbent). A disadvantage of this method is that the adsorbent solubilizes with time.

Immunoadsorbents can also be used for quantitative determination of antigens. Direct coupling of antibodies on insoluble supports does not usually give good results, since the adsorbents thus produced have a very low capacity. It is more convenient to exploit the capacity of an insoluble complex consisting of a fixed antigen conjugated with its antibody to retain new portions of antigen from the solution [14, 16, 75]. Such an "adsorbent-antibody" can be represented schematically as: (cellulose-antigen):antibody.

One gram adsorbent-antibody is able to retain up to 40-50 mg antigen. Drizlikh and Gurvich [16] developed a method for antigen determination on this basis. The fixed antigen is first treated with immune serum and then washed free of unbound proteins. Then the antigen solution is added to the adsorbent-antibody thus obtained. After a second washing, the protein is eluted. The quantity of added antigen will be equal to the difference between the protein contents in the sample and control eluates (adsorbent-antibody + control antigen or another adsorbent-antibody + the test antigen). The results obtained are given in Table 4.

The determination is much simpler if the antigen is radioactively labeled. This eliminates the background and sharply in-

TABLE 4. Determination of the Amount of Antigen Bound to an Adsorbent-Antibody [16]

Adsorbent-antibody to horse γ-globulin, mg	Added to adsorbent		Protein eluted from adsorbent, mg	Increase in antigen protein, mg	Specific increase per mg adsorbent, mg	Nonspecific increase per mg of adsorbent, mg
	0.85% NaCl solution, ml	protein, mg				
2	1.0	–	432	–	–	–
2	–	0.25 (ovalbumin)	442	10	–	5
2	–	0.25 (horse γ globulin)	506	74	32	–

creases sensitivity to 0.005-0.0001 μg protein depending on the counter used.

Olovnikov [25] proposes the following indirect method for quantitative determination of antigens, including nonprecipitating antigens. First, the sample antigen solution is added to a known quantity of antibodies. Then the amount of antibody still capable of binding to the fixed antigen is determined, and the amount of antigen originally in the solution is calculated from a calibration curve. By this method, up to $2\text{-}4 \cdot 10^{-7}$ moles can be determined.

Thus, if several immunoadsorbents are available in the laboratory, both antibodies and antigens can be quantitatively determined.

EQUILIBRIUM DIALYSIS FOR THE QUANTITATIVE MEASUREMENT OF THE HAPTEN – ANTIBODY REACTION

Low molecular weight haptens with known structure used as antigen determinants are frequently of considerable advantage in immunochemical work. The primary advantage of haptens lies in the fact that their reaction with a corresponding antibody can be accurately and quantitatively determined. In contrast to many other antigen–antibody reactions (for example, precipitation, ag-

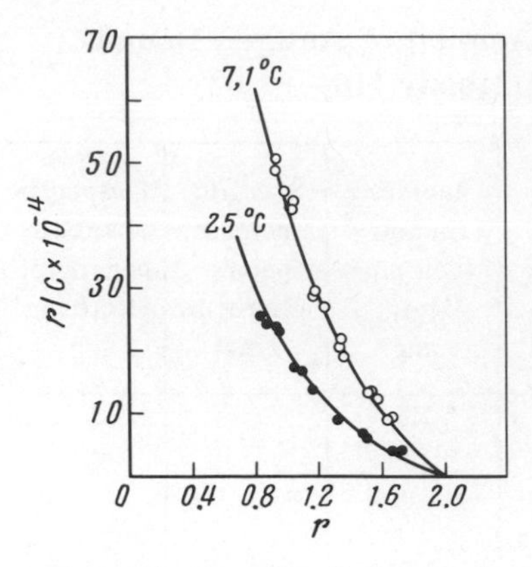

Fig. 4. Binding of D−Ip hapten dye by its specific purified antibody in equilibrium dialysis [54]. The points are experimental and the curves are theoretical. r is moles of bound hapten per mole antibody, and c is the concentration of free hapten.

glutination), the addition of a univalent hapten to an antibody solution generally produces no visible changes. Hence, if the hapten–antibody reaction can be quantitatively measured, then it is possible to give a direct, accurate characterization of the primary antigen–antibody interaction.

The most widely used method for studying the hapten–antibody reaction is equilibrium dialysis. A known quantity of antibody solution is placed in a cellophane bag, and the bag is then immersed in a hapten solution of known concentration. If the hapten concentration in the solution outside the bag after dialysis is known, the number of hapten molecules coupled with each antibody molecule can be calculated.

The procedure for the determination is, briefly, the following [51, 52]: Cellophane tubes manufactured by the Visking Co. (USA) are usually used for the dialyzing membrane. These tubes have the advantage of a high homogeneity. Before the experiment, boiling of the tubes in water is recommended. A measured amount of antibody solution is placed in the prepared bag. To reduce nonspecific adsorption of hapten, pure antibodies, or at least a γ-globulin fraction, should be used, since albumins have an especially high adsorption capacity. Dialysis is carried out in an incubator at a temperature which is kept constant throughout the experiment (for example, 25 ± 0.01°), and is terminated when the hapten concentration within and outside the surface of the membrane are the same. The difference between the initial hapten concentration and the concentration after dialysis will be the amount of hapten bound by antibody (taking into account adsorption on the membrane, which is determined in a special control experiment).

The foregoing can be elucidated by an example taken from the investigations of Eisen and Karush [42], who studied the interaction of purified rabbit antibody specific to p-arsanilic acid with a dye, namely, the homologous hapten dye p-(hydroxyphenylazo)phenylarsonic acid. One ml of a 0.15% antibody solution in 0.01*M* phosphate buffer (pH 7.4) and 0.16*M* NaCl was placed inside a cellophane dialysis bag and dialyzed against the same volume of hapten in the same buffer solution at 29°. In a control experiment, it was found that adsorption on the dialysis bag was 11% of the free dye concentration. Calculations were carried out as follows: in one of the samples, the initial hapten concentration in the outside solution, i.e., outside of the bag, was $3.96 \cdot 10^{-5}M$ whereas after dialysis the hapten concentration was $1.18 \cdot 10^{-5}M$. On the average, the concentration in the solution after dialysis was $2.36 \cdot 10^{-5}M$. Consequently, the total amount of hapten bound was $1.60 \cdot 10^{-5}M$, including $1.35 \cdot 10^{-5}M$ with the antibody. Thus 1.45 mole hapten is bound to 1 mole antibody. Usually, a number of samples are prepared with different hapten concentrations and then the value r/c is plotted against r, where r is the moles of bound hapten per mole antigen, and c is the concentration of free hapten. Figure 4 shows one such curve. The equilibrium constant of the hapten–antibody reaction can be calculated from the data obtained [54]. As will be discussed below, several important properties of antibodies have been characterized with the aid of equilibrium dialysis. For example, the exact antibody valence was ascertained, and antibody heterogeneity was described in detail (see Chapter III). In addition, thermodynamic calculations for the antigen–antibody reaction were carried out on the basis of this method [2].

If the hapten is polarographically active, its free concentration can be polarographically determined, since the antibody-bound hapten is polarographically inactive [67a, 78]. In contrast to equilibrium dialysis, by this method the antibody–hapten interaction can be characterized in a few minutes, and not only whole antibody molecules, but also their subunits can be studied since the latter may pass through the dialysis membrane by virtue of their small size. Zikan [78] showed that results of quantitative polarographic determination of antibody were in good agreement with results obtained by quantitative precipitation.

reaction of 0.02 μg ribonuclease (in 0.1 ml) with its specific antibody was detected [47]. The theoretical foundations for the use of fluorescence polarization in immunochemistry have been laid out by Dandliker [41].

Thus, at the present time, there are accurate immunochemical methods available for quantitative antibody determination. Those methods in which immunoadsorbents, i.e., antigens bound to insoluble supports, are used are especially preferred. With immunoadsorbents, determinations can be carried out quickly under strictly standard conditions without complicated equipment. In addition, they make possible the detection and quantitative determination of univalent nonprecipitating antibody fragments, which is very important in the study of antibody structure (cf. Chapter IV).

It is commonly believed that immunochemical methods in which the last step involves protein determination are not as sensitive as immunological methods of antibody determination. If, however, a radioactivity count is the final step (i.e., if radioactive antibodies are used), the sensitivity of immunoadsorbent methods is increased considerably. Data are given below on instances in which, with the aid of immunoadsorbents and highly efficient counters, it was possible accurately to determine extremely small amounts of antibodies, comparable to the amounts detected by immunological methods. The possibility of strictly quantitative and hypersensitive determinations of radioactive antibodies is very important in the study of antibody biosynthesis (cf. Chapter V). The sensitivities of various methods of antibody determination are given below [36, 45, 59, with additions].

Method	Antibody nitrogen, μg
Specific precipitation	
Qualitative	
Ring test	3-5
In gel	5-10
Radioimmunoelectrophoresis	0.0001
Quantitative	
Kjelldahl micromethod	20
Nephelometric	20
Biuret method	20

Colorimetric:	
Lowry method	4
Lowry micromethod	0.08
Method with bromsulfalein	0.02-0.2

Augmentation of the amount of antibody protein in the immunosorbent

Colorimetric	
Lowry method	2
Lowry micromethod	0.04
Radioactive antibodies	
End-window counter	0.001
Scintillation counter	0.0001
Agglutination of bacteria	
Qualitative	0.05-1.0
Quantitative	10-20
Hemagglutination	
Active (blood group)	0.1-0.2
Passive (adsorbed antigens)	0.003-0.006
Hemolysis	
Active (lytic antibodies and complements)	0.0002-0.03
Passive (adsorbed antigen)	0.3
Complement fixation	0.1
Flocculation test (for syphilis)	0.2-0.5
Anaphylaxis	
Passive (guinea pigs)	30
Passive (uterine muscle)	0.01
Local passive (skin of guinea pigs)	0.003
Local passive (skin of rats)	3
Prausnitz–Küstner reaction	0.01
Arthus phenomenon	10
Neutralization of toxin (rabbit skin)	0.3

LITERATURE CITED

1. G. I. Abelev, "Modification of a method of precipitation in agar for comparison of two antigen–antibody systems," Byul. Eksperim. Biol. i Med. 49(3):118 (1960).
2. W. C. Boyd, Introduction to Immunochemical Specificity, Wiley, New York (1962).
3. O. E. Vyazov, B. V. Konyukhov, and M. L. Lishtvan, "Capillary micromethod for the precipitation reaction in agar," Byul. Eksperim. Biol. i Med. 47(5):117 (1959).
3a. V. S. Gostev and M. A. Shagunova, "Quantitative reaction of specific binding of nitrogen by protein antigens adsorbed on dermatol and paper," Byul. Eksperim. Biol. i Med. 44(10):121 (1957).

4. P. Grabar and P. Burtin, Immunoelectrophoretic Analysis, American Elsevier, New York (1964).
5. A.E. Gurvich, "Determination of protein antigens by paper chromatography," Biokhimiya 20(5):550 (1955).
6. A.E. Gurvich, "Quantitative determination of the antibody content with the aid of protein antigens adsorbed on paper," Biokhimiya 22(6):1028 (1957).
7. A.E. Gurvich, "Use of antigens on an insoluble support to determine absolute quantities of antibodies," in: Current Problems of Immunology, Meditsina, Moscow (1964), p. 71.
8. A.E. Gurvich, "Isolation of pure antibodies and determination of absolute quantities of antibodies in serums with the aid of antigens linked to cellulose," in: Contemporary Methods in Biochemistry, Meditsina, Moscow (1964), p. 73.
9. A.E. Gurvich and M.V. Ispolatovskaya, "Study of antibody content in immune horse serum to *Clostridium oedematiens* with the aid of precipitation on paper or antigens adsorbed on paper," Byul. Eksperim. Biol. i Med. 45(5):79 (1958).
10. A.E. Gurvich, M.V. Ispolatovskaya, and K.N. Myasoedova, "Determination and isolation of antidiphtheria antibodies with the aid of antigens linked to cellulose," Vopr. Med. Khim. 7(1):55 (1961).
11. A.E. Gurvich and R.B. Kopner, "Quantitative determination of antibody content by precipitation on paper," Lab. Delo 4(2):23 (1958).
12. A.E. Gurvich, O.B. Kuzovleva, and A.E. Tumanova, "Use of immunoadsorbent suspensions for determination of absolute antibody contents," Biokhimiya 27(2):246 (1962).
13. A.E. Gurvich and E.V. Sidorova, "Study of inhibition of antibody biosynthesis," Biokhimiya 29(4):556 (1964).
14. A.E. Gurvich and G.I. Drizlikh, "Use of antibodies on an insoluble support for specific detection of radioactive antigens," Nature 203(4945):648 (1964).
15. A.I. Gusev and V.S. Tsvetkov, "Contribution to the technique for setting up a reaction of microprecipitation in agar," Lab. Delo 7(2):43 (1961).
16. G.I. Drizlikh and A.E. Gurvich, "Use of antibodies linked to antigen on an insoluble support for specific extraction of various proteins," Biokhimiya 29(6):1054 (1964).
17. L.A. Zil'ber, Foundations of Immunology [in Russian], Medgiz, Moscow (1959).
18. L.A. Zil'ber and G.I. Abelev, Virology and Immunology of Cancer [in Russian], Medgiz, Moscow (1962).
19. E.A. Zotikov, R.M. Urinson, and L.P. Poreshina, "Sensitive method for detection of weak antibodies," Patol. Fiziol. i Eksperim. Terapiya 7(4):71 (1963).
20. A.P. Konikov, "Modification of the complement fixation reaction for quantitative determination of antigens and antibodies," Zh. Mikrobiol., Epidemiol., Immunobiol. (1):57 (1953).
21. A.P. Konikov and I.A. Tarakhanova, "Theory and practice of quantitative reactions of precipitation and complement fixation," in: Contemporary Problems of Immunology, Institute for Experimental Medicine, Leningrad (1959), p. 180.
22. R.S. Nezlin, "Certain problems of isolation and fractionation of pure antibodies," Biokhimiya 24(3):521 (1959).

23. R. S. Nezlin, "Determination of small quantities of protein and amino acids with the photoelectric colorimeter FEK-M," Vopr. Med. Khim. 8(3):316 (1962).
24. R. S. Nezlin, A. E. Gurvich, and V. I. Tovarnitskii, "Determination of the content of antibodies specific for viruses and rickettsia with the aid of antigens linked to paper," Vopr. Virusol. 4(2):150 (1959).
25. A. M. Olovnikov, "Use of immunoadsorbents for determination of the absolute quantities of haptens and antigens," Biokhimiya 29(4):680 (1964).
26. A. M. Olovnikov, "Preparation of immunoadsorbents as suspensions of benzidineprotein complexes and their use for determination of antibodies by agglutination," Vopr. Med. Khim. 10(5):538 (1964).
27. A. M. Olovnikov, "Semicondensed suspended antibody—immunoadsorbent and its use in the agglutination reaction for determination of antigen content," Dok. Akad. Nauk SSSR 158(5):1202 (1964).
28. V. A. Parnes, "Ouchterlony's reaction of specific precipitation in agar," Byul. Eksperim. Biol. i Med. 44(11):117 (1957).
29. V. I. Safonov, "Serological methods for the investigation of proteins and their prospective use in the biochemistry and physiology of plants," Usp. Biol. Khim. 6:304 (1964).
30. B. N. Sofronov, "On the mechanism and use of Boyden's reaction," Lab. Delo 10(9):542 (1964).
31. I. A. Tarkhanova and A. Ya. Kul'berg, "Study of the serological activity of papain-cleaved antibodies in the complement fixation reaction," Byul. Eksperim. Biol. i Med. 54(8):65 (1962).
32. R. S. Tatarinov, "On the use of a method of precipitation on paper to evaluate immune reactions of serum proteins in syphilis," Byul. Eksperim. Biol. i Med. 47(6):83 (1959).
33. N. I. Khramkova and G. I. Abelev, "Range of sensitivity of a method of precipitation in agar," Byul. Eksperim. Biol. i Med. 52(12):107 (1961).
34. M. A. Shagunova, "Complement fixation reaction in a 50% titer with protein antigens adsorbed on paper and its use in depleting anti-cancer sera," Byul. Eksperim. Biol. i Med. 45(3):122 (1958).
34a. A. E. Essel, Indirect Hemagglutination Reaction [in Russian], Meditsina, Leningrad (1965).
34b. R. Backhausz, Immunodiffusion and Immunoelectrophoresis, Akademiai Kiado, Budapest (1967).
35. D. Boroff and J. Fitzgerald, "Fluorescence of the toxin of *Clostridium botulinium* and its relation to toxicity," Nature 181(4611):751 (1958).
36. W. C. Boyd, Fundamentals of Immunology, Wiley, New York (1956).
37. S. V. Boyden, "Approaches to the problem of detecting antibodies," in: Mechanisms of Hypersensitivity, Churchill, London (1959), p. 95.
38. D. H. Campbell, J. S. Garvey, N. E. Cremer, and D. H. Sussdorf, Methods in Immunology, W. A. Benjamin, New York (1964).
39. R. R. A. Coombs, "Red cell-linked antigen test," in: Immunological Methods, Blackwell, Oxford (1962), p. 397.
40. A. J. Crowle, "Interpretation of immunodiffusion tests," Ann. Rev. Microbiol. 14:161 (1960).

41. W. B. Dandliker, H. C. Schapiro, J. W. Muduski, R. Alonso, G. A. Feigen, and J. R. Hamrick, "Application of fluorescence polarization to the antigen–antibody reaction. Theory and experimental method," Immunochemistry 1(3):165 (1964).

42. H. N. Eisen and F. Karush, "The interaction of purified antibody with homologous hapten. Antibody valence and binding constant," J. Am. Chem. Soc. 71(1):363 (1949).

42a. V. Ghetie and V. Micusan, Immunochemical Analysis [in Rumanian], Edit. Academii, Bucharest (1966).

43. D. Glick, R. A. Good, L. J. Greenberg, J. J. Addy, and N. K. Day, "Measurement of precipitin reactions on the millimicrogram protein-nitrogen range," Science 128(3339):1625 (1958).

44. H. C. Goodman and J. Bozicevich, "Recently developed techniques for the use of inert particles to detect antigen–antibody reactions," in: Immunological Methods, Blackwell, Oxford (1962), p. 93.

45. P. Grabar, "Comparison between quantitative immunological methods and the results of their application," Atti Congr. Intern. Microbiol. 2:169 (1953).

46. L. Gyenes and A. H. Sehon, "The use of toluene-2,4-diisocyanate as a coupling reagent in the passive hemagglutination reaction," Immunochemistry 1(1):43 (1964).

47. E. Haber and J. C. Benett, "Polarization of fluorescence as a measure of antigen–antibody interaction," Proc. Nat. Acad. Sci. USA 48(11):1935 (1962).

47a. J. Haimovich and M. Sela, "Inactivation of poly-*dl*-alanyl bacteriophage T4 with antisera specific towards poly-*dl*-alanine," J. Immunol. 97(3):338 (1966).

48. J. D. Hawkins, "Some studies on the precipitin reaction using a turbidometric method," Immunology 7(3):229 (1964).

49. M. Heidelberger and F. E. Kendall, "A quantitative theory of the precipitin reaction. III. The reaction between crystalline egg albumin and its homologous antibody," J. Exptl. Med. 62(5):697 (1935).

50. A. A. Hirata and D. H. Campbell, "The use of a specific antibody adsorbent for estimation of total antibody against bovine serum albumin," Immunochemistry 2(2):195 (1965).

51. T. R. Hughes and I. M. Klotz, "Analysis of metal-protein complexes," in: Methods of Biochemical Analysis, Vol. 3, Wiley, New York (1956), p. 266.

52. E. A. Kabat, Experimental Immunochemistry, C. C Thomas, Springfield (1964).

53. E. A. Kabat and M. Heidelberger, "A quantitative theory of the precipitin reaction. V. The reaction between crystalline horse serum albumin and antibody formed in the rabbit," J. Exptl. Med. 66(2):229 (1937).

54. F. Karush, "The interaction of purified antibody with optically isomeric haptens," J. Am. Chem. Soc. 78(21):5519 (1956).

55. M. Libich, "Immunochemical titration of antigens and antibodies in an electric field: titration of antigens in complex systems and titration of individual determinant groups," Immunology 4(2):164 (1961).

56. N. R. Ling, "The coupling of protein antigens to erythrocytes with difluorodinitrobenzene," Immunology 4(1):49 (1961).

57. O. H. Lowry and O. A. Bessey, "The adaptation of the Beckman spectrophotometer to measurements on minute quantities of biological materials," J. Biol. Chem. 163(3):633 (1946).

58. F. C. McDuffie and E. A. Kabat, "A comparative study of methods used for analysis of specific precipitates in quantitative immunochemistry," J. Immunol. 77(3):193 (1958).

58a. O. Mäkelä, "Assay of antihapten antibody with the aid of hapten-coupled bacteriophage," Immunology 10(1):81 (1966).

59. J. R. Marrack, "Sensitivity and specificity of methods of detecting antibodies," Brit. Med. Bull. 19(3):178 (1963).

60. C. Mathot, A. Rothen, and J. A. Casals, "A new sensitive method for detecting immunological reactions," Nature 202(4938):1181 (1964).

61. P. H. Maurer and D. W. Talmage, "The effect of complement in rabbit serum on the quantitative precipitin reaction," J. Immunol. 70(2):135 (1953).

61a. Methods in Immunology and Immunochemistry (ed.: M. W. Chase and C. A. Williams), Academic Press, New York (1967).

62. O. Ouchterlony, "Gel-diffusion techniques," in: Immunological Methods, Blackwell, Oxford (1962), p. 55.

63. J. Oudin, "Immunochemical analysis of human serum and its fractions. II. Qualitative and quantitative analysis of the fraction soluble in two-thirds saturated ammonium sulphate," J. Immunol. 81(5):376 (1958).

64. Z. Ovary, "Passive cutaneous anaphylaxis," in: Immunological Methods, Blackwell, Oxford (1962), p. 259.

65. R. A. Patterson, W. O. Weigle, and F. J. Dixon, "Elimination of circulating serum protein antigens as a sensitive measure of antibody," Proc. Soc. Exptl. Biol. and Med. 105(2):330 (1960).

66. H. J. Rapp, "The nature of complement and the sign of a complement fixation test," in: Immunological Methods, Blackwell, Oxford (1964), p. 1.

67. A. Roberts and F. Haurowitz, "Quantitative studies of the bis-diazotized benzidine method of hemagglutination," J. Immunol. 89(3):348 (1962).

67a. H. Schneider and A. H. Sehon, "Determination of the lower limits for the rate constants of a hapten—antibody reaction by polarography," Trans. N. Y. Acad. Sci. 15:25 (1961).

68. A. H. Sehon, "Hemagglutinating factors in allergic sera," in: Allergology, Pergamon Press (1962), p. 301.

69. A. B. Stavitsky, "Hemagglutination and hemagglutination-inhibition reactions with tannic acid and bis-diazotized benzidine—protein-conjugated erythrocytes," in: Immunological Methods, Blackwell, Oxford (1962), p. 363.

70. J. Sterzl and J. Kostka, "Procedure for determination of the minimal quantities of antibodies," Bratislav. Lekarske Listy 43(2):110 (1963).

71. D. W. Talmage, "The measurement of antibody," in: Allergology, Pergamon Press (1962), p. 282.

72. D. A. J. Tyrrell, "Demonstration of antibodies to viruses," in: Immunological Methods, Blackwell, Oxford (1962), p. 313.

73. S. F. Velick, C. W. Parker, and H. N. Eisen, "Excitation transfer and the quantitative study of the antibody hapten reaction," Proc. Nat. Acad. Sci. USA 46(11):1470 (1960).

74. R. J. Weiler, D. Hofstra, A. Szentivanyi, R. Blaisdell, and D. W. Talmage, "The inhibition of labeled antigen precipitation as a measure of serum γ-globulin," J. Immunol. 85(2):130 (1960).
75. R. R. Williams and S. S. Stone, "Application of antigen–antibody reactions on supporting media to the purification of proteins. I.," Arch. Biochem. Biophys. 71(12):377 (1957).
76. Y. Yagi, K. Engel, and D. Pressman, "Quantitative determinations of small amounts of antibody by use of solid absorbents," J. Immunol. 85(5):736 (1962).
77. Y. Yagi, P. Maier, and D. Pressman, "Immunoelectrophoretic identification of guinea pig anti-insulin antibodies," J. Immunol. 89(5):736 (1962).
78. J. Zikan, "Polarographic determination of the extent of association of antibodies against the 2,4-dinitrophenyl group with hapten," Collection Czech. Chem. Commun. 31:4260 (1966).

Chapter II

Isolation of Pure Antibodies

The possibility of isolating pure antibodies is of interest in several respects. Pure preparations are, first of all, necessary for a detailed study of the properties and structure of these highly specific serum proteins. The availability of pure antibodies considerably facilitates immunological investigations, for example, the study of antibody–antigen interaction. Finally, it should be recalled that antibodies are widely used as highly specific reagents, for example, to detect antigen in tissues, and also as effective therapeutic agents against many infectious diseases. In any case, pure antibody preparations are obviously more convenient to use than the presently employed immune serums or their fractions, which contain considerable quantities of nonspecific proteins in addition to antibody.

In the last three decades, a great variety of methods have been proposed for the isolation of antibodies. Some of these are based on physicochemical or chemical fractionation of serum proteins (nonspecific methods), while others involve extraction of antibody from the sum total of serum proteins with the aid of a specific antigen (specific methods). However, antibody preparations obtained by these methods often contain a considerable quantity of protein impurities.

In the following, only those studies in which the isolation of highly purified antibody preparations is described are discussed. Special attention is given to recently developed methods which employ synthetic immunoadsorbents, i.e., antigens bound to insoluble supports.

NONSPECIFIC METHODS

In fractionation of immune serum proteins by chemical or physicochemical methods, antibody is recovered in one of the globulin fractions, usually γ-globulins. By these methods (precipitation with neutral salts or organic solvents, electrophoresis, etc.) preparations in which the ratio of antibody to nonspecific protein is significantly higher than in the initial immune serum can be obtained [4, 51, 87, 146].

As there are no appreciable differences in physicochemical or chemical properties between the antibodies and globulins of a given fraction, it is usually impossible to obtain pure antibody by these methods.

However, highly purified antibodies to pneumococcal polysaccharides can be isolated from hyperimmune horse serum by this method. These antibodies are found only in the euglobulin fraction of horse serum, i.e., in the fraction insoluble in distilled water. By fractionation with EtOH, Felton [64] obtained preparations in which approximately 80% of the protein was precipitated by the corresponding polysaccharides.

Chow and Goebel [53], and later Northrop and Goebel [120] successfully used a 0.2*M* solution of potassium phthalate (pH 3.6) to extract nonspecific proteins from the euglobulin fraction of the serum of horses immunized with polysaccharides. After adding this solution, the nonimmune proteins were precipitated, and the resultant supernatant was a pure antibody solution, in which 90-100% of the protein reacted with antigen. It is very difficult to explain this highly specific effect of phthalate. Possibly, this anion interacts separately with the binding sites of antipolysaccharide antibodies, and the resultant complex under certain conditions is more soluble than the nonspecific γ-globulins [87].

However, other investigators have not been able to obtain such good results with nonspecific methods. The antibody concentration in preparations obtained in this way depends chiefly on the antibody content in the extracted fraction, and although it is several or even tens of times higher than in the initial serum, it is still very seldom higher than 50%.

Nonspecific methods, especially those employing chemical fractionation of serum proteins, are not particularly difficult to use, and for this reason are employed in those cases where a large quantity of partly purified antibody is necessary. The use of ethanol for separation of serum γ-globulin for therapeutical purposes may be mentioned as an example [29]. In the Soviet Union, a procedure developed by Kholchev [32, 33] is used for the production of a γ-globulin preparation. The chief steps in the procedure are as follows:

First, ethanol is added to the serum to a concentration of 25% at pH 6.8–7.0 with the temperature reduced to −5°. The precipitate, which contains impurities (mainly β-globulin) in addition to γ-globulins, is dispersed in a 25-fold volume of 0.01*M* NaCl at pH 5.1. Ethanol is added again to a 17% concentration whereby the salt concentration is reduced to 0.007*M*. Under these conditions at reduced temperature, β-globulins and fats are precipitated. γ-Globulin, which remains in solution, is precipitated at pH 7.2 at an ethanol concentration of 25%. The yield is 8-10% of the total quantity of protein in the serum.

Under laboratory conditions, it is convenient to employ chromatography on ion-exchange resins, in particular, on DEAE cellulose, to separate the major part of immunoglobulins (γG-globulins). In this method, the serum is first dialyzed against phosphate buffer (pH 6.3, 0.0175*M*), and then passed through a column filled with DEAE cellulose equilibrated with this same buffer [95, 107]. As the serum passes through the column, all the serum proteins are retained by the resin, while γ-globulins (γG) emerge with the buffer solution. It is expedient first to precipitate the globulin fraction with sulfate, then to dissolve the precipitate, and, after dialysis, subject the solution to chromatography on resin.

If DEAE-Sephadex in chloride form is used to extract γG-globulins [36, 41], the procedure is even simpler, as preliminary dialysis of the serum is then unnecessary. Briefly, the procedure involves the following:

DEAE Sephadex (A-50 coarse) is swelled in H_2O and treated successively with 0.5*N* NaOH and 0.5*N* HCl solution, after which it is equilibrated with cold 0.01*M* phosphate buffer, pH 6.5. Then to 10 g DEAE Sephadex is added 50 ml human serum, the thick mixture is stirred periodically for one hour in the cold, transferred to a filter, and after filtering by suction washed four times with 25 ml of the same buffer. To the filtrate is added a fresh portion of DEAE Sephadex (10 g) and the mixture is agitated for one hour in the cold with a magnetic mixer. The suspension is filtered off, the DEAE Sephadex is washed with 200 ml buffer (10-20 ml), and the filtrate, containing almost all the serum γG-globulin, is immediately neutralized to pH 7.5 with a (1*M*) K_2HPO_4 solution. The yield is 610 mg (97%). According to our experiments, about 400 mg γG-globulin can be recovered from 50 ml rabbit serum.

It is more convenient to employ the "molecular sieve" principal to isolate macroglobulins (γM-globulins). Thus, Franek [66] used a column of granulated agar to isolate pig γM-globulin. When serum was passed through the column, macroglobulins emerged in the very first fractions. To separate γM-globulins from other serum proteins, the dextran gel Sephadex S-200 is even more effective. This gel retains proteins with a molecular weight less than 200,000 [25, 65]. Maseev [113] used this method. The polyacrylamide gels Bio-gel R, some brands of which retain proteins with a molecular weight less than 300,000 (Bio-gel R-200 and R-300), also show some promise.

A method has also been developed for isolation of γA-globulins from human serum [84]. These globulins are separated from γG-globulins by fractionation on carboxymethyl cellulose, and from trace amounts of other proteins (chiefly α-globulins and albumins) by chromatography on TEAE cellulose. A shortcoming of this method, as well as one of its modifications [62a], is the low yield and the possibility of contamination of the preparation by other proteins. Skvaril and Brummelova [135] developed a method for isolation of γA-globulins from the third ethanol fraction of placenta serum. First, ceruloplasmin is separated with DEAE-Sephadex and the bulk of the protein is precipitated with 0.5*M* zinc acetate. The final purification is carried out by fractionation on Sephadex C-200.

If required, all three of the principal types of immunoglobulins can be extracted from the same serum sample, for example from human serum [148], by the above-mentioned methods.

There are also several other well-known methods for isolation of γ-globulins with rivanol [8, 85], and with concentrated glycine solutions [126].

SPECIFIC METHODS

To separate antibody from other serum proteins it is very convenient to exploit their capacity to react with a specific antigen. The procedure then usually involves the following principal steps: 1) isolation of the antigen–antibody complex formed after addition of the antigen to the serum, and washing it free from nonspecific serum proteins; 2) dissociation of the specific complex; 3) separation of the dissociated antibody from the antigen. Each of these steps is examined below in detail.

Isolation of the Antigen – Antibody Complex

The specific complex formed by the antigen–antibody reaction may be either soluble or insoluble in aqueous solutions. Isolation of the complex is feasible only in the latter case. An example of such a complex is the specific precipitate formed after addition of antigen to serum containing precipitins. After centrifugation and repeated washing (usually with 0.85% NaCl and H_2O) the precipitate is usually sufficiently purified [45]. Its content of nonspecific protein impurities is not more than 2-4% [10]. To avoid contamination of the precipitate with complement proteins, they are removed beforehand by adding a heterologous immune antigen–antibody complex (decomplementation).

Isolation of the specific complex is uncomplicated if the antigen itself is insoluble. Thus, to remove the specific antibody from serum, cells of animal or bacterial origin may be used. After the cells are added to the serum, centrifugation alone is sufficient to extract the antibody–antigen complex. Research conducted on the isolation of antipneumococcal antibody with the aid of pneumococcal cells [105] and antitissue antibody with the aid of tissue from different animal organs [102, 147a] can be cited as examples.

The relative simplicity of isolation of a specific complex when the antigen is insoluble has served as one of the points of departure for the development of methods for binding of antigens to insoluble supports, whereby any soluble antigen can be rendered insoluble (see below). It should be mentioned that nonprecipitating antibody, as well as products of incomplete proteolytic antibody cleavage which are capable of reacting with antigen [18, 119], can be isolated only with the aid of insoluble antigens.

Dissociation of the Antigen – Antibody Complex

After isolation of the specific complex and the removal of nonspecific proteins comes the second stage – dissociation of the antibody from the antigen. This purpose is served by the reversibility of the antigen–antibody reaction when the medium conditions are altered. Usually the antigen – antibody bond is broken by changing the temperature, ionic strength, or pH of the medium.

Temperature Change. The antigen–antibody reaction is exothermic and evolves considerable heat [45]. Hence, it may be assumed that if the temperature is raised the reaction will be reversed. And indeed, if the antigen–antibody complex was formed at 0°, when the temperature is raised to 40-60° some of the antibody is dissociated and enters solution. Antibodies which agglutinate bacteria [91] and erythrocytes [72, 103] have been isolated in this way.

Although only a small portion of the conjugated antibody is dissociated from the antigen by heating, some investigators have nevertheless been able to isolate enough antibody to carry out detailed physicochemical and immunochemical studies. For example, Gordon's study on the isolation and properties of cold hemagglutinins from human serum is well known. Korngold and Pressman [102, 124], adsorbed ^{131}I-labeled antibody from immune antitissue sera on tissue from liver and other organs. The labeled antibodies were then dissociated by heating and used to determine the localization of antigens in tissue.

Nevertheless, it must be noted that antibody dissociation by heating is not to be preferred because of the low antibody yield and, in addition, the possibility that the antibodies may be damaged.

Change in Ionic Strength. A portion of antipneumococcal antibodies can be dissociated from antigen by placing the specific precipitate or agglutinate in 15% NaCl at 37°. Heidelberger and Kabat [83] used this method to isolate antibodies to several types of pneumococci from the sera of different animals; 60-100% of the protein in the antibody preparations was precipitated by the corresponding polysaccharide. Davis et al. [56] also used a 15% NaCl solution to isolate Wassermann antibodies from the serum of syphilitic patients.

It must be pointed out that when concentrated NaCl solutions are used, only a portion of the antipolysaccharide antibodies are dissociated from the antigen. The yield of antibody is negligible if the antigen is a protein [122]. Elution with a 5.5*M* solution of potassium iodide in 0.05*M* Tris buffer (pH 8.2) has proved to be effective, however. The quantity of antibody separated from insoluble antigen with this solution was the same as by elution with a buffer with pH 2.2.

Changes in pH of Medium. This method for dissociating the antigen–antibody complex is, perhaps, the most widely used. The specific complexes are stable only in the near-neutral pH zone. Thus, complexes of rabbit antibody with ovalbumin are stable at pH 6.25–8.45 [100], complexes containing bovine serum albumin are stable at 4.5–7.5 [133], and complexes with human serum albumin are stable at pH 5–9 [136]. If the pH is increased or decreased, the specific bonds begin gradually to break. This is true of antibody complexes with protein antigens as well as with polysaccharides.

Many investigators have used alkali solutions to dissociate antibody. Highly pure preparations of antibodies to pneumococcal polysaccharides have been obtained using NaOH [54]. Kabat [92] used $Ba(OH)_2$ for the same purpose, and Sternberger and Pressman [137] used $Ca(OH)_2$ in a method which they developed for isolating antibodies to proteins.

Alkali dissociation has some definite disadvantages, however, precisely because of the high pH necessary. In a study of the effect of alkali on a precipitate containing ovalbumin it was shown [101] that at pH 11.0 only one electrophoretic peak appears in spite of dissolution of the precipitate. At pH 11.7, about 17% of the antibody is dissociated. Total dissociation was achieved only at pH 12.3. However, a medium with such high alkalinity damages antibody. According to the same investigators, treatment of immune globulins with alkali at pH 12.3 for two hours caused an increase in the amount of precipitate formed when antigen was added. On the other hand, after 24-hr exposure of γ-globulins to the same pH, the amount of precipitate decreased. Heidelberger and Pedersen [83a] obtained data on possible damage to antibodies in alkaline media. They studied antipneumococcal polysaccharide antibodies isolated with alkali in an ultracentrifuge, and found that even a brief treatment with $Ba(OH)_2$ in the cold resulted in decomposition of the antibody molecules. These data were confirmed in a further study of the sedimentation properties of pure antibodies of the same type at different pH values. Even a 10-minute treatment of the antibodies at pH 12.4 led to total decomposition of the antibody molecules. At pH 10.9 a reduction in the molecular size as a result of cleavage was also observed [92].

Use of acids is more efficient since dissociation of the specific complex takes place at pH values at which antibody damage is negligible. At pH 3, 75% of the antibody is split off pneumococcal cells, while at pH 2.3, 86% splits off [105]. According to electrophoretic data, total dissociation of the bovine serum albumin complex with a specific antibody occurs at pH 2.4 [133]. For a complex of human serum albumin with its specific antibody the total dissociation takes place at pH 2.7 [136]. Acid treatment of antibody at these pH values has almost no effect on their capacity to react with antigen. In a sedimentation study, Singer [132] demonstrated the complete reversibility of dissociation of a bovine serum albumin complex at pH values <4.5. As a result of acid treatment at pH 2 for 24 hr at 0°, the precipitating capacity of antibody was reduced by 3%. It should be mentioned that acids, nevertheless, cause changes in the molecular structure of antibodies which affect their aggregation capacity. Thus, Kabat [92] showed that the sedimentation coefficient of antibodies treated for 10 minutes at pH 3.41 and 1.44 was somewhat higher than that of antibodies treated at pH 4-9. Notwithstanding, acid dissociation is used at the present time in the majority of investigations on isolation of pure antibody by virtue of its simplicity and the relatively slight damage it causes to antibodies.

Other Methods for Dissociation of a Specific Complex. Kleinschmidt and Boyer [101] studied the effect of several substances on the stability of the specific antigen—antibody complex. Precipitates containing ovalbumin were completely dissolved in 0.3*M* KCl or 2*M* urea. However, electrophoresis detected only 5% antibody dissociation. Our data also indicates that urea dissociates only a small quantity of antibody; thus, 1*M* urea eluted only 17% of the amount of antibody eluted at pH 3.2 from human albumin bound to cellulose [24]. Even a 6*M* urea solution dissociated only 20% of the quantity of antibody eluted by a solution at pH 3 from a bound antigen [40]. However, 8*M* urea solutions can be used to isolate antibodies specific to haptens as well as to proteins [67a, 135a]. First, the precipitate is dissolved in 8*M* urea, then the antibodies are separated by passing the solution through DEAE cellulose or carboxymethylcellulose in an 8*M* urea solution. In the experiments of Kleinschmidt and Boyer [101] dodecylsulfate solutions did not dissociate antibody from ovalbumin, although they inhibited precipitation in this system. However, by using 0.05-0.2% solutions of this detergent, Cebra et al. [52] detected the dissociations of active papain fragments of antibodies from ovalbumin by ultracentrifugation. Aqueous solutions of CO_2 (0.035*M*, pH 5.0) can be used to dissociate specific complexes. Precipitates containing polysaccharides as well as protein antigens [147] are dissolved in these solutions, whereby some of the antibody molecules are split off. This method is probably one of the gentlest of the methods presently known, but it is unsuited for isolation of a large quantity of antibody because of the low yield (15-25% for antibody to ovalbumin).

Dissociation of antigen from its complex with antibody by means of negatively charged polyelectrolytes is a method of particular interest. Experiments were conducted on the dissociation of influenza virus from a complex with its specific antibody by adding polymethacrylate at pH 5.4-6.5 [89]. Approximately 30% of the antibody was brought down in the precipitate with the polymer, while the virus remained in solution. This effect can be accounted for by the fact that the antibody and virus are oppositely charged in this pH range. Consequently, the positively charged polymethacrylate is bound selectively to the antibody. Antibody was isolated from bovine serum albumin with polystyrene sulfonate in a similar manner. However, as it is very difficult to separate antibody from the newly formed antibody—polymer complex, the antibody yield in the aforementioned experiments were relatively low [87].

Antibodies to low-molecular-weight substances — haptens — can be split from conjugated proteins by placing the specific complex in a hapten solution. Thus, Karush purified antibodies to phenyl-(p-azobenzoylamino)-acetate and p-azophenyl-β-lactoside [96, 97, 128]. A method for isolation of the antibodies specific for the 2,4-dinitrophenyl and 2,4,6-trinitrophenyl groups [63, 67, 107a] has been described in detail.

Dissociation of specific complexes with the aid of hapten solutions is very convenient since the entire process of antibody isolation can be carried out at pH values in the neutral range. However, in this procedure only those antibodies bound less strongly with antigen can pass into solution from the specific precipitate. Another complication arises from the fact that it is extremely difficult to liberate the antibody from trace quantities of hapten bound with the antibody. Thus, in the aforementioned experiments of Farah et al. [63], in which radioisotope-labeled hapten was used, it was shown that about 10% of the binding sites of antibody molecules in the purified preparations may be blocked by hapten. Froese and Sehon [68] also point out this difficulty.

Separation of Antibody from Antigen after Dissociation of a Specific Complex

Separation of Antibody from Soluble Antigen. After dissociation of a specific complex, antibody must be separated from antigen. However, this is very intricate when the antigen is soluble. Nevertheless, sometimes both components can be separated if advantage is taken of differences in the properties of antibodies and antigens.

If the antigen molecular weight differs considerably from that of the antibody, separation can be accomplished by ultracentrifugation. Isliker and Strauss [89] isolated pure anti-influenza antibody in this way by sedimentation of the influenza virus after dissociation of the complex. The neutralizing capacity of 0.1 mg protein of the antibody preparation in inhibition of hemagglutination corresponded to a titer of 1:200.

In several cases, antibodies and antigens differing in size can be separated by gel filtration in acid solutions in which the complex is dissociated. Thus, Givol et al. [69] were able to separate the two components of a lysozyme–antilysozyme complex dissolved at pH 1.8 from each other by filtration through Sephadex G-75. A small amount of antigen (0.06% of the protein in the preparation) remained in the resultant antibody preparation; this was detected by using ^{131}I-labeled lysozyme. Even larger antigens can be separated from antibodies with a molecular weight of 160,000 by using Sephadex G-100 and G-200 [80]. Tarkhanova [30a] isolated antiovalbumin antibody in this manner.

The principal methodological complication encountered when such methods are used is that it is extremely difficult to achieve complete dissociation of the specific complex. In a study of dissociation of lysozyme, conjugated with p-aminohippurate, from its specific antibody, it was found that none of the methods employed (heating, 1*M* acetic acid, 2% sodium laurylsulfate, or excess hapten) could yield antibody preparations which would be completely free of trace quantities of antigen [42]. Hence, if highly purified antibody preparations are required, it is better to use antigen labeled with isotopes or with fluorescein isothiocyanate [42] to facilitate its assay [42].

Some investigators treated a specific complex with proteolytic enzymes to isolate antibodies to bacterial toxins [26, 125, 149], utilizing the fact that antitoxins are less susceptible to proteases than toxins. Northrop [26] obtained very good results in isolating antibodies specific for diphtheria toxin. The toxin–antitoxin complex was treated with a trypsin solution which resulted in separation of 30-60% of the antibodies. The isolated antibody preparations were highly purified. One of the fractions obtained from fractionation of antitoxin by means of ammonium sulfate was capable of crystallization.

However, it must be taken into account that only active fragments of molecules, and not the native molecules, are isolated by this method. In an ultracentrifuge investigation of a diphtheria antitoxin isolated with trypsin, the sedimentation coefficient was 5.3 S, while that of an antitoxin isolated by heating a toxin–antitoxin complex in an acid medium was 6.8 S [26]. Consequently, trypsin treatment also causes partial hydrolysis of antitoxin, resulting in a decrease in its molecular weight. In this treatment, presumably only that part of the molecule containing no binding sites for the reaction with the toxin is separated, since the purified antitoxin retains its full neutralizing capacity. An attempt to employ proteolytic digestion for isolation of antibody to gelatin was successful; in the case in question, the complex was digested with collagenase [38]. However, an attempt to isolate antibody to polytyrosyl gelatin by the same method yielded preparations containing approximately 3% antigen impurity due to incomplete digestion. These impurities, which contained tyrosine residues, can be removed if the antibody solution is acidified and passed through

Sephadex G-100 [70]. After digestion of the dextran–antidextran precipitate with the enzyme dextranase, antibody preparations were isolated; 80-90% of the protein in these preparations was precipitated with antigen. However, complete extraction of oligosaccharides involves considerable difficulties [93, 94, 142].

Campbell and Lanni [49] precipitated the antibody to ovalbumin after dissociation of the complex with half-saturated ammonium sulfate. The precipitate was isolated and dissolved in a physiological solution. About 90% of the protein in the solution was precipitated with antigen. Antibody yield was 70-90%.

If a specific precipitate is dissolved in an acid medium and the solution is immediately neutralized, the bulk of the antibodies form with the antigen a new complex which is brought down in the precipitate. However, after separating the precipitate, a small quantity of free antibody remains in the supernatant. In this way, Abelev and Avenirova [1] isolated precipitins to specific antigens of mouse liver and hepatoma.

Isolation of antibody is easier if the antigen is poorly soluble in an acid medium. After acidification of the medium, the specific complex is dissociated and the antigen is precipitated. Thus, for example, antibodies specific for ureases were isolated [140].

Proteins linked with haptens via a diazo bond, sometimes referred to as azoproteins, are also poorly soluble in acids. This property was exploited by Haurowitz et al. for isolation of antibodies to azoproteins [82] and to native proteins [81]. In the first case, the precipitate formed after addition of azoprotein to serum was dissolved in a 5% NaCl solution in HCl or phosphoric acid (for example, in a 5% NaCl solution in 0.1*N* HCl). After centrifugation, up to 90% of the proteins in the neutralized supernatant was precipitated with the appropriate azoprotein. In the second case, rabbits were immunized with pseudoglobulins, and antibody was precipitated from the immune serum by azo derivatives of these same proteins. Then the antibody–azoprotein complex was dissociated in 0.1-1.0*N* HCl or 0.1*M* phosphoric acid. The resultant preparations contained 36-55% protein precipitable by azo-antigen.

The reduced solubility of azoproteins at low pH values was exploited by Nisonoff and Pressman [118] to isolate antibody specific for the p-aminobenzoyl group. Antibody–azoprotein complexes were dissociated by buffer solutions at pH 3.3. After centrifugation, one-third of the antibodies originally in the precipitate was found in the supernatant. The purity of the preparations obtained was 69-98%.

For isolation of antibodies to haptens, some investigators have used polyhapten substances which are insoluble in an acid medium. Antibodies from the serum of a rabbit immunized with bovine serum proteins bound with p-azophenylarsonic acid were precipitated by Campbell et al. [48] by a substance of the following structure:

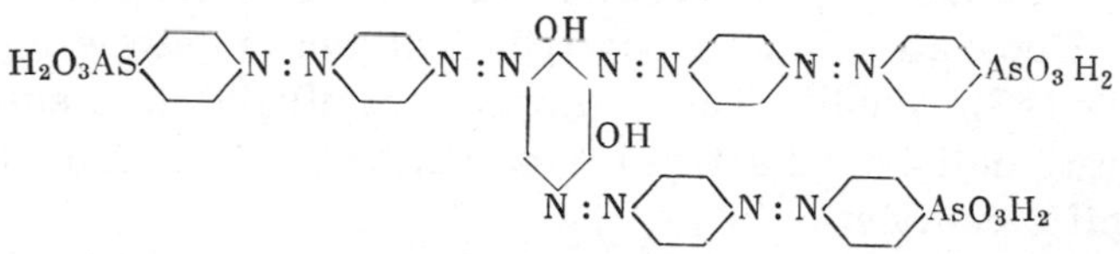

$$H_2O_3AS\langle\rangle N:N\langle\rangle N:N\,\overset{OH}{\wedge}\,N:N\langle\rangle N:N\langle\rangle AsO_3H_2$$
$$|\;\;|OH$$
$$N:N\langle\rangle N:N\langle\rangle AsO_3H_2$$

The precipitate thus obtained was dissociated with a solution of sodium arsanilate after which the pH of the solution was adjusted to 3.2. The antibody yield was 84-97%, and the purity of the preparations was 87-98%. A polyhapten with a somewhat different structure was used by Epstein et al. [62] to isolate antibody from the serum of a rabbit immunized with horse serum protein treated with diazotized arsanilic acid.

In addition to the aforementioned methods, which are suited for isolation of individual types of antibodies, two other general methods for obtaining pure antibodies to soluble protein antigens have been proposed. In both, chemically modified antigens are used to precipitate antibodies from immune serums.

Sternberger and Pressman [137] coupled a protein antigen with diazotized p-aminophenylarsonic acid, or p-aminobenzoic acid. Such treatment had no effect on the capacity of antigen to react with antibodies. After formation of the complex between modified antigen and antibody, the pH was adjusted to 12.5 by addition of $Ca(OH)_2$. Then the antigen was precipitated with calcium aluminate. These investigators succeeded in isolating the antibodies to bovine serum albumin and ovine serum albumin in preparations of 88-100% purity. The antibody yield was 30-40% of the quantity bound by the antigen.

Singer et al. [134] thiolated protein antigens with N-acetyl-D,L-homocysteinethiolactone. After precipitation of antibody with the antigen thus treated, the specific precipitate was dissolved at pH 2.4. Then a substance whose molecule contained two mercury atoms — 3,6-bis-(acetoxymercurimethyl)-dioxane — was added to the solution. This compound bound the antigen molecules via an S—Hg bond, whereby the antigen was quickly precipitated while the bulk of the antibodies remained in solution. Antibodies to ovalbumin, bovine serum albumin, and RNA-ase were isolated by this method. No less than 90% of the protein in the antiovalbumin preparation reacted with the antigen.

Separation of Antibody from Insoluble Antigens. Separation of antibody from antigen is quite simple if the latter is insoluble. Thus, simple centrifugation is sufficient to separate cell-bound antigen from antibody after dissociation of the specific complex.

However, the use of cellular antigens to isolate antibody has several disadvantages: In the first place, the antibody preparations may be contaminated with soluble proteins extracted from the cells during the preparation; in the second place, it must be taken into account that the cell membrane usually contains several antigens. Hence, by using cells, the sum total of antibodies to all antigens can be isolated, although it is impossible to isolate the antibody to any one of them.

Campbell and Lanni [49] used denatured ovalbumin as an insoluble antigen. After adsorption of the antibodies from immune serum, the specific complex was dissociated at pH 3.2. After separation of antigen by centrifugation, almost all antibodies adsorbed from the serum were found in the supernatant.

However, only some of the antibodies can be extracted with denatured antigens. After adsorption with these antigens, a portion of the antibodies reacting with native antigen always remains in the serum. This restricts the use of denatured antigens for isolation of antibody [47].

Day et al. [57] used fibrin powder to purify antifibrin antibodies labeled with radioactive iodine. After elution by heating at 64°, the antibodies were used to localize fibrinogen in tumor tissue in rats.

Protein can easily be rendered insoluble with bis- or tetradiazobenzidine. The insoluble aggregates thus obtained can probably be used not only to determine antibody and antigen, but also for their isolation [28, 58].

Insoluble protein polymers of albumin or globulin were obtained by Pressman in the following way [121]: first, mercaptosuccinyl groups were added to the protein. Then the modified proteins were polymerized with tris-[1-(2-methyl)-aziridinyl]-phosphinoxide. By binding hapten to such an insoluble polymer, the adsorption capacity of the resultant adsorbent with respect to antihapten antibody was one gram antibody per gram adsorbent. The quantity of antibody bound per gram adsorbent was less (0.4-0.7 g) for bovine albumin linked via bis-diazobenzidine. Antibodies isolated with such adsorbents had a high purity. However, nonspecific adsorption on adsorbents of this type was considerable.

Avrameas and Ternink polymerized protein antigens and antibodies with ethylchloroformate. The polymers were insoluble even in 6*N* HCl. The adsorption capacity was higher if polymerization was carried out at acid pH values, and was usually 0.4-0.7 g antibody per gram adsorbent. Polymers containing immune globulins retained their adsorption capacity for specific antigens. The use of formaldehyde for formation of polymer antigen complexes with a high adsorption capacity has also been described [113a]. Cross-linking of proteins through disulfide bridges can be similarly used for production of immunoadsorbents [136a].

In consideration of the advantages to be derived from isolation of antibody with insoluble antigens, methods for rendering soluble antigens insoluble have been developed. The usual procedure is to fix the antigens in some way to insoluble supports. Methods of protein fixation for the purpose of preparing these synthetic immunoadsorbents merit consideration in all cases where it is necessary to render protein insoluble without loss of biological activity. Hence, the data accumulated up to the present on synthetic immunoadsorbents deserve a separate, detailed discussion.

ISOLATION OF ANTIBODIES WITH THE AID OF FIXED ANTIGENS (Synthetic immunoadsorbents)

Antigens Adsorbed on an Insoluble Support

In the preparation of immunoadsorbents, the capacity of certain substances to adsorb proteins, including protein antigens, on their surface may be utilized. There are quite a number of substances known on which proteins can be adsorbed without loss of their immunological activity; some of these are: ferric and aluminum oxides, kaolin, carbon [35, 44], cholesterol, collodion [61, 117], alizarine [2], bismuth gallate [3], cellulose treated with chitosan [139, 152], polystyrene latex [131], etc. Antigens adsorbed on kaolin [115] and charcoal [143] have also been used to isolate antibody.

When paper impregnated with a protein solution is dried, some of the protein molecules remain firmly adsorbed to it and cannot be eluted by a stream of liquid. Gurvich [9] employed antigens bound in this way to entrap antibodies for assay purposes. Antigens adsorbed on cotton cloth and paper have also been used by Gostev et al. [5-7]. However, after drying only a small quantity of protein remains firmly bound to the paper, so that this is not an efficient way to prepare immunoadsorbents for preparative isolation of pure antibody.

Antigens adsorbed on Pyrex glass have been successfully used to isolate antibody [141]. A small quantity of ovalbumin can be fixed on this type of glass and still retain its capacity to react with antibody. For preparative purposes, a column filled with small glass beads of 44 μ diameter was used. One gram of beads firmly adsorbed 0.8 g ovalbumin.

However, immunoadsorbents prepared by adsorption of antigen on an insoluble support have several essential shortcomings. In the first place, they usually have a marked capacity for nonspecific adsorption of proteins from immune serum so that there is always the possibility that some of these proteins will then be eluted with the antibodies to contaminate the resultant preparations. In the second place, it is difficult to bind firmly all the antigen molecules by means of adsorption. For example, in the above-mentioned experiments [141], only ovalbumin was firmly adsorbed by glass. A certain amount of other fixed proteins (human and bovine serum albumins, bovine serum globulin) was washed out into the eluate by exposure to a solution with pH 3.0. Thirdly, the capacity of this type of immunoadsorbent, i.e., the quantity of antibody bound per unit weight of the immunoadsorbent, is usually relatively small.

Antigens Linked to an Insoluble Support via a Chemical Bond

Antigens linked to an insoluble support via chemical bonds have considerable advantages. For example, the relatively mild means used to split the antibody–antigen bond do not detach the antigen. Moreover, an insoluble support with a minimal capacity for nonspecific adsorption of protein can be employed. And finally, by chemical binding an immunoadsorbent with a very high capacity can sometimes be obtained.

Several investigators have used fixed haptens to isolate antihapten antibodies. Landsteiner and van der Scheer [104] attached diazotized aromatic amines to erythrocyte stroma. After such treatment, the stroma were used to adsorb antibodies from the serum of rabbits immunized with azoproteins.

In a similar fashion, Eisen and Karush [60] obtained preparations of antibodies to p-arsanilic acid whereby 90% of the protein was precipitated by erythrocyte stroma chemically bound with hapten. Later, Karush used haptens bound to fibrinogen [96, 97] to purify antibodies specific for phenyl-(p-azobenzoylamino)-acetate and β-lactoside.

Lerman [106] used powdered cellulose esterified with resorcinol as an insoluble support. Diazotized p-aminophenyl-azo-p-phenylarsonic acid was attached to the treated cellulose to yield an immunoadsorbent capable of extracting antibody from immune rabbit serum. One ml packed moist powder entrapped up to 2 ml antibody protein. This investigator used the immunoadsorbent thus obtained for isolation and chromatography of pure antibody.

A diazo bond is most often used [23, 94, 129, 130] to bind

protein antigens. This has the following advantages: it is known that several amino acid groups, such as the phenol groups of tyrosine, the imidazole groups of histidine, the ε-amino groups of lysine, the SH groups of cysteine, the indole groups of tryptophan, the guanidine groups of arginine and the amino groups of glycine [86, 144], and possibly also free COOH groups [108], react with diazo compounds. These groups are usually contained in proteins, by virtue of which the latter are able to react with substances having a free diazo group. It is, therefore, not surprising that various protein antigens can be coupled to insoluble supports via a diazo bond. For example, human and horse serum albumins, horse γ-globulin, casein, vicilin, glycinin, soluble tissue proteins, antigens of various viruses, bacterial toxins [11-15, 20, 22], insulin [39], certain allergens [109], etc., have been bound to cellulose in this way. The use of a diazo bond is advantageous for yet another reason: namely, diazonium salts are able to react with proteins over a wide range of external conditions. Thus, the amount of albumin linked to paper via a diazo bond is approximately the same at pH values from 6 to 9, and at temperatures from 0° to 37° [20]. It is important that this binding can take place at temperatures and pH values at which proteins are not denatured and retain their capacity to react with specific antibodies.

Compounds containing acid chloride or isocyanate groups are able to react with proteins. In several instances, these groups have been used to attach proteins to insoluble supports [16, 46, 87].

In 1964-1965 several reports were published on the use of H_2O-soluble carbodiimides with the structure $R-N=C=N=R^1$ to bind proteins and haptens; these compounds are capable of reacting with a large number of functional groups (carboxyl, amino, thiol, phosphate, etc.) [71, 154, 156]. In [155], carboxymethylcellulose was used as the insoluble support, and dicyclohexylcarbodiimide was used for binding, which presumably took place via an amide bond between the antibody amino group and the carboxyl group of carboxymethylcellulose:

$$\text{Cellulose}-O-CH_2-CO-OH+RNH_2 \rightarrow$$
$$\text{Cellulose}-O-CH_2-CO-NH-R+H_2O.$$

Bovine serum albumin, human γ-globulin, hemocyanin, dinitrophenyllysine, and p-(p-aminophenylazo)-phenylarsonic acid [154a,

155] were bound in this way, while in other experiments 6-trichloromethylpurine [153], tyramine, and p-arsanilic acid [150] were bound.

Erythrocyte stroma or other insoluble proteins can be used as insoluble supports for fixation of protein antigens. In such cases, the intermediate "bridge" will be a compound with at least two diazo groups, one of which reacts with the support, while the other reacts with the antigen. Pressman et al. [123] bound protein to erythrocyte stroma via bis-diazotized benzidine. Coombs [55] bound protein antigens to erythrocytes via tetra-diazotized benzidine and then used them to assay nonprecipitating antibodies. In principal, any bifunctional compound reacting with proteins can be used for this purpose (cf. Chapter I).

Antigens fixed on cellulose, ion-exchange resins, and polystyrene have proven to be the most effective for antibody isolation. These compounds are relatively resistant to various external factors. Since a column technique can be used for the isolation of antibody by antigens fixed on these polymers, it is possible to simplify considerably and to automate the process, and to study processes of antibody adsorption and elution by collecting the liquid emerging from the column in fractions.

Cellulose-Based Immunoadsorbents. Campbell et al. [50, 109] described an immunoadsorbent based on powdered cellulose. First the cellulose was treated with p-nitrobenzylchloride, and the p-nitrobenzylcellulose was successively reduced and diazotized, the result being a diazo group bound to cellulose. Then an antigen solution (bovine serum albumin) was added to the cellulose. After incubation, the cellulose was washed free of excess antigen. The final dried immunoadsorbent contained 1.5% protein. After treatment with immune serum, the antibody which had combined with the fixed antigen was eluted at pH 3.2. Approximately 90% of the protein in the antibody eluate was precipitated when the antigen was added. Subsequently, commercial p-aminobenzylcellulose was also used to isolate antibodies to various antigens [109]. It should be mentioned, however, that Campbell et al. also reported these immunoadsorbents to have a considerable nonspecific adsorption capacity: tenths of a μg protein per ml serum appeared in the eluate. An immunoadsorbent based on carboxymethylcellulose which was later used by these same investigators had only one-tenth the nonspecific protein adsorption capacity of the former adsorbent [155].

Talmage et al. [145] employed antigen bound to cellulose by means of p-nitrobenzylchloride to obtain ^{131}I-labeled antibody preparations with a high specific activity.

Another procedure for linking protein to cellulose was first developed by Gurvich [11, 15] and involves the following: as Kursanov and Solodkov [19] showed, tetra-substituted ammonium salts with the structure $[R-O-CH_2-NR'_3]Cl$ react extremely readily with cellulose and form a stable chemical bond with it. These investigators synthesized pyridine salts of methylchloride esters, in particular, N-(m-nitrobenzyloxymethyl)-pyridine chloride (alkyl halide) with the structure

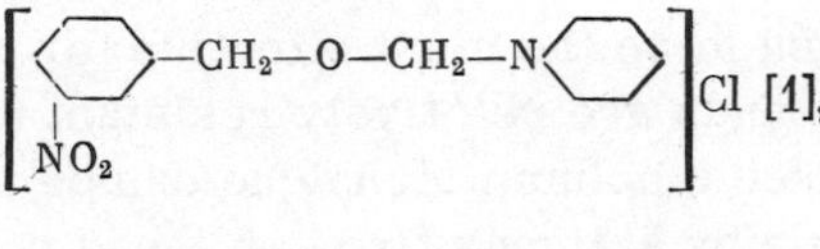

$$\left[C_6H_4(NO_2)-CH_2-O-CH_2-NC_5H_5 \right] Cl \; [1], \qquad (I)$$

which is an aromatic analog of substances of the type indicated above. At 125° the alkyl halide residue (I) becomes bound to cellulose and by subsequent reduction and diazotization a diazonium salt firmly bound to cellulose can be obtained. The sequence of the reactions in this process is shown below.

$$[I] \quad \left[C_6H_4(NO_2)-CH_2-O-CH_2-NC_5H_5 \right] Cl +$$

$$\downarrow 125^\circ$$

$$C_6H_4(NO_2)-CH_2-O-CH_2-O-\text{cellulose}$$

$$\downarrow 15\% \; Na_2S_2O_4, \; 60^\circ$$

$$C_6H_4(NH_2)-CH_2-O-CH_2-O-\text{cellulose}$$

$$\downarrow 1.5\% \; NaNO_2 + 5\% \; HCl, \; 0-4^\circ$$

$$C_6H_4(N{\equiv}N^+ \, Cl^-)-CH_2-O-CH_2-O-\text{cellulose}$$

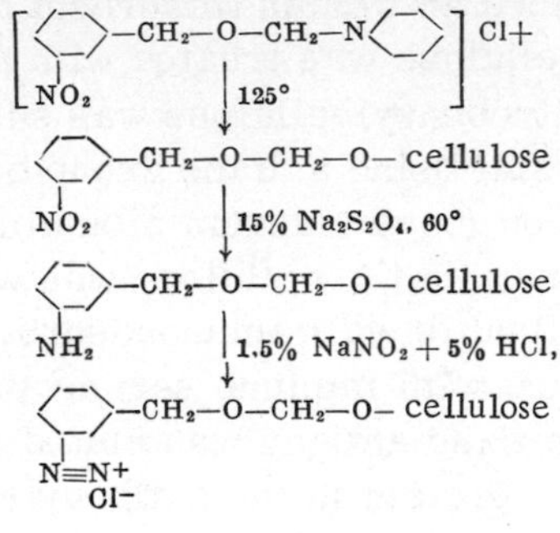

With the aid of the alkyl halide (I), the synthesis of which is described in several places [11, 19, 94], proteins have been attached to paper, cellulose powder, and to minute cellulose particles in suspension.

The quantity of antibody which can be retained by an immunoadsorbent is dependent chiefly on the number of antigen protein molecules fixed to the surface of the particles of the insoluble support. This in turn depends on several factors. In the first place, it is restricted to a certain degree by the number of diazo groups on the cellulose, which is itself a function of the amount of unreacted molecules of alkyl halide (I). However, the quantity of bound antigen protein increases only to a certain limit with increase in the concentration of alkyl halide (I). Thus, the maximum binding of protein to paper discs is reached at an alkyl halide (I) concentration of 8% in solution, although the number of free diazo groups continues to increase with increase in the alkyl halide concentration [20]. It was not possible to attach more than 10-50 mg protein per gram adsorbent dry weight [11, 13, 24] to powdered cellulose of large particle size. This limit was not due to a deficit of diazo groups, since after attachment of the protein the adsorbent reacted with β-naphthol via these groups, producing an intense bright red color. Another limiting factor was the total cellulose particle surface, increase of which augmented the adsorbent capacity. Surface area was increased by using smaller cellulose particles.

Gurvich, Kuzovleva, and Tumanova [14] reprecipitated the treated cellulose powder at the aminocellulose stage in a copper–ammonium solution to reduce the particle size. As a result, the adsorbent capacity was increased tenfold to 0.3-0.5 g antibody per gram adsorbent and higher. Shakhanina [34] developed another procedure for obtaining a suspension of minute particles of adsorbent [34] in which the cellulose is first treated with alkali and then with carbon disulfide. The cellulose is then salted out of the resultant solution by a large volume of water.

It is of course more advantageous to use such a high-frequency immunoadsorbent as a suspension of small particles to isolate antibody. The advantages of such an adsorbent have also been noted by other investigators [40, 68a, 116]. The general aspects of preparation of an adsorbent in suspension are given below.

The cellulose powder to be used as the insoluble support should be in the form of small particles and be readily soluble in a copper–ammonium solution. To 50 g powder is added 100 ml 1-2% alkyl halide (I) in a 0.35-0.7% solution of sodium acetate and the mixture is blended carefully. Then the thick paste formed is applied in a thin layer to a glass plate and dried thoroughly at 125°, periodically mixing and breaking up lumps. At this stage, it is

TABLE 5. Extraction of Anti-Influenza Antibody from Immune Serum with the Aid of a Fixed Virus [22]

Fraction	Protein concentration, %	Titer in hemagglutination in inhibition
I	5.74	1:64
II	5.79	1:128
III	6.18	1:128
IV	6.32	1:128
VII	—	1:4096
Serum before passing through the column	7.34	1:4096

very important that the entire mass of the powder is thoroughly dried and heated. Then the powder is washed free of unbound alkyl halide (I) with water, carefully dried, and washed thrice with benzene, after which the powder cake is transferred to a beaker. After all the benzene has evaporated, the powder is treated with freshly prepared sodium hydrosulfite (15% aqueous solution) and incubated while stirring occasionally at 50-60° for 30 minutes. Then it is washed with H_2O and extracted three times with a 30% acetic acid solution. The aminocellulose thus obtained can be kept for an extended period. To prepare a suspension of small cellulose particles, the aminocellulose powder is dissolved in a Schweizer copper–ammonium solution (10 g powder in 200 ml concentrated NH_4OH, 15 g copper hydroxide, 3.3 g sucrose, and 130 ml water), an additional 600 ml concentrated NH_4OH is added, and the entire solution is decanted into a large volume of water. After a certain period the solution becomes cloudy and very small particles of cellulose begin to precipitate. To accelerate precipitation, the solution may be slightly acidified by adding drops of sulfuric acid while stirring vigorously. On the following day, the liquid is decanted and the brown precipitate is neutralized with a 10% solution of sulfuric acid (until it becomes white) and washed free of excess acid with water.

For diazotization, up to 5% HCl is added to the aqueous suspension, and, after cooling, up to 1.5% sodium nitrate is added. The mixture is stirred vigorously on an ice bath for 30 minutes, then quickly washed with cold water until the pH is neutral, and a 2% solution of antigen protein is added. After 18 hr of mixing the immunoadsorbent is washed free of unbound protein. The bulk of the protein is bound during the first 30 minutes [20]. The immunoadsorbent can be kept for a long time in the cold as a suspension in the presence of antibiotics. It should be washed several times with a physiological solution before use.

TABLE 6. Quantity of Precipitins to Horse Albumin in Eluates [13]

Total protein content in eluate, μg	Addition of antigen, μg	Amount of precipitate, μg	Antibody content μg	Precipitins, %
223	25	232	207	93
213	35	263	228	107
2912	300	3300	3000	103
4510	350	3920	3570	79
90	6	46	40	44

It is not difficult to ascertain that proteins bound in this way to cellulose are capable of specific adsorption of corresponding antibodies from serum. Thus, if a serum containing two antibodies is passed through a column containing the immunoadsorbent, only heterologous antibody will be found in the emerging fractions, while antibody specific to the fixed antigen will be retained in the column [13]. An example of specific antibody adsorption from immune serum is given in Table 5, showing results of an experiment in which antiinfluenza rabbit serum was passed through a column containing a fixed influenza virus.

Solutions with pH 2.5–3.0 are used to dissociate the coupled antibodies. If columns are employed in conjunction with fraction collectors, acidified 1% NaCl solutions can be used for simultaneous isolation and fractionation of antibody (Fig. 6).

For preparative isolation of antibody, whether with columns or by washing, the adsorbent and elution acid buffer solutions should be used as these quickly detach the coupled antibodies, as is seen in Fig. 6 [21]. The antibody-saturated adsorbents can be kept for some time without suffering losses in subsequent elution.

Antibody preparations isolated with immunoadsorbents have a high degree of purity. Table 6 gives results of a determination of the quantity of precipitins in eluates from an immunoadsorbent.

The purity of the preparations can be verified by electrophoretic and sedimentation studies [13]. The precipitation properties of antibodies are not altered by extraction. Figure 7 gives precipitation curves of antibodies before extraction from serum and in eluates from immunoadsorbents.

were bound by a reaction between acid chloride groups of the resin and the reactive groups of protein (e.g., an amino group). By this method, erythrocyte stroma, human serum albumin, and influenza A virus were bound. All these antigens retained their complex-forming capacity with antibody. Antibodies of blood groups (anti-A and anti-B) were eluted with acid or alkaline buffers, potassium periodate, D-galactose, D-lactose, and N-acetyl-D-glucosamine. The eluates contained 75-90% of the antibody coupled with antigen. Antibodies to albumin and influenza virus were eluted with solutions having pH 3.6.

Isliker [16] points out another method for fixation in resins in which sulfonated aromatic amines are bound to anion-exchange resins. After diazotization, protein molecules were capable of conjugation with resins via a diazo bond.

As Isliker and Strauss noted [89], the chief disadvantage of immunoadsorbents prepared on ion-exchange resin supports is their small capacity. In addition, there is always the possibility that some of the bound antigen will be split off the resin during antibody extraction. Antigen (influenza virus) forms the most stable conjugate with the resin KHE 64 (COCl), evidently by means of peptide bonds between the resin and virus.

In recent years, successful experiments on the preparation of immunoadsorbents on a polystyrene base have been carried out. First the polystyrene is nitrated, and then reduced to polyamino-polystyrene, which can be diazotized to yield polyisocyanate polystyrene. In the first case, the diazo groups react with protein, while in the second the isocyanate groups react.

Grubhofer and Schleith [74, 75] were among the first to demonstrate the possibility of fixation of proteins on polystyrene. These investigators conjugated amylase, pepsin, RNA-ase, and carboxypeptidase via a diazo bond. The proteins retained their enzymic activity after fixation. In the same way, bovine serum albumin was fixed in the ratio of 13 g to 1 g polymer. These investigators calculated that this amount of protein occupied an area of 3 m^2; 1 g polystyrene had approximately the same surface area. Subsequently, several investigators fixed protein antigens and enzymes [75, 110-112, 151] to polystyrene. In all cases, the bound proteins retained their specific activity. Sehon [76-78, 129, 130] used antigens bound to polystyrene by a diazo bond to isolate pure antibody. Bovine and human serum albumins, and human serum γ-globulin were used as antigens. All these proteins were capable of extracting their specific precipitins from rabbit serum and the antibody was dissociated by dilute HCl (pH 3). The isolated antibodies were capable of reacting with antigen to form both soluble and insoluble complexes in an excess of antigen. Allergens of ambrosia pollen were fixed on polystyrene. The resultant immunoadsorbents provided total extraction of skin-sensitizing and blocking antibodies from the serum of humans sensitive to this pollen. However, the antibodies to the allergens could not be eluted at pH 3. The exact nature of the bond by which these allergens are linked is not completely clear. It is possible that a considerable portion of the allergens are simply adsorbed on the polymer. Thus, even more allergens are fixed on polyaminostyrene than on polydiazo-compounds [127].

The preparation of immunoadsorbents on the basis of polystyrenes involves certain methodological difficulties, for example, in the reduction of polynitropolystyrene to a polyamino compound. Immunoadsorbents based on a polymer with a relatively small number of amino groups have a considerable nonspecific adsorption capacity, and the antibodies they adsorb are only partially eluted [77].

In spite of the shortcomings indicated, antigens linked to polystyrene have been successfully used for various purposes. For example, the rheumatic factor was fractionated on such an adsorbent [121a].

Olovnikov [27] developed a procedure for preparation of immunoadsorbents on a Soviet-produced nonisotactic emulsion polystyrene.

Kent and Slade [98, 99] used an interesting method for preparation of immunoadsorbents based on a linear styrene polymer. After nitration, reduction, and diazotization, protein (γ-globulin, extracted from immune rabbit serum) was added to the polymer. The protein molecules reacted with diazo groups and formed cross-links with the polymer chain. In some instances, the newly formed cross-linked polymer contained up to 60-70% protein. In other experiments, mercury atoms were added to the linear polymer, and the SH groups of the protein molecules reacted with Hg to form cross-links with the chains. Fixed rabbit antibodies to *Yersinia pestis* and to egg albumin, and horse atibodies to *Bacillus anthracis* were coupled to their corresponding antigens. However, these immunoadsorbents had a considerable specific adsorption capacity.

Immunoadsorbents can be prepared by polymerizing any substance in an antigen solution, whereby some of the antigen molecules are mechanically bound by the newly formed polymer. Such experiments were conducted with polyacrylamide [43].

The dextran gel Sephadex can be used as a base for preparation of immunoadsorbents. Thus, antibodies to gonadotropin were fixed to modified Sephadex (isothiocyanatephenoxyhydroxypropylsephadex) and the adsorbent was used to determine gonadotropin [157].

As is evident from the foregoing, there are many methods presently available for isolating pure antibodies. Some of these methods yield quite large quantities of very pure antibody preparations suitable for different types of analysis. If soluble antigens are used, methods employing synthetic immunoadsorbents, i.e., antigens fixed on insoluble supports by means of a chemical bond, are most suitable. Of the presently known types of immunoadsorbents, antigens linked to very small cellulose particles [14] enjoy essential advantages for isolation of antibodies: they are quite simple to prepare, and have a high capacity and low nonspecific adsorption.

LITERATURE CITED

1. G. I. Abelev and Z. A. Avenirova, "Isolation of precipitins to specific antigens of liver and hepatoma of mice," Vopr. Onkol. 6(6):57 (1950).
2. A. K. Adamov, "Properties of antibodies fixed on particles and their prospective use in microbiology, VI," Zh. Mikrobiol. Epidemiol. i Immunobiol. 42(2):100 (1965).
3. I. K. Babich, "On the early and rapid diagnosis of infectious diseases. A new adsorbent of microbe haptens," Zh. Mikrobiol. Epidemiol. i Immunobiol. No. 7, p. 24 (1949).
4. A. V. Beilinson, N. N. Orlova, K. L. Shakhanina, T. A. Vitokhina, G. V. Chistoserdova, and A. I. L'vova, "Purification and concentration of polyvalent serums against influenza by a fractional salting-out method," Vopr. Virusol. 5(2):140 (1960).

5. V. S. Gostev, The Chemistry of Specific Immunity [in Russian], Medgiz, Moscow (1959).
6. V. S. Gostev and D. G. Grigor'yan, "Method for a serological study of desoxyribonucleoproteins," Byul. Eksperim. Biol. i Med. 45(1):122 (1958).
7. V. S. Gostev and N. A. Shagunova, "Quantitative reaction of specific nitrogen fixation by protein antigens adsorbed on dermatol and paper," Byul. Eksperim. Biol. i Med. 44(10):121 (1957).
8. T. L. Gubenko and V. I. Smirnova, "Quantitative rivanol—alcohol method for γ-globulin determination," Ukr. Biokhim. Zh. 35(5):747 (1963).
9. A. E. Gurvich, "Determination of protein antigens by paper chromatography," Biokhimiya 20(5):550 (1955).
10. A. E. Gurvich, "Ratio between synthesized specific and nonspecific serum γ-globulins," Vopr. Med. Khim. 1(2):169 (1955).
11. A. E. Gurvich, "Preparation of pure antibodies and determination of the absolute quantities of antibody in serums with the aid of antigens fixed on cellulose," in: Contemporary Methods in Biochemistry [in Russian], Meditsina, Moscow (1964), p. 73.
12. A. E. Gurvich, M. V. Ispolatovskaya, and K. N. Myasoedova, "Determination and isolation of anti-diphtheria antibodies with the aid of antigens fixed on cellulose," Vopr. Med. Khim. 7(1):55 (1961).
13. A. E. Gurvich, R. B. Kapner, and R. S. Nezlin, "Isolation of pure antibodies with the aid of antigens fixed on cellulose and the study of their properties," Biokhimiya 24(1):144 (1959).
14. A. E. Gurvich, O. B. Kuzovleva, and A. E. Tumanova, "Preparation of protein—cellulose complexes (immunoadsorbents) as suspensions capable of combining with large quantities of antibodies," Biokhimiya 26(5):934 (1961).
15. A. E. Gurvich and R. S. Nezlin, "Method for extracting antibodies by adsorption to antigens chemically linked to cellulose with subsequent elution," Biokhimiya 26(5):934 (1961).
16. G. Isliker, "Purification of antibodies with the aid of antigens linked to ion-exchange resins," in: Ion-Exchange Resins in Medicine and Biology [Russian translation], IL, Moscow (1956), p. 230.
17. S. A. Kibardin and T. A. Nikolaeva, "Adsorption chromatographic method for isolation of pure γ-globulin," Zh. Mikrobiol. Epidemiol. i Immunobiol. 42(1):70 (1965).
17a. T. S. Kotova, "Preparation of an immunoadsorbent on a cellulose base," Lab. Delo, No. 2, p. 93 (1966).
17b. O. B. Kuzovleva, A. E. Gurvich, Z. A. Rogovin, and A. D. Virnik, "Synthesis of new immunoadsorbents based on cellulose ether and 4-β-hydroxyethylsulfanyl-2-aminoanisole," Vopr. Med. Khim. 12(1):106 (1966).
18. A. Ya. Kul'berg and I. A. Tarkhanova, "Isolation of the specific binding site of antibody from papain-fermented antiserum with antigen fixed on cellulose," Byul. Eksperim. Biol. i Med. 50(11):76 (1960).
19. D. N. Kursanov and P. A. Solodkov, "New method for obtaining colored cellulose derivatives," Zh. Prikl. Khim. 16(11-12):351 (1943).
20. R. S. Nezlin, "Effect of different conditions on chemical fixation of proteins to paper," Biokhimiya 24(2):301 (1959).

21. R. S. Nezlin, "Certain problems in isolation and fractionation of pure antibodies," Biokhimiya 24(3):521 (1959).
22. R. S. Nezlin, "Isolation of antiinfluenza antibodies with the aid of influenza virus fixed on cellulose," Dokl. Akad. Nauk SSSR 131(3):676 (1960).
23. R. S. Nezlin, "Contemporary methods of isolation of pure antibodies," Usp. Sovrem. Biologia 52(4):19 (1961).
24. R. S. Nezlin, "Isolation of pure antibodies and the study of their properties," Dissertion, Akad. Med. Nauk SSSR, Moscow (1961).
25. R. S. Nezlin and L. M. Kul'pina, "Separation of serum proteins according to molecular weight with the aid of the dextran gel Sephadex G-200," Vopr. Med. Khim. 10(5):543 (1964).
26. J. Northrop, M. Kunits, and R. Kherriot, Crystalline Enzymes, Columbia University Press, New York (1948).
27. A. M. Olovnikov, "Preparation of immunoadsorbents based on emulsion non-isotactile polystyrene," Lab. Delo 8(8):31 (1962).
28. A. M. Olovnikov, "Preparation of immunoadsorbents in the form of suspensions of benzidine–protein complexes and their use for determination of antibodies by agglutination," Vopr. Med. Khim. 10(5):538 (1964).
29. N. A. Ponomareva and A. S. Nechaeva, Gamma-Globulin [in Russian], Meditsina, Moscow (1965).
30. Z. A. Rogovin, Sun Tun, A. D. Virnik, and N. Ya. Khvostenko, "Synthesis of new derivatives of cellulose and other polysaccharides, XIX," Vysokomolekul. Soedin. 4(4):571 (1962).
30a. I. A. Tarkhanova, "Use of chromatography on Sephadex G-200 to isolate antibodies from a specific precipitate," Vopr. Med. Khim. 12(1):108 (1966).
31. F. Franek and R. S. Nezlin, "Study of the role of different peptide chains of antibody in the antigen–antibody reaction," Biokhimiya 28(2):193 (1963).
32. N. V. Kholchev, "γ-Globulin (methods of isolation and standardization)," Doctoral dissertation, Akad. Med. Nauk SSSR, Moscow (1965).
33. N. V. Kholchev and L. I. Kolesnikova, "Procedure for isolation of a purified γ-globulin fraction," Zh. Mikrobiol. Epidemiol. i Immunobiol., No. 5, p. 6 (1947).
34. K. L. Shakhanina, "Isolation of large quantities of diphtheria and tetanus antibodies," Vopr. Med. Khim. 10(6):619 (1964).
35. W. A. Engel'gardt, "On the effect of antiphenolase in absorbed state," Biochem. Z. 148(5-6):463 (1924).
36. O. Aalund, J. W. Osebold, and F. A. Murphy, "Isolation and characterization of ovine γ-globulins," Arch. Biochem. Biophys. 109(1):142 (1965).
37. R. E. Amelunxen and A. A. Werder, "Quantitative considerations of influenza virus neutralization. I. Number of antibody molecules per virus particle," J. Bacteriol. 80(2):271 (1960).
38. R. Arnon and M. Sela, "Isolation of antibodies to gelatin from antigen–antibody complex by proteolysis," Science 132(3419):86 (1960).
39. E. R. Arquilla and J. Finn, "Genetic differences in antibody production to determinant groups on insulin," Science 142(3590):400 (1963).
40. J. E. Bata, L. Gyenes, and A. H. Sehon, "The effect of urea on antibody–antigen reactions," Immunochemistry 1(4):289 (1964).

41. J. S. Baumstark, R. J. Laffin, and W. A. Bardawil, "A preparative method for the separation of 7S-γ-globulin from human plasma," Arch. Biochem. Biophys. 108(3):514 (1964).

41a. M. M. Behrens, J. K. Inman, and W. E. Vannier, "Protein—cellulose derivatives for use as immunoadsorbents: preparation employing an active ester intermediate," Arch. Biochem. Biophys. 119(1-3):411 (1967).

42. J. C. Benett and E. Haber, "Studies on antigen conformation during antibody purification," J. Biol. Chem. 238(4):1362 (1963).

43. P. Bernfeld and J. Wan, "Antigen and enzymes made insoluble by entrapping them into lattices of synthetic polymers," Science 142(3593):678 (1963).

43a. F. Borek, Y. Stupp, and M. Sela, "Formation and isolation of rabbit antibodies to a synthetic antigen of low molecular weight," J. Immunol. 98(4):739 (1967).

44. J. Böszöremenyi, "A simple method for absorbing heterologous antibodies from precipitating sera," Nature 181(4619):1344 (1958).

45. W. C. Boyd, Fundamentals of Immunology, Wiley, New York (1956).

46. H. Brandenberger, "Methods for linking enzymes to insoluble carriers," Rev. Ferment. Inds. Aliment. 11(5):237 (1956).

47. D. H. Campbell, "Some problems in the detection, isolation, and purification of antibodies," in: Immunity and Virus Infection, Wiley, New York (1959), p. 139.

48. D. H. Campbell, R. H. Blaker, and A. B. Pardee, "The purification and properties of antibody against p-azophenylarsonic acid and molecular weight studies from light scattering data," J. Am. Chem. Soc. 70(7):2496 (1948).

49. D. H. Campbell and F. Lanni, "The chemistry of antibodies," in: Amino Acids and Proteins, C. C. Thomas, Springfield (1951), p. 649.

50. D. H. Campbell, E. Luescher, and L. S. Lerman, "Immunological absorbents. I. Isolation of antibody by means of a cellulose—protein antigen," Proc. Nat. Acad. Sci. USA 37(9):575 (1951).

51. J. Cann, "The separation and purification of antibodies," in: Immunity and Virus Infection, Wiley, New York (1959), p. 100.

52. J. J. Cebra, D. Givol, and E. Katchalski, "Soluble complexes of antigen and antibody fragments," J. Biol. Chem. 237(3):751 (1962).

53. B. F. Chow and W. F. Goebel, "The purification of the antibodies in type 1 antipneumococcus serum and the chemical nature of the type specific precipitin reaction," J. Exp. Med. 62(2):179 (1935).

54. B. F. Chow and H. Wu, "Isolation of immunologically pure antibodies," Science 84(2179):316 (1936).

55. R. R. A. Coombs, "Methods for the detection of incomplete antibodies and their diagnostic importance," Schweiz. Z. Allgem. Pathol. Bacteriol. 17(4):424 (1954).

56. B. D. Davis, D. H. Moore, E. A. Kabat, and A. Harris, "Electrophoretic, ultracentrifugal, and immunochemical studies on Wasserman antibodies," J. Immunol. 50(1):1 (1945).

57. E. D. Day, J. A. Planinsek, and D. Pressman, "Localization in vivo of radioiodinated anti-rat-fibrin antibodies and radioiodinated rat fibrinogen in the Murphy rat lymphosarcoma and in other transplantable rat tumors," J. Nat. Cancer Inst. 22(2):413 (1959).

58. S. DeCarvalho, A. J. Lewis, H. J. Rand, and J. R. Uhrick, "Immunochromatographic partition of soluble antigens on columns of insoluble diazo-γ-globulins," Nature 204(4955):265 (1964).
59. P. Delorme, M. Richter, S. Grant, and B. Rose, "Studies on the uptake of reagin by polyaminostyrene—ragweed complexes," Can. J. Biochem. Physiol. 40(4):519 (1962).
60. H. N. Eisen and F. Karush, "The interaction of purified antibody with homologous hapten. Antibody valence and binding constant," J. Am. Chem. Soc. 71(1):363 (1949).
61. M. Eisler, "On the effect of absorbents on antigens and their antibodies," Therapie 115(6):441 (1958).
62. S. J. Epstein, P. Doty, and W. C. Boyd, "A thermodynamic study of hapten antibody association," J. Am. Chem. Soc. 78(14):3306 (1956).
62a. J. L. Fahey, E. C. Franklin, R. S. Nezlin, and T. Webb, "Antibody purification on insoluble adsorbents and purification and characterization of immunoglobulin," Bull. World Health Organ. 35(2):779 (1966).
63. F. S. Farah, M. Kern, and H. N. Eisen, "The preparation and some properties of purified antibody specific for the 2,4-dinitrophenyl group," J. Exp. Med. 112(6):1195 (1960).
64. D. L. Felton, "The use of ethyl alcohol as a precipitant in the concentration of antipneumococcus serum," J. Immunol. 21(5):357 (1931).
65. P. Flodin and J. Killander, "Fractionation of human serum proteins by gel filtration," Biochim. Biophys. Acta 63(3):403 (1962).
66. F. Franek, "Isolation and some molecular characteristics of pig macroglobulin," Collection. Czech. Chem. Commun. 27(12):2808 (1962).
67. F. Franek, L. Simek, and O. Kotynek, "Antidinitrophenyl antibodies in pigs and bulls," Folia Microbiol. 10(6):335 (1965).
67a. M. Freedman, L. Slobin, J. Robbins, and M. Sela, "Soluble antigen—antibody complexes as intermediates in the purification of antibodies in 8 molar urea," Arch. Biochem. Biophys. 116:82 (1966).
68. A. Froese and A. H. Sehon, "Problems involved in the purification of antihapten antibodies by the use of stroma—hapten conjugates," Can. J. Biochem. Physiol. 39(6):1067 (1961).
68a. R. G. G. Gallop, B. T. Tozer, J. Stephen, and H. Smith, "Separation of antigens by immunological specificity. Use of cellulose-linked antibodies as immunoadsorbents," Biochem. J. 101(3):711 (1966).
69. D. Givol, S. Fuchs, and M. Sela, "Isolation of antibodies to antigens of low molecular weight," Biochim. Biophys. Acta 63:222 (1962).
70. D. Givol and M. Sela, "Isolation and fragmentation of antibodies to polytyrosyl gelatin," Biochemistry 3(3):444 (1964).
71. T. L. Goodfriend, L. Levine, and G. D. Fasman, "Antibodies to bradykinin and angiotensin: a use of carbodiimides in immunology," Science 144(3624):1344 (1964).
72. H. C. Goodman, "Chromatographic fractionation and characterization of antibody based on the thermal dissociation of the specific antigen—antibody complex," Nature 193(4813):350 (1952).

73. R. S. Gordon, "The preparation and properties of cold hemagglutinin," J. Immunol. 71(4):220 (1953).

74. N. Grubhofer and L. Schleith, "Modified ion-exchangers as specific absorbents," Nature 40(19):508 (1953).

75. N. Grubhofer and L. Schleith, "The coupling of proteins to diazotized polyaminostyrene," Hoppe-Seylers Z. Physiol. Chem. 297(2):108 (1954).

76. L. Gyenes, B. Rose, and A. H. Sehon, "Isolation of antibodies on antigen–antibody conjugates," Nature 181(4621):1465 (1958).

77. L. Gyenes and A. H. Sehon, "Preparation and evaluation of polystyrene–antigen conjugates for the isolation of antibodies," Can. Biochem. Physiol. 38(11):1235 (1960).

78. L. Gyenes and A. H. Sehon, "The use of polystyrene conjugates for the removal of antibodies from sera of allergic individuals," Can. J. Biochem. Physiol. 38(11):1249 (1960).

79. A. Hässig and E. Lüscher, "On the binding of dissolved antigens to corpuscular surfaces," Schweiz. Z. Allgem. Pathol. Bacteriol. 17(4):440 (1954).

80. E. Haber, "Recovery of antigenic specificity after denaturation and complete reduction of disulfides in a papain fragment of antibody," Proc. Nat. Acad. Sci. USA 52(4):1099 (1964).

81. F. Haurowitz, L. Etili, and S. Tunc, "Purification of antiprotein antibodies," Bul. Soc. Chem. Biol. 30(1-2):220 (1948).

82. F. Haurowitz, Sh. Tekman, M. Bilen, and P. Schwerin, "The purification of azoprotein antibodies by the dissociation of specific precipitates," Biochem. J. 41(2):304 (1947).

83. M. Heidelberger and E. A. Kabat, "Quantitative studies on antibody purification. II. The dissociation of antibody from pneumococcus specific precipitates and specific agglutinated pneumococci," J. Exp. Med. 67(2):181 (1938).

83a. M. Heidelberger and K. O. Pedersen, "The molecular weight of antibodies," J. Exp. Med. 65(3):393 (1937).

84. J. F. Heremans, In: Molecular and Cellular Basis of Antibody Formation, Academic Press, New York (1965), p. 277.

85. J. Hořejši and R. Smetana, "The isolation of γglobulin from blood-serum by rivanol," Acta Med. Scand. 155(1):65 (1956).

86. A. N. Howard and F. Wild, "The reactions of diazonium compounds with amino acids and proteins," Biochem. J. 65(4):651 (1957).

87. H. C. Isliker, "The chemical nature of antibodies," Advan. Protein Chem. 12:387 (1957).

88. H. Isliker and P. H. Strauss, "Use of the ion exchange resins for the purification of antibodies," Federation Proc. 13(13):236 (1954).

89. H. C. Isliker and P. H. Strauss, "The purification of antibody to PR8 influenza A virus," Vox Sanguinis 4(3):196 (1959).

90. A. T. Jagendorf, A. Patchornik, and M. Sela, "Use of antibody bound to modified cellulose as an immunospecific absorbent of antigens," Biochim. Biophys. Acta 78(3):516 (1963).

91. C. E. Jenkins, "The release of antibody by sensitized antigens," Brit. J. Exp. Pathol. 27(2):111 (1946).

92. E. A. Kabat, "The molecular weights of antibodies," J. Exp. Med. 69:103 (1939).

93. E. A. Kabat, "Purification of anti-dextran," Science 120:782 (1954).

94. E. A. Kabat, Experimental Immunochemistry, 2nd ed., C. C Thomas, Springfield (1964).

95. M. A. Kapusta and D. Halberstam, "Preparation of pure 7S γ-globulin," Biochim. Biophys. Acta 93(3):657 (1964).

96. F. Karush, "The interaction of purified anti-β-lactoside antibody with haptens," J. Am. Chem. Soc. 79(13):3381 (1957).

97. F. Karush and R. Marks, "The preparation and properties of purified anti-hapten antibody," J. Immunol. 78(4):296 (1957).

98. L. H. Kent and J. H. R. Slade, "Immunochemically active cross-linked polystyrene preparations," Biochem. J. 77(1):12 (1960).

99. L. H. Kent and J. H. R. Slade, "Immunochemically active cross-linked polystyrene," Nature 183(4657):325 (1959).

100. W. J. Kleinschmidt and F. D. Boyer, "Interaction of protein antigens and antibodies. I. Inhibition studies with egg albumin—anti-egg albumin system," J. Immunol. 69(3):247 (1952).

101. W. J. Kleinschmidt and P. D. Boyer, "Interaction of protein antigens and antibodies. II. Dissociation studies with egg albumin—anti-egg albumin precipitates," J. Immunol. 69(2):257 (1952).

102. L. Korngold and D. Pressman, "The in vitro purification of tissue localizing antibodies," J. Immunol. 71(1):1 (1953).

103. K. Landsteiner and C. Miller, "Serological studies on the blood of the primates. II. The blood groups in anthropoid apes," J. Exp. Med. 42(6):853 (1925).

104. K. Landsteiner and J. van der Scheer, "On the cross-reaction of immune sera to azoproteins," J. Exp. Med. 63(3):325 (1936).

105. K. H. Lee and H. Wu, "Isolation of antibody from agglutinate of type I pneumococcus by treatment with acid," Proc. Soc. Exp. Biol. Med. 43(1):65 (1940).

106. L. S. Lerman, "Antibody chromatography on an immunologically specific adsorbent," Nature 172(4379):635 (1953).

107. H. B. Levy and H. A. Sober, "A simple chromatographic method for preparation of γ-globulin," Proc. Soc. Exp. Biol. Med. 103(1):250 (1960).

107a. J. R. Little and H. N. Eisen, "Preparation and characterization of antibodies specific for the 2,4,6-trinitrophenyl group," Biochemistry 5(11):3385 (1966).

108. A. V. Luisada and H. Sobotka, "Coupling of proteins with radioactive p-diazobenzoic acid and its methyl ether," Immunochemistry 2(2):127 (1965).

109. A. Malley and D. H. Campbell, "Isolation of antibody by means of an immunologically specific adsorbent," J. Am. Chem. Soc. 85(4):487 (1962).

110. G. Manecke, "Reactive polymers and their use for the preparation of antibody and enzyme resins," Pure Appl. Chem. 4(2-4):507 (1962).

111. G. Manecke, "On serologically active protein resins and enzyme resins," Nature 51(2):3 (1964).

112. G. Manecke, S. Singer, and K. E. Gillert, "Serologically specific adsorbents," Nature 45(18):440 (1958).

113. R. Maseyeff (Maseev), J. Gombert, and J. Joselin, "Method of preparation of serum β_2-macroglobulin," Immunochemistry 2(2):177 (1965).

113a. J. C. Metcalfe, H. F. Marlow, and A. S. V. Burgem, "Immunoadsorbents of high capacity," Nature 209(5028):1142 (1966).

114. H. Metzger and H. Edelhoch, "Purification of anti-thyroglobulin antibodies," Nature 193(4812):275 (1962).

115. K. Meyer and A. Pic, "On the isolation of antibody by fixation to an antigen adsorbent system and consecutive dissociation," Ann. Inst. Pasteur 56(4):401 (1936).

116. N. R. Moudgal and R. R. Porter, "The use of antigen—cellulose suspensions for the isolation of specific antibodies," Biochim. Acta 71(1):185 (1963).

117. S. Mudd, B. Lucke, M. McCutcheon, and M. Strumia, "On the mechanism of opsonin and bacteriotropin action. VI. Agglutination and tropin action by precipitin sera. Characterization of the sensitized surface," J. Exp. Med. 52(3):313 (1930).

118. A. Nisonoff and D. Pressman, "Heterogeneity and average combining constant of antibodies from individual rabbits," J. Immunol. 80(6):417 (1958).

119. A. Nisonoff, F. C. Wissler, and D. L. Woernley, "Properties of univalent fragments of rabbit antibody isolated by specific adsorption," Arch. Biochem. Biophys. 88(2):241 (1960).

120. J. H. Northrop and W. F. Goebel, "Crystalline pneumococcus antibody," J. Gen. Physiol. 32(6):705 (1949).

121. K. Onoue, Y. Yagi, and D. Pressman, "Immunoadsorbents with high capacity," Immunochemistry 2(2):181 (1965).

121a. I. Oreskes, "The use of immunoadsorbents for the fractionation of rheumatoid factor and rabbit anti-human γG-antibodies," Immunology 11(5):489 (1966).

122. J. Oudin and P. Grabar, "Quantitative study of the homologous ovalbumin—antibody precipitating system of rabbits. II. Solubility of specific precipitates in a concentrated saline solution," Ann. Inst. Pasteur 70(1-2):7 (1944).

123. D. Pressman, D. H. Campbell, and L. Pauling, "The agglutination of intact azo-erythrocytes by antisera homologous to the attached groups," J. Immunol. 44(2):101 (1942).

124. S. Pressman and L. Korngold, "Localizing properties of antiplacenta serum," J. Immunol. 78(2):75 (1957).

125. C. G. Pope and M. Healey, "The preparation of diphtheria antitoxin in a state of high purity," Brit. J. Exp. Pathol. 20(3):213 (1939).

126. J. Porath and N. Ui, "Chemical studies on immunoglobulins. I. A new preparative procedure for γ-globulins employing glycine-rich solvent systems," Biochim. Biophys. Acta 90(2):324 (1964).

127. M. Richter, P. Delorme, S. Grant, and B. Rose, "Studies on the uptake of ragweed pollen allergens by polyaminostyrene," Can. J. Biochem. Physiol. 40(4):471 (1962).

127a. J. B. Robbins, J. Haimovich, and M. Sela, "Purification of antibodies with immunoadsorbents prepared using bromoacetyl cellulose," Immunochemistry 4(1):11 (1967).

128. J. H. Rockey, N. R. Klinman, and F. Karush, "Equine antihapten antibody. I. 7S β_{2A}- and 10S γ_1-globulin components of purified antilactoside antibody," J. Exp. Med. 120(4):589 (1964).

129. A. H. Sehon, "Antibody—antigen reactions model system for the specific interactions of biological macromolecules," Pure Appl. Chem. 4(2-4):483 (1962).

130. A. H. Sehon, "Physicochemical and immunochemical methods for the isolation and characterization of antibodies," Brit. Med. Bull. 19(3):183 (1963).

131. J. M. Singer, I. Oreskes, F. Hutterer, and J. Ernst, "Mechanism of particulate carrier reactions. V. Adsorption of human γ-globulin to 0.2 micron diameter latex particles and their agglutination by rheumatoid factor," Ann. Rheumatic Diseases 22(6):424 (1963).

132. S. J. Singer, "Physical-chemical studies of the nature of antigen—antibody reactions," J. Cellular Comp. Physiol., Vol. 50, Suppl. 1, p. 51 (1957).

133. S. J. Singer and D. H. Campbell, "Physical-chemical studies of soluble antigen—antibody complexes. IV. The effect of p^H on the reaction between bovine serum albumin and its rabbit antibodies," J. Am. Chem. Soc. 77(13):3504 (1955).

134. S. J. Singer, J. E. Fothergill, and J. R. Shainoff, "A general method for the isolation of antibodies," J. Am. Chem. Soc. 82(3):565 (1960).

135. F. Skvaril and V. Brummelova, "Isolation of γ_{1A}-globulin from the ethanol fraction III of placental serum," Collection. Czech. Chem. Commun. 30(8):2886 (1965).

135a. L. J. Slobin and M. Sela, "Use of urea in the purification of antibodies," Biochim. Biophys. Acta 107(3):593 (1965).

136. R. F. Steiner, "Reversible association processes of globular proteins. IX. The effect of pH and electrolytes upon an antigen—antibody combination," Arch. Biochem. Biophys. 55(1):235 (1955).

136a. J. Stephen, R. G. G. Gallop, and H. Smith, "Separation of antigens by immunological specificity. Use of disulfide-linked antibodies as immunoadsorbents," Biochem. J. 101(3):717 (1966).

137. L. A. Sternberger and D. A. Pressman, "A general method for the specific purification of antiprotein antibodies," J. Immunol. 65(1):65 (1950).

138. A. J. L. Strauss, P. G. Kemp, W. E. Vannier, and H. C. Goodman, "Purification of human serum γ-globulin for immunological studies: γ-globulin fragmentation after sulfate precipitation and prolonged dialysis," J. Immunol. 93(1):24 (1964).

139. S. S. Stone and R. R. Williams, "Application of antigen—antibody reactions on supporting media for the purification of proteins. II. The removal of a subtilisin-like enzyme from carboxypeptidase preparations by anti-subtilisin on cellulose," Arch. Biochem. Biophys. 71(2):386 (1957).

140. J. B. Sumner and J. S. Kirk, "On the chemical nature of urease," Hoppe-Seylers Z. Physiol. Chem. 205(6):219 (1932).

141. G. B. Sutherland and D. H. Campbell, "The use of antigen-coated glass as a specific adsorbent for antibody," J. Immunol. 80(4):294 (1958).

142. O. Swinford, Jr., R. Hoene, S. Quelch, and D. Samsell, "Purification of antibodies. I. Dextranase purification of antibody precipitated from type II antipneumococcal rabbit serum by dextran," J. Allergy 30(5):433 (1959).

143. H. Svensson, Arkiv. Kemi 17:1 (1943).

144. M. Tabachnik and H. Sobotka, "Azoproteins. II. A spectrophotometric study of the coupling of diazotized arsanilic acid with proteins," J. Biol. Chem. 235(4):1051 (1960).

145. D.W.Talmage, G. R. Baker, and W. Akeson, "The separation and analysis of labelled antibodies," J. Infect. Diseases 94(2):199 (1954).

146. J. R. Thurston, M. S. Rheins, and E. V. Buehler, "A rapid method for recovering serologically active globulins by sodium sulfate precipitation," J. Lab. Clin. Med. 49(4):647 (1957).

147. B. T. Tozer, K. A. Cammack, and H. Smith, "Dissociation of serological complexes of ovalbumin and hemoglobin using aqueous carbon dioxide," Nature 182(4636):668 (1958).

147a. O. Tsuzuku, Y. Yagi, and D. Pressman, "Highly specific lung localizing antibodies," J. Immunol. 99(1):1 (1967).

148. J. P. Vaerman, J. F. Heremans, and C. Vaerman, "Studies of the immunoglobulins of human serum. I. A method for the simultaneous isolation of the three immune globulins from individual small serum samples," J. Immunol. 91(1):7 (1963).

149. A. J. Van Triet, "Enzymatic purification of antidiphtheria and antitetanus sera obtained from sheep," Brit. J. Exp. Pathol. 40(6):559 (1959).

150. W. E. Vannier, W. P. Bryan, and D. H. Campbell, "The preparation and properties of a hapten–cellulose antibody adsorbent," Immunochemistry 2(1):1 (1965).

151. T. Webb and C. Lapresle, "Study of the adsorption from polystyrene-human serum albumin conjugates of rabbit anti-human serum albumin antibodies having different specificites," J. Exp. Med. 114(1):43 (1961).

152. R. R. Williams and S. S. Stone, "Application of the antigen–antibody reactions on supporting media to the purification of proteins. I," Arch. Biochem. Biophys. 71(2):377 (1957).

153. H. H. Weetall and N. Weliky, "Immunoadsorbent for isolation of purine-specific antibodies," Science 148(3674):1235 (1965).

154. H.H.Weetall and N. Weliky, "New cellulose derivatives for the isolation of biologically active molecules," Nature 204(4961):896 (1964).

154a. H. H. Weetall and N. Weliky, "The coupling of biologically active substances to the insoluble polymers: antibody on cellulose," Biochim. Biophys. Acta 107(1):150 (1965).

155. N. Weliky, H. H. Weetall, R. V. Gilden, and D. H. Campbell, "The synthesis and use of some insoluble immunologically specific adsorbents," Immunochemistry 1(3):219 (1964).

156. N. Weliky and H. H. Weetall, "The chemistry and use of cellulose derivatives for the study of biological systems," Immunochemistry 2(4):293 (1965).

157. L. Wide and J. Porath, "Radioimmunoassay of proteins with the use of Sephadex coupled antibodies," Biochim. Biophys. Acta 130(1):257 (1966).

163. M. [illegible] and H. [illegible], [illegible] "[illegible] the coupling of [illegible] antibodies [illegible]," J. [illegible] Chem. [illegible] (1966).

164. [illegible], and [illegible], "The separation and [illegible] of labelled antibodies," [illegible] (1966).

165. [illegible] and [illegible], "[illegible] globulin [illegible] specific [illegible]," [illegible] (1967).

166. [illegible], A. [illegible], and [illegible], "[illegible] complexes of ovalbumin and [illegible] using [illegible]," Nature [illegible] (1969).

167. [illegible] and D. [illegible], "[illegible]," J. Immunol. [illegible] (1971).

168. [illegible], [illegible], and [illegible], "[illegible] of the [illegible]: A method for the simultaneous isolation of the three [illegible] from [illegible]," J. Immunol. [illegible] (1968).

169. [illegible], "[illegible] purification of [illegible] and antibodies [illegible] obtained from [illegible]," [illegible] (1968).

170. W. [illegible], and D. [illegible], "[illegible] the preparation and properties of a [illegible] cellulose [illegible]," [illegible] (1968).

171. L. Wofsy and [illegible], "Study of the [illegible] of [illegible] serum albumin conjugates [illegible] antibodies having different specificities," J. [illegible] (1967).

172. [illegible] and [illegible], "[illegible] of antibodies [illegible] for the purification of [illegible]," J. Immunol. [illegible] (1971).

173. [illegible] and [illegible], "[illegible] adsorbent for isolation of [illegible] specific antibodies," Science [illegible] (1962).

174. H. H. Weetall and [illegible], "New [illegible] derivatives for the isolation of biologically active molecules," [illegible] (1968).

175. H. H. Weetall and [illegible], "The coupling of biologically active substances to the [illegible] with [illegible]," [illegible] (1968).

176. [illegible] and [illegible], "The [illegible] and use of some [illegible] with specific [illegible]," Immunochemistry [illegible].

177. [illegible] and H. H. Weetall, "The chemistry [illegible] derivatives for the study of biological systems," [illegible].

178. [illegible], "[illegible] of [illegible] and the use of [illegible] to [illegible]," [illegible] (1970).

Chapter III

Properties of Antibodies

Antibodies are distinguished from other biologically active proteins, for example, enzymes, by their pronounced heterogeneity [75, 226]. The same antibody binding site, i.e., that part of the molecule which is responsible for specific conjugation with an antigen, can be found in molecules with distinctly different properties. This can primarily be attributed to the fact that the same antibody to a given antigen can belong to a different type of immunoglobulin in different animal species [67, 78, 120, 196]. Before investigating the properties of antibodies, it is necessary to examine the various types of immunoglobulins currently known to be present in animal serums, and to elucidate their mutual differences.

THE CHIEF TYPES OF IMMUNOGLOBULINS

According to a definition of a special committee of the World Health Organization [5, 169], the concept immunoglobulin comprises proteins of animal origin which have a definite antibody activity, as well as proteins similar to them in chemical structure, and consequently in antigenic specificity. In humans, in particular, this group includes proteins exhibiting no antibody activity, for example, myeloma proteins, Bence-Jones proteins, and immunoglobulin subunits found in the organism.

Immunoglobulins are found in other fluids and tissues as well as in plasma, for example, in the urine, spinal fluid, lymph nodes, and spleen.

In 1964 this same committee worked out a classification for human immunoglobulins [5, 169], which at present includes three chief types: γG, γA, and γM immunoglobulins, but other types

TABLE 7. Content of Different Types of Immunoglobulins in Human Serum (in mg%)

Immuno-globulin	Mean [56]	Range of variations [56]	Mean [81]	Standard (mean square) deviation [81]
γ G	1194	760--2000	1210	120
γ A	395	118--1065	280	70
γ M	119	64--380	120	35
Type K	--	--	800	200
Type L	--	--	450	120

TABLE 8. Immunoglobulin Levels in Various Diseases (in mg%) [56]†

Disease	γG	γA	γM
Hypogammaglobulinemia in men, congenital, sex-linked	130	<40	<15
γA-type myeloma	260	> 3800	10
γG-type myeloma	900	57	61
Lymphoma:			
Hodgkin's disease	1100	417	77
Solitary myeloma and chronic lymphatic leukemia	800	269	50
Acute leukemia	2000	1802	118
Chronic lymphatic leukemia	620	125	51
Lymphosarcoma	600	48	42
Macroglobulinemia	900	59	800
Systemic lupus erythematosus	1300	507	29
Laennec's cirrhosis	2000	< 04	590
Posthepatic cirrhosis	1600	857	400
Cirrhosis	620	2102	160
Chronic pneumonitis	1800	853	48
Vasculitis	1100	909	102
Fever, eosinophilia	1400	507	272

*Data from a study of the serums of individual patients are given.

also exist. The contents of the different types of immunoglobulins in serum differ (Table 7) and can vary considerably during illness (Table 8).

The different types of immunoglobulins can be partially separated by fractionation on DEAE cellulose. This method has been

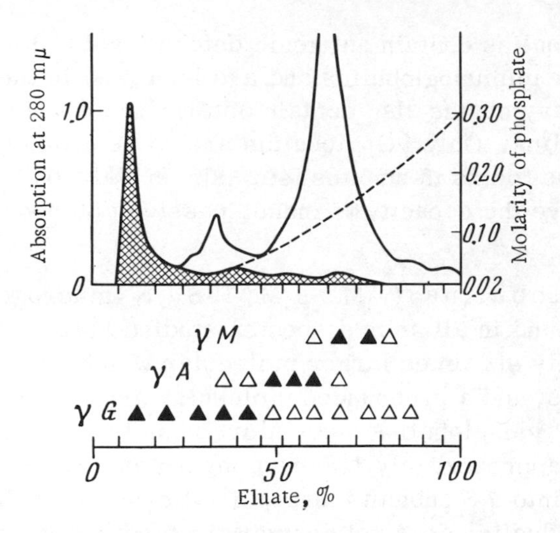

Fig. 8. Fractionation of serum proteins on DEAE cellulose. The shaded area is the distribution of γ-globulins in zonal electrophoresis. The distribution of the different types of immunoglobulins in the fractions was determined by gel-diffusion with specific antiserums. The solid triangles are fractions with the highest content of the given immunoglobulin; the plain triangles are fractions with a low content of the same protein [75].

developed in detail by Fahey et al. [75, 79, 99]. In Fig. 8 it can be seen that with the given method of elution, only γG-globulins emerge initially, followed by γA- and γM-globulins with a small admixture of γG-globulins. Macroglobulins can also be isolated by density-gradient centrifugation and by gel filtration through Sephadex G-200 (cf. Chapter II).

γG-Globulins (γ, 7Sγ, 6.6Sγ, γ_2, γ_{SS}) comprise the bulk of serum immunoglobulins [186]. This immunoglobulin has been the most thoroughly studied. It has a molecular weight of about 150,000, and contains about 2.5% carbohydrates. In electrophoresis at alkaline pH it is the slowest of all serum proteins. The molecule is built up of two light polypeptide chains and two heavy chains with molecular weights of 20,000 and 50,000, respectively. Both the heavy and light chains possibly consist of two different smaller chains.

γG-Globulins contain antigenic determinants which are common to other immunoglobulins and are localized in the light chains, but their heavy chains also contain determinants which are peculiar to γG-globulins. Only γG-globulins are transmitted from the mother to the foetus in humans [46, 58]. In addition, only γG-globulins have the capacity to induce passive cutaneous anaphylaxis [176].

γM-Globulins (γ_1M, β_2M, 19 Sγ, γ-macroglobulins) have been found in all animal species studied [142]. They have an approximately six times larger molecular weight (900,000) than γG-globulins, and a pronounced molecular asymmetry. In electrophoresis, γM-globulins move more rapidly than γG-globulins, and contain approximately 10% carbohydrates. When reduced, they break down into 7 S subunits and in most cases lose their activity. However, reduction does not destroy the binding site since the subunits are able to reassociate with restoration of activity [128]. Sometimes the subunits themselves exhibit a capacity to react with antigen (cf. Chapter VI). In more thorough reduction the molecule breaks down into a heavy and light chain, the latter containing the common antigenic determinants and the heavy chains containing those which are peculiar to γM-globulins.

γA-Globulins (β_2A; γ_1A) have been identified in humans [118] and in many species of animals, for example, rabbits [173, 173a], rats [34, 42], mice [33, 82], and birds [67, 102]. Their molecular weight is somewhat higher than that of γG-globulins (about 170,000), and in some cases is between the weights of γG- and γM-globulins. Proteins of this type are polymers of 7 S subunits [231a]. γA-Globulins have been found in various secretions (e.g., sputum, colostrum). In rabbit colostrum, secretory γA-globulin has a molecular weight of 370,000. This molecule probably consists of four pairs of heavy and light chains, and in addition contains a T-component, with a molecular weight of about 50,000, which is made up of one or two peptide chains synthesized in gland cells [52c]. The carbohydrate content in γA-globulin is somewhat less than 10%.

It has now been demonstrated that γA-globulins can contain antibody [34, 82, 120, 127, 144, 196]. Certain antibodies found in the serum of allergic patients, in particular, skin-sensitizing antibodies [88, 119, 197, 203] and blood group antibodies [189, 192], probably belong to this class of immunoglobulin.

Other Serum Immunoglobulins. The aforementioned types of immunoglobulins do not exhaust all the proteins of this class which can be found in animal serum. Recently, several proteins have been discovered, the properties of which preclude their classification into one of these three main groups of immunoglobulins.

Thus, protein with a sedimentation constant of 7 S, but with the antigen properties of γM-globulins, has been found in horse serum [200]. Karush et al. [196] discovered a protein with a sedimentation constant of 10 S in preparations of pure horse antibody to β-lactoside hapten. This antibody had antigenic properties distinct from that of γG-, γA-, and γM-antibodies.

The properties of still another horse immunoglobulin (T) have been studied in detail by Porter [236a]. This protein has several properties in common with γG-globulin, but differs from the latter in its higher carbohydrate content, the unique antigen properties of its heavy chains, and the unusual way in which it is cleaved by papain.

An immunoglobulin with peculiar antigen properties (γD) was identified not long ago in human serum with the aid of antibody to myeloma protein isolated from a patient. On the average, normal human serum contains about 0.03 mg/ml of this immunoglobulin. In addition to its individual antigenic determinants it also possesses the K- and L-types [199]. Ishizaka et al. [126a] described still another human immunoglobulin – γE, which contains characteristic heavy chains. This immunoglobulin contains antibody of the reagin type.

Microglobulins in Plasma and Urine. In normal humans and animals, proteins which have antigen properties akin to those of immunoglobulins (Fab and Fc fragments) and a molecular weight lower than that of γG-globulins can be found in plasma and urine. Various investigators have given quite dissimilar descriptions of the properties of purified proteins of this type, although the disparity is quite possibly due to the different methods of isolation. Thus, Berggård and Edelman [41] maintain that human plasma microglobulins consist of light immunoglobulin chains, and indeed, the low molecular weight proteins which they investigated had antigen properties common to light chains, the same electrophoretic mobility in a starch gel, and a similar molecular weight

(25,000). Moreover, their solubility when heated was similar to that of light chains and Bence-Jones protein (low molecular weight proteins found in the urine of myeloma patients [115]). These microglobulins were precipitated on heating to 48-60° and dissolved with further rise in temperature.

According to Schmid et al. [87, 126, 127a, 220], human serum contains low-molecular weight proteins with the electrophoretic mobility of γ-globulins, but with other chemical properties than the light chains of immunoglobulins. They form two groups according to molecular weight: 3 S γ_1-globulins (molecular weight 25,000) and 2 S γ_2-globulins (molecular weight 14,000). Their relationship to immunoglobulins is still uncertain.

A low-molecular-weight globulin is found in the plasma of new-born pigs. Franek, Riha, and Sterzl [91, 92] demonstrated that the principal component of pig globulins was only partially identical in its antigen properties with γ-globulins of adult pigs, and did not have antibody activity. The sedimentation constant of this heterogeneous protein was 5.1 S. Its component was relatively low at 40-80 μg/ml.

Investigations with electrophoresis and gel filtration have shown that low molecular weight globulins of urine are heterogeneous [62, 107, 218]. In some cases a fraction with a molecular weight of 10,000-13,000 [11, 12, 107, 236] was isolated. Urine microglobulins are similar to Bence-Jones protein in solubility and antigen properties [94a, 218, 221], and may also belong to either the K or L type. The urine of some patients may contain two types of microglobulins at the same time: namely, unpolymerized light chains with a molecular weight of about 17,000, and Bence-Jones protein which contain these chains, plus an additional antigen determinant, and which have a molecular weight which is twice as high [65a]. It has recently been shown that urine microglobulins also contain a very small quantity of protein which in its antigen properties is akin to the heavy chains of γG-globulins [107], in particular, to their Fc fragments [40a].

A very interesting discovery, made by several investigators, is the capacity of low molecular weight urine globulins to react with antigen. Evidence of this capacity was provided by inhibition of gel-diffusion [12], the virus-neutralizing activity [107], and even the capacity to precipitate with antigen in a gel [193]. It is difficult to explain this property, and all the more in that it sometimes places the very antibody activity of microglobulins in doubt [198]. It would be a very important step to establish whether these proteins are the cleavage products of serum globulins, or whether they are independently synthesized units. The only partial antigenic identity with Fab fragments is to a certain extent evidence against the proteolytic origin of microglobulins. Early experiments [94, 235], in which elimination of a radioactive label in the urine, after injection of ^{131}I-labeled γ-globulin into the blood, was studied, gave indications that microglobulins could be products of serum protein metabolism. However, a recent repetition of the same experiments showed that the radioactive label content in the microglobulin fraction was negligible whereas in the γG-globulin fraction it was very high [218].

SIZE AND SHAPE OF MOLECULES

The size and shape of molecules of antibodies and other immunoglobulins has been determined by various methods. A simple inspection of the f/f_0 coefficients, which characterize molecular asymmetry, will show that γG-globulins, and especially γM-globulins, are asymmetric in aqueous solutions (Table 9). Oncley et al. [172] determined dimensions of 235 × 44 Å for γG-globulins by sedimentation, viscosimetric, and diffusion studies. Values obtained by double light diffractions of flow were quite similar: the length of the human γG-globulin molecule was 230 Å [68].

Boyd [44] determined the size of molecules of several antibodies from physicochemical characteristics taken from the literature. The dimensions of γM antibodies were 930-981 Å × 47 Å for unhydrated, and 882-928 × 55 Å for hydrated molecules (0.2 g water/g protein). However, for γG-antibodies the length varied considerably (from 169 to 363 Å for the unhydrated molecule). Results of a study of γG-globulins made by scattering X-rays at a small angle are very interesting. According to the data of Austrian investigators [140], the length of the γG-globulin molecule is 237-240 Å, while the two transverse axes are 19 and 57 Å. The calculated molecular weight of the protein was 159,000 and in shape the molecule resembled an elliptical cylinder. The molecule was also found to contain a dense nucleus which com-

TABLE 9. Sedimentation and Diffusion Properties of Pure Antibodies and γ-Globulins

Protein	Sedimentation constant, s_{20}	Diffusion constant D_{20}, $cm^2/sec \cdot 10^{-7}$	Molecular weight	f/f_0	Reference
Rabbit					
γG-globulin	6.5	–	187,600 ± 1400	–	54
"	–	–	145,000 ± 5000	1.47	167
"	6.79	4.66	137,000	1.34	49
"	–	–	170,000 ± 2000	–	208
γM-globulin	–	–	850,000-900,000	–	144a
Antibodies to:					
polysaccharides	–	–	–	–	–
pneumococci	7.0	7.23	157,000	1.4	131
azophenylarsanilate	–	–	158,000-192,000 (light dispersion)	–	74
azophenylarsanilate	–	–	158,000 (light dispersion)	–	50
azobenzoate	6.9 and 21 (less than 2%)	–	–	–	163
horse serum albumin	6.1	–	–	–	4
ovalbumin	6.56	–	–	–	227
polytyrosyl gelatin	6.5	–	–	–	96
lysozyme	6.56	–	–	–	95
γA-globulin	7.2	–	173,000	–	52b
γA-globulin from colostrum	–	–	370,000	–	52b
Human					
γG-globulin	7.1	3.84	177,000	1.5	131
"	6.53	–	–	–	39
"	6.3	–	–	–	238
γM-globulin	17.9	1.75	890,000	1.92	158a
Antibodies to pneumococcal polysaccharides	7.4	3.6	195,000	1.5	131

TABLE 9 (Continued)

Protein	Sedimentation constant, s_{20}	Diffusion constant D_{20}, $cm^2/sec \cdot 10^{-7}$	Molecular weight	f/f_0	Reference
Human (Cont)					
Cold hemagglutinins	18.0	–	–	–	101
γA-globulin from serum and ascites fluid	6.9	–	–	–	226a
γA-globulin from sputum and colostrum	11.4	–	–	–	226a
Horse					
γG-globulins	7.75(25°)	4.8	151,000	–	177
Antibodies:					
antitetanus (γ, G, and T)	7 and 10 (10-20%)	–	180,000	–	210
to β-lactoside	6.5-6.8 and 9.4-9.5	–	–	–	196
to pneumococcal polysaccharides	19.3	1.8	920,000	2.0	131
diphtheria antitoxin	6.8	–	–	–	17
Cattle					
γG-globulin	7 and ~10 (5-15%)	–	–	–	55
Antibodies to pneumococcal polysaccharides	18.1	1.69	910,000	2.0	131
Pig					
Antibodies to pneumococcal polysaccharides	18.0	1.64	930,000	2.0	131
γM-globulins	19.2	–	870,000	–	90a

TABLE 9 (Continued)

Protein	Sedimentation constant, s_{20}	Diffusion constant D_{20}, $cm^2/sec \cdot 10^{-7}$	Molecular weight	f/f_0	Reference
Monkey					
Antibodies to pneumococcal polysaccharides	6.7	4.08	157,000	1.5	131
Mouse					
γG-globulin	6.7-6.9	–	–	–	82
γA-globulin	7 and 9; 11 and 13	–	–	–	82
γM-globulin	18	–	–	–	82
Guinea pig					
Antibodies to the dinitrophenyl group	6.3-6.5	–	–	–	40

prised 76% of the entire mass, and a less dense shell. The dimensions of the Bence-Jones protein molecule were found to be 21 × 48.3 × 74.8 Å [121].

Electron micrographs of pure antibodies gave quite similar values: 250 × 30-40 Å [106]. Interesting data were obtained from an electron microscope study of antigen–antibody aggregates [86]. In particular, in a study of complexes of polyoma and wart viruses with specific rabbit antigens it was found that the antibody molecules had the shape of cylindrical rods with a long axis of 250-270 Å and a short axis of 35-40 Å [31]. However, electron micrographs of other investigators gave a somewhat different picture [85]. By using a shadow technique, particles with a width of 80-120 Å and a thickness of 34 Å were visible on electron micrographs of γG-globulin, whereas with a negative contrast method slightly asymmetrical particles with a maximum linear dimension of approximately 105 Å were noted. Pictures of antigen–antibody complexes at large antigen excess (ferritin was used as the antigen) showed antibodies with almost no change of shape. However, at equivalent antigen–antibody ratios, the complexes had a different appearance: Antigen molecules were linked by strands with a diameter of 15 Å and a length of more than 200 Å. These investigators conjectured that antibody molecules are able to bend and unbend in the middle where they are most pliant (as if on hinges). In a report published in 1965, Almeida et al. [32] also notes that antibody molecules sometimes appear V-shaped. This model resembles somewhat that of Tanford et al. [167], who, on the basis of physicochemical investigations, hypothesized that the subunits of antibody molecules have a rather high degree of freedom relative to each other. Results of observations made of complexes of polyoma virus with antibodies in an antibody excess indicated a definite flexibility of antibody molecules. In these complexes, the antibody molecules linked to virus particles were kinked and bent [32].

Valentine and Green [231b] obtained clear results in a study of complexes of rabbit antibody with divalent hapten. It was found that in a complex with hapten, antibody has the shape of the letter Y, in which the lower branch corresponds to the papain-cleaved fragment Fc, since it disappears as a result of treatment with pepsin. The Fab fragment (two upper branches) was 60 Å long and 35 Å wide, while fragment Fc had a length and width of 45 Å and 40 Å, respectively.

In spite of the pronounced heterogeneity of many antibody properties, the dimensions of γG-globulin and antibody molecules do not vary markedly. Thus, Edsall and Foster [68] found that human γG-globulin behaved as if all molecules were of the same length, at least in dilute solutions. In a special study, Pain [178] showed that horse γG-globulin in gel filtration through Sephadex – G-200 gave a very symmetrical peak, and that the molecular weight of protein from different parts of the main peak was the same at 150,000.

MOLECULAR WEIGHT

The molecular weight of an antibody depends on the class of immunoglobulins to which it belongs [113]. Table 9 summarizes values obtained in a determination of the molecular weight of antibodies and immunoglobulins. These determinations were carried

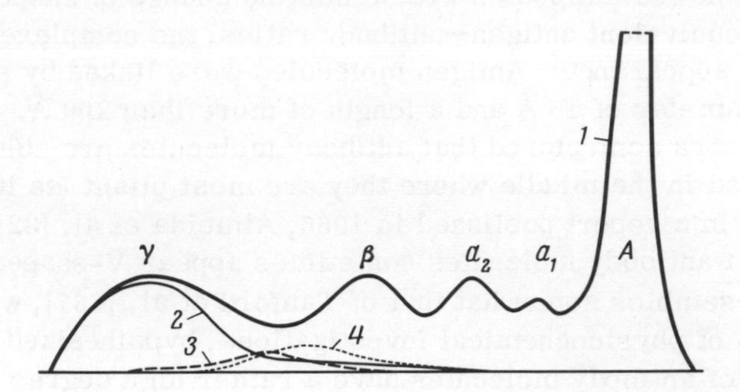

Fig. 9. Distribution of the principal types of immunoglobulins in zonal electrophoresis. Content: 1) protein; 2) γG-globulins; 3) γM-globulins; 4) γA-globulins [75].

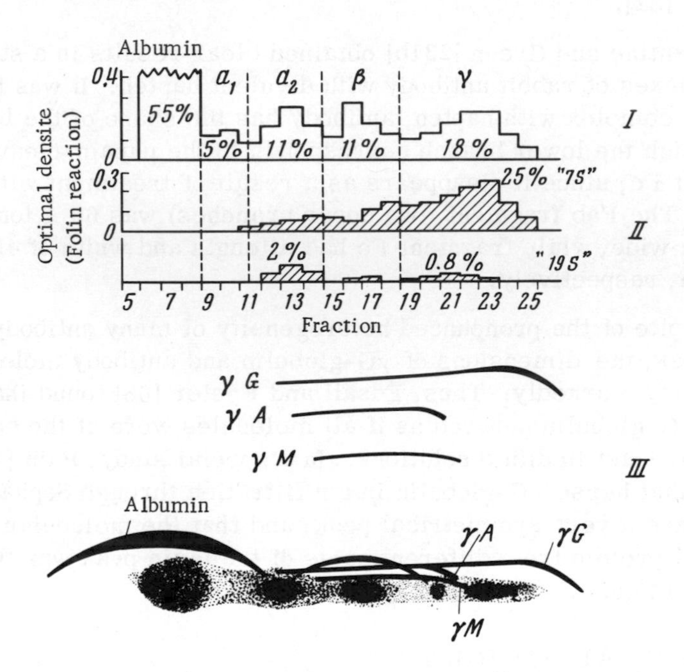

Fig. 10. Sedimentation (II), electrophoretic (I), and immunoelectrophoretic (III) studies of serum immunoglobulins [99, 232a]. The contents of the various fractions are expressed in %.

out mainly by a sedimentation method. The majority of the preparations had a sedimentation constant either somewhat less than 7 S (γG), or about 19 S (γM). In some instances, an additional peak with a sedimentation constant of 10 S was observed; this corresponded to a molecule twice as large as the γG-globulin molecule. In some cases, this peak may have been due to aggregation of γG-globulins. However, Karush [196] discovered antibodies with a sedimentation constant around 9.5 S among pure horse antibodies specific for β-lactoside. These antibodies had their own characteristic antigenic properties. It appears that antibodies with a molecular weight around 400,000 exist, at least in horses.

A certain quantity of molecules with these same sedimentation properties are sometimes found in preparations of γA-immunoglobulins. Some investigators have discovered antibodies with sedimentation constants between 7 S and 19 S [223]. For example, after density-gradient ultracentrifugation of human serums or their individual fractions, certain isohemagglutinins with sedimentation constants of 9-15 S were found in the protein fractions [197]. In the same way it was demonstrated that skin-sensitizing antibodies to highly purified glucagon had a sedimentation constant around 8-11 S. The same was true of skin-sensitizing antibodies of rats [42]. This type of antibody is possibly a γA-immunoglobulin.

The molecular weight of γG-globulins (cf. Table 12) varies from 140,000 to 190,000 according to different determinations. The variation is dependent on the animal species, and is presumably (see Porter [49, 186]) attributable to experimental error. In any event, some recent experiments yielded data which tended toward lower values: 151,000 [177], 137,000 [49], amd 145,000 [152]. It is possible that the presence of a small amount of aggregates in γG-globulin preparations will raise the values for the molecular weights of these proteins [178].

ELECTROPHORETIC PROPERTIES

Since Tiselius and Kabat [225] made the first electrophoretic investigation of antibody properties, considerable research has been done in this area. Practically all the available electrophoretic methods have been employed; namely, in a Tiselius apparatus, on paper [222], in starch block [159, 217], in starch gel [214], in agar

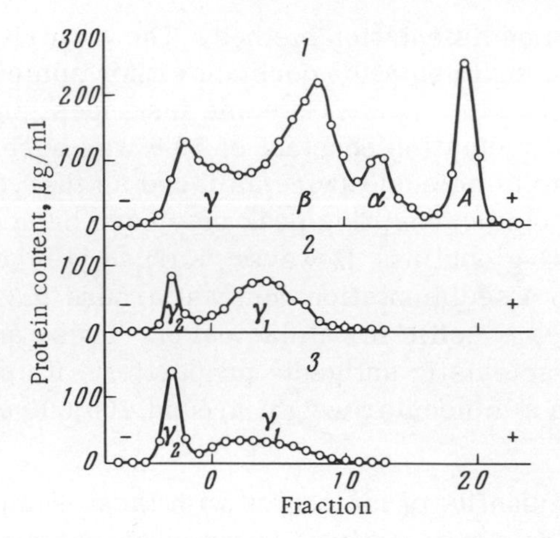

Fig. 11. Electrophoresis on starch of antitoxins isolated from the plasma of hyperimmune horses by means of immunoadsorbents, and of horse serum proteins [28]. 1) Horse serum proteins; 2) diphtheria antitoxin; 3) tetanus antitoxin; A) albumins; α, β, γ, γ_1, and γ_2, corresponding globulin fractions.

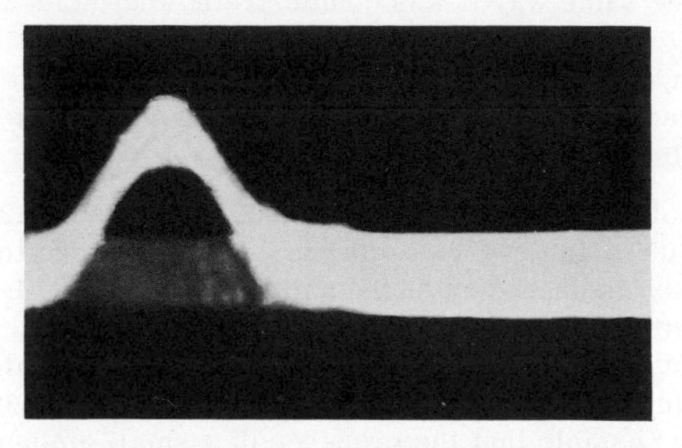

Fig. 12. Electrophoresis in a Tiselius apparatus of antibody to horse serum albumin isolated from rabbits serum by means of an immunoadsorbent [4].

gel; and also combinations with immunological methods, such as Grabar's immunoelectrophoretic method and electrophoresis plus precipitation [3]. The principal results of such studies are given in Figs. 9 and 10.

As is the case with other properties, the electrophoretic mobility of a given antibody is primarily determined by the class of immunoglobulins to which it belongs. γG-antibodies are slower at alkaline pH values than γA- and γM-antibodies. The peak of macroglobulin antibodies lags somewhat behind the peak of γG-antibodies. These data were obtained primarily from the study of human proteins, although other mammals exhibit a similar picture. When horses are immunized with toxins and certain other antigens, so-called T(γ_1) globulins appear. These have a mobility intermediate between γ- and β-globulins (Fig. 11).

The electrophoretic properties of γG-antibodies have been the most thoroughly studied. In an electrophoretic study in a Tiselius apparatus and in studies by other methods, pure preparations of these antibodies give flat peaks with a wide base or wide bands in zone electrophoresis, indicative of their considerable microheterogeneity (Fig. 12). It was found [77] that the electrophoretic mobility of the entire γG-globulin molecule is determined by the mobility of the Fab fragment, which is obtainable by papain hydrolysis, and in particular, by the heavy chain section of this fragment (Fd fragment). Other data indicate that fast and slow γG-globulins, as well as various chromatographic fractions, have different amino acid compositions [84, 151]. Thus, the heavy chain section in which the specific antibody determinant is most likely localized (cf. Chapter IV) shows considerable variations in its properties even for the same antibody.

It has recently been shown that immunoglobulins and antibodies of the γG type can be further fractionated, especially by electrophoresis in agar. This fractionation was especially distinct in experiments with guinea pig proteins [171]: two additional fractions, a fast and a slow, designated as γ_1 and γ_2, were visible on electrophoregrams. Similar fractions were found in mice [83] and humans [240, 241]. Horse antibodies were even found to contain three components [135]. It is worthy of note that these two additional fractions in mice had different biological properties [170].

Different antibodies from the same animal species may have somewhat dissimilar electrophoretic properties. Thus, the partial fractionation of two different antibodies by electrophoresis convection has been reported [51]. An electrophoretic study under

TABLE 10. Electrophoretic Properties of Immunoglobulins and Pure Antibodies

Substance	Buffer pH	Mobility in $cm^2/sec\text{-}V \cdot 10^5$	Reference
Rabbit			
γG-globulin	8.6	1.5-1.68	186
γG-globulin	8.2	1.4	4
Antibodies to:			
ovalbumin	8.2	1.64	4
horse serum albumin	8.2	1.05-1.22	4
ovalbumin	8.6	1.2	227
pneumococcal polysaccharides	7.72	0.95	225
Horse			
T-globulin	8.5-8.6	1.8-2.6	212
Antibodies to pneumococcal polysaccharides	7.72	1.7	225
Cattle			
Antibodies to pneumococcal polysaccharides	7.72	1.31	225
Pig			
Antibodies to pneumococcal polysaccharides	7.72	1.30	225
Human			
Cold human hemagglutinins	7.4	0.83	101

identical conditions of antibodies differing in specificity disclosed distinct differences in mobility between different antibodies, as well as between antibodies, and γ-globulins. As is evident from Table 10, pure rabbit antibody to horse albumin was somewhat slower than antibody to ovalbumin and nonspecific rabbit γG-globulins [4].

Differences in electrophoretic mobility for antibodies of different specificities are clearly illustrated in Fig. 13 [171], in which

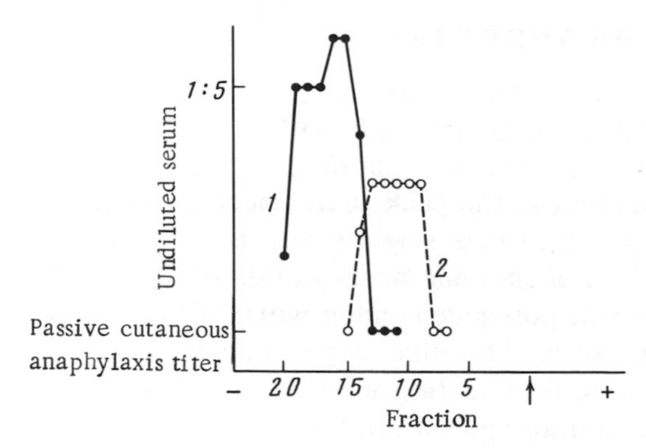

Fig. 13. Localization according to passive cuatneous analphylaxis of anti-DNP and anti-arsanilic antibody in fractions obtained by electrophoresis in agar of serum of a guinea pig immunized simultaneously with bovine DNP-γ-globulin and guinea pig arsanilic acid azo albumin. Antibodies to: 1) DNP groups; 2) arsanilic groups [171]. The arrow indicates the input of the serum. Each fraction is an eluate from a 2-mm gel strip.

results are given of experimental determinations of the passive cutaneous anaphylaxis activities of two antibodies in fractions obtained by agar-electrophoresis of the serum of a guinea pig immunized with two different haptens. As is seen, the anti-DNP antibody moved further from the input point than the antibody to arsanilic acid. The evidence of these experiments, which demonstrated the differing electrophoretic mobility of antibodies, can be explained either by differences in the folding of antibody chains, or by differences in the amino composition of the antibodies. The aforementioned data showing differences in amino acid composition of different γG-globulin fractions favor the latter possibility.

Denaturation of γ-globulins results in an increase in their electrophoretic mobility. Similar changes were brought about by heat treatment and by treatment with different chemical agents [18, 24, 27]. Evidence has been published indicating the presence of antibodies in the α-globulin fraction [219]. Data from isoelectric point determinations are summarized by Alberty [30, 186]. Titration of hydrogen ions by Weltman and Sela [237] showed that these ions are more or less evenly distributed in the γ-globulin molecule.

OPTICAL PROPERTIES

Ultraviolet Absorption. Like other proteins, γ-globulins and antibodies have a broad ultraviolet absorption maximum near 280 mμ. For human, bovine, and horse γ-globulin and horse T-globulin, the peak is at approximately 278 mμ, while for bovine T-globulin it is slightly beyond 280 mμ [211]. The absorption maxima of immune precipitates of antibodies to albumins and pneumococcal polysaccharides were in the vicinity of 290 mμ [69]. Troitskii et al. [26] discovered in a study of beef γ-globulin that absorption of this protein increased somewhat, and that the absorption maximum shifted slightly toward shorter wavelengths after treatment with 8*M* urea. At 280 mμ the absorption of 1% solutions of rabbit and human γG-globulins was $E^{1 cm} = 13.5$ and 13.6 [63], respectively, and rabbit γG-globulins had $E^{1 cm} = 13.8$ [167] at 278 mμ. For rabbit γA-globulin absorption was $E^{1 cm} = 13.5$ [52b] at 280 mμ, while one human pathological macroglobulin had $E^{1 cm} = 11.85$ [158a] at the same wavelength.

Optical Rotatory Dispersion. Data obtained in a study of optical rotatory dispersion, i.e., the wavelength dependence of specific rotation, are of considerable interest since they can give some idea of the configuration or secondary structure of protein molecules [8, 24, 25, 216a, 231]. γ-Globulin preparations from different animal species were studied by this spectropolarimetric method.

The principal result of these studies was to establish the dissimilarities between the properties of γ-globulins and of these proteins which are definitely known to have helical structures. Thus, the dispersion constant λ_c in the equation expressing the wavelength dependence of specific rotation (Drude equation) has a value of 240-280 mμ for helical proteins, while for γ-globulins it is considerably lower (205-217 mμ). This constant is dependent on the configuration of the protein in solution. Usually, the Moffit equation in which one coefficient (b_o) is proportional to the content of α-helix in the protein [2] is used to evaluate spectropolarimetric data. It has been found that for proteins with a definite helical configuration (for example, myosin, hemoglobin), this coefficient has high values (up to $-630°$ at 100% helicity) while for γ-globulins it is equal to zero. Other data indicate that immunoglobulins differ from proteins and from polypeptides lacking an ordered conformation [130].

It should be mentioned that spectropolarimetric data are far from unambiguous, and their interpretation is very difficult. Some investigators, in particular Doty et al. [100] and Jirgensons [129], believe that their spectropolarimetric characteristics indicate the absence of α-helical sections in γ-globulins. The latest experiments of Jirgensons, in which he studied optical rotatory dispersion in the far ultraviolet region, also provided no definite evidence of helixes [130]. The absence of the α-helix can be explained by the high proline content in γ-globulin, for this amino acid is known to hinder formation of helical structure. Valine and disulfide bonds are also not particularly favorable for the formation of these structures. Some data might be explained by assuming the simultaneous presence of equal quantities of right and left helices. However, the evidence gathered in a study of the behavior of γ-globulin during denaturation seems to be against this assumption [100, 129].

Another group of investigators consider very plausible the explanation that right-handed α-helix exists together with a compensating quantity of β-structures in both γG- and γM-globulins [24, 47, 48], and it is pointed out that theoretically there are no obstacles for formation of the α-helix in γ-globulin in as much as it is found in 2-chlorethanol [47, 239]. This point of view is adhered to by Troitskii, Okulov, and Kiryukhin [26] who used an expanded Moffit-Yang equation for the analysis of spectropolarimetric data. These investigators are of the opinion that β-structures are nearer to the center of the molecule, and that α-helical sections make up its periphery. The space in between is occupied by a disordered structure, which explains the special sensitivity to proteases.

Judging from recent data [1], the infrared absorption bands (amide I, amide A, and amide B) of rabbit γG-globulin in solution differ from the infrared bands of horse hemoglobin which contains 75% α-helix, and are very similar to the band of ribonuclease and chymotrypsin, in which were found a high content of β-structures by X-ray diffraction analysis. In contrast, the infrared spectra of γG-globulin differ slightly from ribonuclease spectra; this is apparently attributable to the peculiar type of β-structure in immunoglobulin [130].

In a study of the rate of hydrogen–deuterium exchange it was found that rabbit γG-globulin has a high percentage of ordered conformation [1, 100]. The combined data on H–D exchange and infrared spectroscopy give evidence for the existence of both β-structures and a fixed random coil in γG-globulin. Data on small-angle scattering of X-rays also indicate the existence of two regions with different electron densities [140]. A study of H–D exchange and infrared spectra [1] also disclosed that isolated light chains of rabbit γG-globulin are ordered in solution, and contain both the above-mentioned structures characteristic for whole molecules of these proteins.

Here it should be noted that hydrophobic interactions make a considerable contribution to the maintenance of the native structure of immunoglobulins [18a].

Differences between antidinitrophenyl rabbit antibody and nonspecific immunoglobulins were discovered by an ultraviolet rotatory dispersion method. These differences were due to the different properties of Fab fragments [216a]. This is essentially the only case where differences in properties between a given antibody and other immunoglobulins of the same type have been detected by physicochemical methods.

Fluorescence Polarization. Studies on fluorescence polarization were carried out on γG-globulins to which a fluorescent dye had been coupled [237a, 242]. According to calculations based on various physicochemical characteristics (sedimentation, viscosity, diffusion) the rotational relaxation time is about 220 nsec, assuming that the protein molecule is a compact sphere. However, experimental values were 60 nsec [242]. This indicates that even in the native state the molecule has a considerable internal flexibility, and that some of its subunits are capable of rotation relative to each other.

This latter assumption is supported by a comparison of the properties of the whole γG-globulin molecule with the properties of its papain-cleaved fragments. These fragments proved to have some of the characteristics of typical globular proteins, whereas the native molecule is much less compact [166, 167]. With this in mind, Tanford [167] worked out a model (Fig. 31) for the structure of γG-globulin which in his opinion was not less probable than a model specifying a rigid structure (cf. Chapter IV). It can be seen that the individual subunits are conjugated in such a way as to give them a quite considerable mutual degree of freedom. As noted above, electron-microscope observations showed a similar picture [231b].

The independence of the rotation of subunits of IgG should not affect the behavior of the molecule as such in the solution. In particular, if the linear arrangement of subunits in the IgG molecule corresponds to the minimum free energy, this molecule will behave in solution as a rod or elongated elipsoid, as substantiated by hydrodynamic and other investigations. On the other hand, the flexibility of bonds between subunits makes it possible for the IgG molecule to change its form under the action of comparatively weak forces, such as those due to the interaction between antibody and antigen.

CHEMICAL PROPERTIES

The principal types of immunoglobulins resemble glycoproteins in composition.

Amino Acid Composition. The amino acid composition of immunoglobulins and antibodies has been determined many times by different methods, for example, by paper chromatography and with ion-exchange resins in automatic amino acid analyzers [21, 58, 89, 116, 139, 213]. Table 11 presents a summary of data from an analysis of normal immunoglobulins by Porter [53, 58, 63]. As is stressed by Cohen and Porter [58], as compared with other proteins immunoglobulins have a relatively high content of hydroxyamino acids and dicarboxylic amino acids; in particular, the proline content is higher than in other globular proteins. Hydroxylysine has not infrequently been found in certain antibodies (Table 13).

Whether differences exist between two antibodies with different specificities is a question of fundamental importance. Smith et al. [213] and Horowitz et al. [89] were not able to detect any differences between different rabbit antibodies in their experiments. However, Koshland et al. [138, 139] showed that the amino acid composition may vary slightly among different antibodies. Table 12 presents the results of their experiments, which are carried out with exceptional precision.

In the first series of experiments, differences in amino acid composition were found between antibodies to a negatively charged

TABLE 11. Composition of Normal Immunoglobulins

Amino acid	Grams of amino acid residue/100 g protein				
	rabbit γG[63]	human γG[63]	human γM[63]	horse γG[58]	horse (T-globulin) [58]
Lysine	5.76	7.06	4.91	6.77	6.50
Histidine	1.73	2.44	1.98	2.58	2.57
Arginine	4.42	4.02	4.75	3.34	3.02
Aspartic acid	8.08	7.77	6.95	7.25	7.31
Theronine	10.37	7.04	6.17	8.31	7.13
Serine	8.32	9.13	6.58	9.68	9.29
Glutamic acid	11.05	11.18	9.92	10.27	9.36
Proline	6.79	6.40	4.95	6.02	6.52
Glycine	3.98	3.37	2.91	3.68	3.53
Alanine	3.71	3.29	3.12	3.60	3.28
Valine	8.36	7.92	5.77	8.14	7.74
Methionine	1.13	0.93	1.02	0.78	0.59
Isoleucine	3.49	2.16	2.83	3.14	2.70
Leucine	6.73	7.40	6.09	6.63	6.51
Tyrosine	6.17	5.76	4.01	5.55	4.93
Phenylalanine	4.15	4.07	3.85	3.79	3.47
Cystine (1/2)	2.63	2.07	2.30	2.08	1.90
Tryptophan	2.90	2.63	2.47	2.57	2.47
Carbohydrates	2.40	2.87	12.22	2.40	4.90
Total	--	97.51	92.80	97.80	94.80

hapten (p-azophenylarsonic acid) and antibodies to a positive charged hapten (p-azophenyltrimethylammonium) from the same rabbit. In the first, the arginine and isoleucine contents, and in the second, the aspartic acid and leucine contents were higher. In the second series of experiments [139], antibody to p-azophenylarsonic acid was compared with antibody to β-phenyllactoside. The anti-lactoside antibody was found to contain more aspartic acid, while in the antiarsonic antibody the tyrosine content was higher. The composition of the peptide bonds from the different antibodies also differed [139a].

Since the differences found were beyond the limits of possible error, the question arises as to whether they can be attributed to differences in the amino acid composition of the binding sites of the given antibodies. However, this interpretation conflicts with the following evidence. In the first place, it is known that different γG antibodies exhibit somewhat different electrophoretic mobilities, and distinct electrophoretic fractions of γG-globulin differ

TABLE 12. Amino Acid Composition of Purified Rabbit Antibodies to p-Azophenylarsonic Acid (Anti-Ars), p-Azotrimethylammonium Group (Anti-Amm), and β-Phenyllactoside (Anti-Lac) [138, 139]*

Amino acid	Residues per 160,000 g		Standard error of a single determination		Residues per 160,000 g		Standard error of a single determination	
	anti-Ars	anti-Amm	anti-Ars	anti-Amm	anti-Ars	anti-Lac	anti-Ars	anti-Lac
	First series of experiments [138]				Second series of experiments [139]			
Lysine	69.6	69.4	0.87	0.56	70.0	70.8	0.41	0.84
Histidine	16.4	16.6	0.23	0.16	16.5	16.9	0.45	0.29
Arginine	44.7	42.5	0.46	0.49	44.6	44.6	0.53	0.82
Aspartic acid	106	110	0.75	1.07	105	112	1.41	1.37
Threonine	162	162	3.05	2.53	161	163	2.55	1.77
Serine	151	151	1.90	1.70	148	143	1.90	1.94
Glutamic acid	125	127	2.15	2.00	122	121	2.24	2.35
Proline	109	110	2.74	2.35	110	110	1.27	1.77
Glycine	110	110	1.73	1.50	109	109	1.06	2.03
Alanine	81.1	81.4	0.86	1.65	79.8	77.3	1.46	2.21
Valine	128	128	1.80	1.70	128	129	1.00	1.73
Methionine	13.8	13.5	0.36	0.53	13.5	13.6	0.21	0.30
Isoleucine + alloisoleucine	48.4	46.4	0.93	0.96	48.1	47.2	0.80	1.17
Leucine	89	91	--	--	89	89	---	--
Tyrosine	56.1	56.2	1.07	0.61	56.3	50.9	0.49	0.55
Phenylalanine	44.3	44.9	0.49	0.43	44.2	44.5	0.35	0.61

*In each series of experiments antibodies were isolated from the serum of one rabbit.

in amino acid composition. In the second place, as will be explained in detail later, different antibodies of a heterozygote animal may belong to different genetic types of γ-globulin, which would also explain differences in amino acid composition. In Koshland's first series of experiments, a genetic analysis of the donor animals was not performed.

It is even more difficult to interpret data obtained from a study of the amino acid composition of antibodies from the same human. Table 13 presents data of Kabat [38].

As is evident from the table, appreciable differences, independent of the genetic type, were found in amino acid contents be-

TABLE 13. Differences in Amino Acid Composition Among γG-Antibodies from One Individual [38]

Amino acid	Antibody to Levan*	Dextran*	Teichoic acid*	Substance A†	Normal γG-globulin (Gma + b +)	Pooled γM-globulin
Glycine	101	99.5	120	108	107	105
Valine	132	134	102	115	139	105
Leucine	116	109	110	115	116	98.0
Tyrosine	45.5	56.0	53.3	53.2	55.9	41.0
Arginine	46.4	48.2	58.6	56.1	47.8	55.5
Lysine	93.5	87.5	81.9	79.1	93.2	64.0
Hydroxyllysine	0	3.2	3.7	0	0	4.0
Threonine	113	109	110	127	111	116
Proline	101	101	108	108	107	97.1
(Arginine + Lysine)	140	136	141	135	141	120

*Gm-group of antibodies: a − b −.
†Gm-group of antibodies: a + b −.

tween different antibodies, and between antibodies and γG-globulins. However, as the authors themselves have pointed out, human γG-globulins exhibit a considerable heterogeneity, which is even greater than that of rabbit proteins. Several reasons may be adduced to explain variations in amino acid composition among human antibodies; namely, genetic variations, the presence of a large number of different light chains in each antibody preparation, possible intragroup differences in heavy and light chains, individual antigenic variations in γG-globulins, and finally, differences in binding sites [38]. Thus, although it is probable that the amino acid composition of binding sites varies among different antibodies, this has not yet been proven.

Not all γG-globulin molecules have free sulfhydryl groups. According to Cecil and Stevenson [52a], human γG-globulin has only 0.2, while rabbit γG-globulin has 0.3 free sulfhydryl groups per mole protein.

Terminal Amino Acid Residues. A number of studies have been devoted to determination of the N-terminal amino acids of normal immunoglobulins and antibodies in animals and man. Results are summarized in Tables 14 and 21 (cf. Chapter IV).

TABLE 14. N-Terminal Amino Acids of Immunoglobulins and Antibodies (in moles/mole protein)

Animal	Preparation	Aspartic acid	Glutamic acid	Serine	Alanine	Valine	Leucine	Threonine	Reference
Rabbit	Antibody to ovalbumin	–	–	–	1	–		–	182
	γG-globulin and anti-polysaccharide antibody	0.28-0.52	–	–	~1	–		–	157
Horse	γ-globulin	0.15	0.09	0.09	0.06	0.15	0.17	0.03	157a
	γG(T)-globulin	0.19	0.07	0.16	0.14	0.16	0.14	0.09	157a
	Antipneumococcal antibody	0.23	0.10	0.07	0.44	0.18	0.09	–	157a
	Antipneumococcal antibody	0.23	0.18	0.14	0.38	0.18	0.18	–	157a
Human	γG-globulin								
	Fraction II 1, 2	1.06	1.82	0.1	–	–	–	–	156
	Fraction II, 1, 2	1	~1	0.1-0.2	–	–	–	–	191
	Fraction II 3	1.01	1.06	0.17	–	–	–	–	156
Cattle	γ-globulin	0.12	0.03	0.12	0.11	0.43	–	–	190
	Antibody to ovalbumin	0.04	0.00	0.10	0.10	0.25	–	–	190
Pig	γ-globulin	–	1.32	0.17	0.70	0.10	0.10	0.05	60

TABLE 15. Carbohydrates of Normal γG-Globulins of Different Species of Animals and Man (in g/100 g protein) [20]

Substance	Protein		
	human	rabbit	horse
Fucose	0.210	0.045	0.320
Neuraminic acid ...	0.336	0.170	0.188
Hexoses.........	1.040	0.500	0.920

TABLE 16. Carbohydrates of Normal γG- and γM-Globulins of Humans (in g/100 g protein) [53, 57, 159, 160, 168, 188, 202]

Substance	γG-Globulin				γM-Globulin			
Hexose	1.02	1.22	0.94	1.25	4.80	6.16	4.80	4.90
Fucose	Present	0.29	0.20	0.19	Present	0.74	0.80	—
Hexosamine...	1.20	1.14	1.00	1.26	3.00	3.31	3.80	2.70
Sialic acid....	0.03	0.22	0.20	0.14	1.06	2.01	1.80	1.70
Total	2.25	2.87	2.34	2.65	8.86	12.22	11.20	9.30

The sum total of N-terminal amino acids is usually less than one, but even when it is more than one this value does not reflect the true number of polypeptide chains, which in γG-globulin is at least four. Consequently, a considerable number of N-terminal groups is somehow blocked (cf. Chapter IV). It has been found that the N-terminal residue in heavy chains is pyrrolidone carboxylic acid. There is yet one more important point to be mentioned: different γ-globulin fractions can have different terminal groups. Immunoglobulins are evidently heterogeneous in this respect also.

The C-terminal amino groups have been considerably less studied. Human γ-globulin contains mainly serine and glycine, and trace quantities of alanine, threonine, and glutamic acid [149];

or according to other data, 2 moles glycine, 1 mole arginine, and 0.5 mole each of cysteine and serine [39a]. Four moles of C-terminal amino acids have been found in rabbit protein. According to some investigators these are glycine, serine, threonine, and alanine (2:1:0.5:0.5) [205], while other investigators [39a] have found 2 moles glycine and 1 mole each of arginine and aspargine.

Carbohydrate Component of Immunoglobulins. Each of the principal types of immunoglobulins contains carbohydrates. This component consists basically of hexoses, aminosugars, and sialic acid. The contents of these substances vary according to the origin of the immunoglobulin (Table 15).

The total content of carbohydrates is up to ~2.5% in γG-globulins and ~10% in γM-globulins (Table 16).

According to recent data, the chief carbohydrate component is attached to heavy chains via an aspartic acid residue and is recovered in the Fc fragment after papain digestion. Apparently, one more smaller carbohydrate component also exists (cf. Chapter IV). Glycopeptides isolated from human γG-globulins have a molecular weight of 2300 in some cases [57], and 3000-4000 in others [209]. The molecules of different γ-globulin fractions may have carbohydrate components with different compositions.

The study of the structure of the carbohydrate component is only in its early stages. According to Rozenfel'd and Kostyukovskaya [20], all neuraminic acid and some of the fucose (33% in man, 60-65% in rabbits, and 14-15% in horses) is located at the end of the carbohydrate chain. Human γG-globulin contains 2-acetoamido-2-desoxy-β-D-glucose and two 6-desoxy-L-galactose (L-fucose) residues and one N-neuraminic acid residue as the nonreducing terminal group [57]. Fucose and N-neuraminic acid are linked to the rest of the oligosaccharide via a galactose molecule [57].

VALENCE OF ANTIBODIES

By antibody valence is understood the number of regions (binding sites) on an antibody which are capable of specific interaction with an antigen. It can now be considered as certain that the valence of γG-precipitins is two. This fact can be demonstrated, for example, by a study of soluble specific complexes obtained in a large excess of antigen. By electrophoretic and sedimentation analyses it has been shown that the overwhelming majority of these complexes consist of two molecules antigen and one molecule antibody [207]. Another convincing proof of the bivalence of γG-antibodies is found in results of a study of the binding of haptens by antihapten antibodies carried out by equilibrium dialysis (cf. Chapter I). By extrapolating data on the number of hapten molecules bound to antibody, a value of approximately 2 is obtained. Figure 4 gives an example of such a study conducted by Karush [132]. It is seen that experimentally found values can be plotted on a curve which intersects the horizontal axis at a point corresponding to two hapten molecules per antibody molecule.

The valence of γM-antibodies has not yet been precisely determined. The dissociation of γM-antibodies into active 7 S subunits (cf. Chapter IV) is evidence that these antibodies have a higher valence than γG-antibodies. Indeed, Onoue et al. [172a] found that the valence of macroglobulin antibodies was six (assuming the molecular weight to be 1,000,000). Dissociation into 7 S subunits did not appreciably alter the association constants or the number of binding sites.

It is known that some of the antibodies in the serum of immunized animals do not form a precipitate when antigen is added. It has been thought either that such nonprecipitating antibodies have only one binding site, or that their affinity for antigen is so low that they are not able to form precipitates. However, this explanation can no longer be considered correct. Experiments conducted with nonprecipitating horse antibodies to p-azophenyl-β-lactoside hapten have shown that each antibody molecule combines with two hapten molecules. It was also found that the hapten affinity of these antibodies is even higher than that of γG-precipitins. There are apparently some other reasons, possibly having to do with steric properties, that nonprecipitating antibodies are not capable of forming a precipitate with antigen [136]. So-called incomplete antibodies [14] are also probably bivalent [181].

The question as to whether the binding sites of antibody formed after addition of two haptens have identical specificities is of considerable significance. The evidence collected by several investigators indicates that each antibody molecule formed in this way probably has similar binding sites [70, 109, 165].

ANTIGEN PROPERTIES

The intensive study in recent years of the antigenic properties of immunoglobulins has brought to light the considerable complexity of the antigenic structure of these proteins [80, 191a]. Up to seven different types of antigenic determinants can be found in human immunoglobulins (which have been the most studied) and similar determinant groups are probably present in other animals. Below, the chief types of antigenic determinants will be examined briefly. It must be kept in mind that every type is not necessarily associated with a distinct part of the molecule since it is possible that they overlap in some cases. The antigenic properties are determined primarily by the protein and not the carbohydrate fragment of the molecule.

Determinants Common to All Immunoglobulins. All the immunoglobulins of a given animal species have antigenic properties in common. Two types of determinants have been found in man – type K (type I) and type L (type II) – which are common to all the chief types of immunoglobulins as well as to Bence-Jones proteins and pathological globulins. About 60% of the molecules of normal human γG-globulin belong to type K, while 30% or so belong to type L [153, 154]. Human antibodies usually possess both types of determinants, although their contents vary considerably from antibody to antibody and differ from the total content in the collective globulin fraction. The determinant proportion can even be different among different antibodies in the same person [155].

It has now been definitely established that the determinant types K and L are present in the light chains of all types of immunoglobulins [76].

Determinants Characteristic of the Principal Types of Immunoglobulins. According to recent data, the antigenic determinants characteristic of each type of immunoglobulins are localized in the heavy polypeptide chains of these proteins, i.e., in the γ, α, and μ chains, respectively. Recently, another type of heavy chain, the D-type, has been studied [199]. In all cases, determinations of this type are localized in the Fc papain-cleaved fragments. This is true of mice, rabbit, and guinea pig proteins as well as of human proteins [75].

Determinants of Subgroups of the Principal Types of Immunoglobulins. It is known that subgroups of the principal types of immunoglobulins exist (cf. Chapter IV), and in particular, in the case of γG-globulins of man [37, 66, 105], horses [135], mice [83], and guinea pigs [224]. The differences are also governed by the heavy chains and apparently by that section of these chains which is recovered in the Fc fragment in papain hydrolysis. Subgroups of K and L determinants probably exist also [215].

Allotypical Determinants. The genetically determined subgroups of immunoglobulins are characterized by allotypical determinants. They can be detected by immunization of animals belonging to the same species but having different genetic characteristics with immunoglobulins. Allotypical determinants have been studied in man and many species of animals (for a more thorough treatment of this subject, see Chapter VII).

Determinants Exposed by Recombination of Heavy and Light Chains. Heavy and light polypeptide chains are able to undergo spontaneous recombination to form a whole molecule (cf. Chapter IV), as a result of which determinants hidden in the isolated chains appear. In one instance, the human genotypical determinant Gm 3 is apparently located completely in heavy chains, but is detectable only after combination with a light chain (cf. Chapter IV for more detail). This determinant is located in the Fd fragment. A study of myeloma proteins has also shown that several determinants of the papain-cleaved fragment Fab can be detected only after chain recombination. However, it is not clear if this is due to changes in the configuration of the heavy chain when it combines with the light chain, as is probably the case with the Gm 3 determinant, or with an antibody binding site, or whether these determinants are interchain in nature, i.e., are composed of parts of both chains [104].

Internal Determinants. Immunoglobulins, like other proteins, have antigenic determinants which are revealed only by quite thorough enzymic hydrolysis. Some of these determinants have been detected by Kul'berg, Tarkhanova, and Khramkova [11, 13] by thorough papain hydrolysis of rabbit γ-globulin. Internal determinants have been found in anti-Rh antibodies after pepsin hydrolysis [174]. Tatarinov discovered hidden determinants after prolonged storage of human γ-globulin [22, 23].

Individual Determinants of Immunoglobulins and Antibodies. Many attempts have been made to isolate antibody specific to the binding site of an antibody, i.e., anti-antibodies [10]. Treffers and Heidelberger were not able to detect any differences in antigenic properties between antipolysaccharide γG-antibodies with different specificities. The results were the same whether in experiments with horse antibody or with rabbit antibody [228-230]. Kryzhanovskii, Fontalin, and Pevnitskii [10] obtained similar results in a study of horse antitoxins (tetanus and diphtheria).

However, Gurvich and Olovnikov [6], by using pure antibodies to human and equine albumins, were able to demonstrate differences in antigenic structure both between these antibodies and between antibodies and γ-globulins of rabbit serum. Somewhat earlier, Zil'ber, Gardash'yan, and Avenirova [7] also discovered differences between antibodies and γ-globulins in rabbits by means of a highly sensitive method of sensitization and desensitization of guinea pigs.

Oudin and Michel [175] found unique antigenic properties in antityphus rabbit antibodies which differed from the properties of normal globulins isolated before immunization and from the properties of antipolysaccharide antibodies of the same rabbit.

It is rather difficult to interpret these experiments, and not only because in experiments of this kind it is necessary to take into account the genetic characteristics of the proteins under study. In recent years, individual antigenic properties have been discovered in an antibody isolated from a given individual and in a myeloma globulin from a given patient. Thus, Kunkel et al. [104, 143] detected individual antigenic determinants in four γG-antibodies and three γM-antibodies among the various antipolysaccharide antibodies studied. Antibodies with the same specificity but from different persons did not exhibit a cross-reactivity of these individual antigenic determinants, which were localized in Fab fragments. Kunkel also demonstrated the existence of individual antigenic determinants in myeloma proteins; these were localized in either the heavy or light chains, or appeared only after recombination of heavy and light chains [104].

Thus, some immunoglobulins, including antibodies, may contain individual determinants, although it is not completely clear what role the antibody binding site plays in them.

HETEROGENEITY OF ANTIBODIES

As has been pointed out, antibodies may belong to any of the principal types of immunoglobulins, which differ from each other in a number of properties. This fact alone will occasion a considerable heterogeneity among antibodies to the same antigen. Antibodies of allergic patients exhibit a considerable heterogeneity [1, 203]. However, even antibodies of the same type, for example γG, are far from homogeneous. This is evident, in particular, from the shape of the peaks obtained in antibody electrophoresis: they are flat with a wide base (Fig. 12). There is a considerable body of evidence of antibody heterogeneity; for example, antibodies can be separated into distinct fractions by various methods. Several examples of fractionation of rabbit antibodies, belonging primarily to the class of γG-globulins, are given below.

Antibodies and γ-globulins can be fractionated by partition chromatography [35, 36, 123, 183, 184, 187]. In chromatography of immune globulins, antibodies are recovered in several different fractions (Fig. 14). This has been demonstrated with rabbit antibody to human serum albumin, pneumococcal polysaccharides, and influenza virus. Quantitative differences in the antigen-binding capacity were found among antibodies in the different fractions [123].

Rabbit precipitins can be separated into several fractions on DEAE cellulose under specially selected conditions. In this way, we were able to fractionate antibody to human serum albumin isolated with the aid of immunoadsorbents [16]. In Fig. 15, which illustrates the results of an experiment with 6.6 mg antibody, it is seen that under the given conditions some of the antibody was not retained in the column. Three fractions were eluted with phosphate buffer solutions of differing molarity, after which an additional quantity of protein was eluted by treating the resin with 0.1*N* NaOH. In this way, five antibody fractions were obtained.

Sela et al. [158b, 203a, 204, 204a] have shown that under certain conditions γG-globulins can be split into two fractions on a column containing DEAE Sephadex (Fig. 15A). A study of the distribution of antibodies between these fractions revealed that antibodies to basic (with respect to charge) antigens (losozyme, RNA-ase, trypsin, synthetic polypeptides) are contained chiefly in the second peak, while antibodies to acid antigens were collected

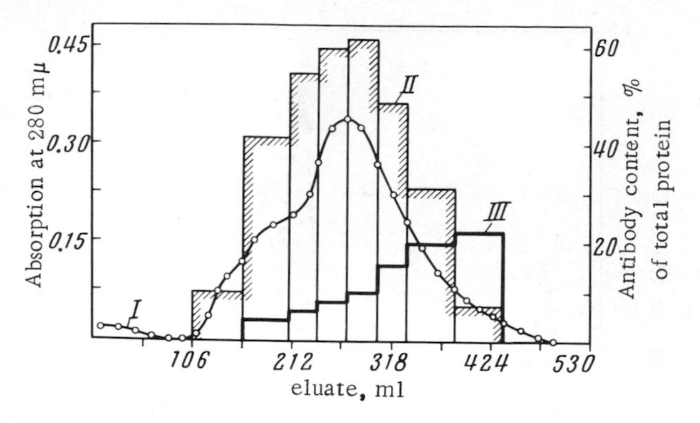

Fig. 14. Partition chromatography of γ-globulin isolated from the serum of a rabbit immunized with ovalbumin and pneumococcal polysaccharides [123]. I) Protein concentration; II) content of antibody of ovalbumin; III) content of antibody to polysaccharide.

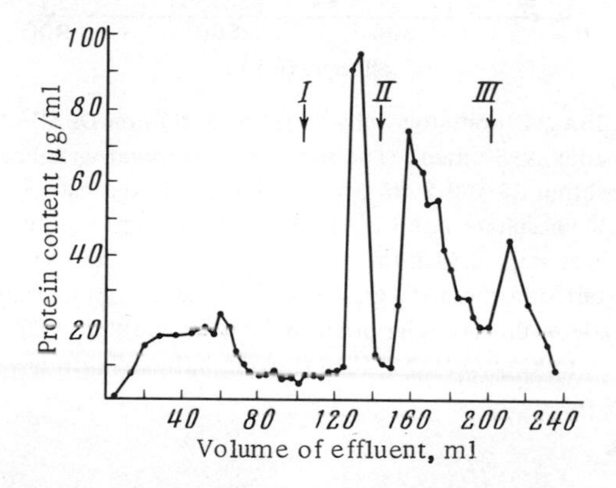

Fig. 15. Fractionation of pure antibodies to human serum albumin on DEAE cellulose. Stepwise addition of eluent solutions: I) 0.02*M* phosphate buffer; II) 0.05*M* Na_2HPO_4 solution; III) 0.1*M* NaCl solution in 0.1*M* Na_2HPO_4 solution.

principally in the first peak. The localization of antibodies to the dinitrophenyl group depended on the charge of the macromolecule carrier of this hapten. These experiments yielded the very interesting and important discovery that antigen determines not only the nature of the binding site, but also exerts an influence on the properties of other sections of the antibody molecule.

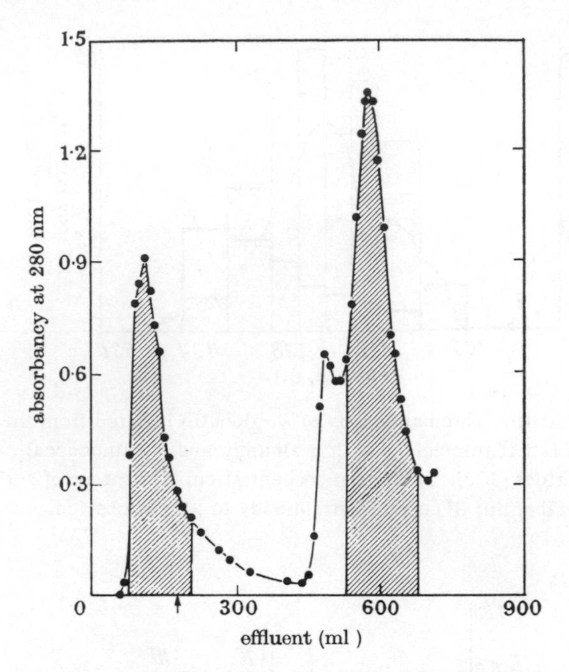

Fig. 15A. Chromatography of rabbit serum on DEAE-Sephadex A-50 med. The serum was chromatographed on column 33 × 2.3 cm after 18 h dialysis against 0.02*M* phosphate buffer, pH 8.0. Elution from the column was started with this buffer. The arrow indicates the start of a molarity gradient obtained by connecting a bottle with 0.3*M* phosphate buffer, pH 8.0 (500 ml) with a mixing chamber with starting buffer (780 ml) [204a].

The same antibody may have either pseudoglobulin or euglobulin properties. Hence, by dialysis against distilled water antibody can be separated into a soluble and an insoluble fraction. For example, in dialysis of rabbit antibody to horse serum albumin (γG-type) 25% protein (eu-antibody) is precipitated, while 75% (pseudo-antibody) remains in solution [4].

Much information is now available on an important aspect of antibody heterogeneity – differences in the antigen-conjugating capacity. Experiments in which this difference was detected were carried out with antibodies to the most frequently used antigens, e.g., proteins, polysaccharides, and proteins conjugated with haptens.

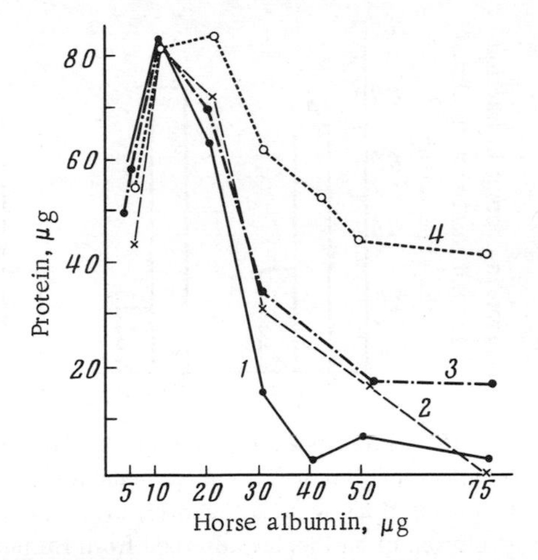

Fig. 16. Dependence of the quantity of precipitate on the amount of horse albumin added in different fractions of anti-horse serum albumin eluted from an immunoadsorbent by 1% NaCl (pH 3.2) [4]. 1) Fraction 11; 2) fraction 12; 3) fraction 13; 4) fraction 14.

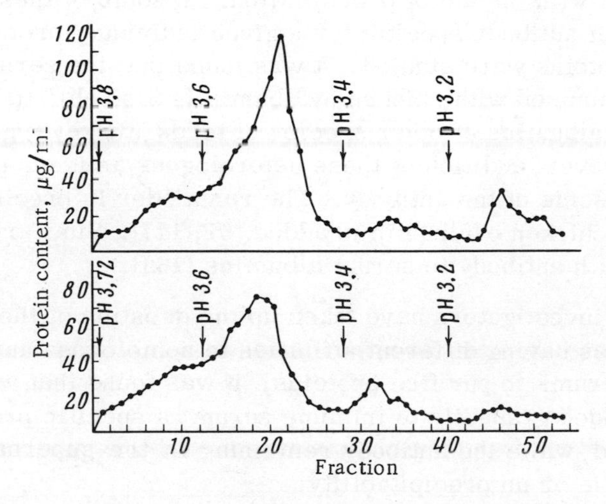

Fig. 17. Elution of antibody to horse albumin from an immunoadsorbent by 1% NaCl solutions of different acidities [4].

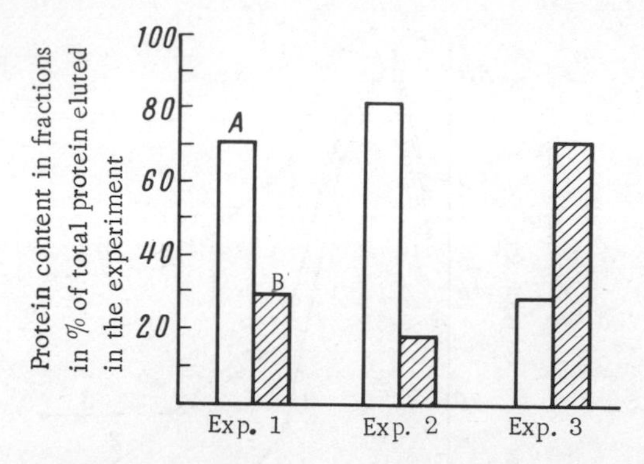

Fig. 18. Readsorption of fractions of pure antibodies obtained by elution from an immunoadsorbent with buffer solutions at pH 3.2 and pH 3.6. Experiment 1) elution of antibodies adsorbed from immune serum; Experiment 2) elution of antibodies adsorbed from fraction A of Experiment 1; Experiment 3) elution of antibodies adsorbed from fraction B of Experiment 1 (unshaded columns – elution at pH 3.6; shaded columns – elution at pH 3.2) [15].

Several investigations of antiprotein antibodies have been carried out with the aid of precipitation. In some of these, cross reactions of antibody specific for a given individual protein with related proteins were studied. It was found that the serum of a rabbit immunized with chicken ovalbumin is also able to react with the ovalbumins of other species of birds (turkeys, geese, ducks, etc.). However, addition of these heterologous antigens precipitates only some of the antibody. The remainder is precipitated only when chicken ovalbumin is added [61, 147]. Similar data were obtained with antibody to serum albumins [162].

Other investigators have taken up the question of the presence of antibodies having different affinities to homologous antigen in immune serums to purified proteins. It was found that when antigen was added gradually to immune serum, a specific precipitate was formed, while the antibody remaining in the supernatant showed little or no precipitability.

Such nonprecipitating antibodies which, however, become precipitable when added to precipitins and antigen, were found in rabbit as well as horse serum [108, 112, 179]. It was also shown in these studies that there was no sharp demarcation between precipitating and nonprecipitating antibodies. Immune serum usually contains a whole variety of antibodies with a wide range of precipitation properties.

The heterogeneity of antibodies to pneumococcal polysaccharides has been studied using similar methods by Heidelberger et al. [110]. Using cross reactions, these investigators were able to demonstrate the presence of two types of antibodies in immune serums to type-specific pneumococcal polysaccharides. Thus, in the serum of a horse immunized by VIII-type (S VIII) pneumococcal polysaccharide, some of the antibodies reacted with both S VIII and III-type pneumococcal polysaccharides (S III). These antibodies could be separated in turn into groups of antibodies which in reaction with S VIII gave a steeply ascending precipitation curve, and antibodies which gave a more gently sloping linear curve. Another antibody fraction reacted only with homologous S VIII, but was also not homogeneous, consisting of subfractions having different reaction capabilities to homologous antigen.

In other experiments [111] it was shown that antibodies to S III in immune serum can be divided into groups of antibodies reacting only with S III, and antibodies reacting with both S III and with the methyl derivative of this antigen.

The heterogeneity of antibodies to polysaccharides can be demonstrated without recourse to cross reactions by fractionation with homologous antigen. If the antigen is added to immune serum in a smaller amount than is necessary to reach the equivalence point, the supernatant remaining after formation of the precipitate will contain antibody having a weaker affinity to the antigen than the precipitated antibodies.

The heterogeneity of antibodies with respect to the antigen-conjugating capacity can be clearly demonstrated by chromatography with immunoadsorbents.

As indicated in Chapter II, elution of antibody from an immunoadsorbent with a 1% NaCl solution (pH 3.2) follows a sloping curve, so that the emergent solution can be collected in fractions. It was found that the antibodies of these fractions differed in their capacity to form a precipitate in the zone of antigen excess. Figure 16 shows precipitation curves of antibodies to horse serum albumin which were successively eluted from an immunoadsorbent [4]. As is evident, the maximum quantity of precipitate was the same in all fractions. However, in an antigen excess, the quantity of precipitate was larger in each succeeding fraction.

The heterogeneity of antibodies with respect to the binding strength with a fixed antigen can be established by successive elution with solutions of different acidities. Figure 17 gives results of an experiment on the elution of antibody to horse albumin by 1% NaCl solutions with successively decreasing pH values [4]. The major portion of the antibody was eluted at pH 3.6, although further elution with solutions at pH 3.4 and 3.2 also freed a certain quantity of protein. These fractions were not the result of some artifact, since in a subsequent adsorption and elution the bulk of the antibody was again eluted at the same pH values as in the previous case. Thus, for example, elution of antibody to serum albumin at pH 3.6, and then at 3.2 yielded fractions the major part of which in a subsequent adsorption on a fixed antigen was again eluted at pH 3.6 for the first fraction, and 3.2 for the second (Fig. 18) [15].

The fractionation can also be carried out by other methods. For example, isoantibodies can be divided into a large number of fractions by successive thermal elution from erythrocytes. By this method, Goodman [97, 98] obtained a large number of anti-A fractions with differing properties.

Schlossman and Kabat [201] describe experiments on fractionation of antibody to dextran, a polysaccharide constructed from isomaltose. This antibody was adsorbed on the dextran gel Sephadex G-25 and eluted successively first with a solution of isomaltose or isomaltotriose, and then with a solution of isomaltohexose. In each case, discrete antibody fractions were obtained. The first of these contained an antibody whose antigen conjugation was inhibited chiefly by low-molecular-weight oligosaccharides. The second fraction contained an antibody whose precipitation with dextran was inhibited by larger molecules. It is evident that in the first case, i.e., in elution with isomaltose or isomaltotriose, the antibody had a smaller binding site than the antibody molecules in the second fraction. Thus, antibodies with different reaction capabilities are formed even to an antigen constructed of identical units (isomaltose).

A considerable number of investigators have studied the heterogeneity of antibodies to conjugated proteins, i.e., proteins coupled with a small chemical group of known structure (hapten).

Landsteiner [145], who was the first to use this type of antigen extensively, and other investigators, demonstrated with the aid of a precipitation reaction that immune serum to conjugated protein contains antibodies which react only with the supporting protein, only with the hapten, or only with the whole protein–hapten complex. This can be observed especially conveniently when immune serum is successively depleted.

Thus, for example, if to the serum of an animal immunized with ovine arsanilazo-globulin is added another arsanilazo protein, a certain amount of antibody will be precipitated. If then to the same portion of serum is added a solution of ovine globulin, the precipitate thus formed will contain antibody which reacts with the carrier protein but not with the hapten. However, subsequent additions of ovine arsanilazo-globulin will yield yet another precipitate containing antibodies which thus can only be precipitated by adding the conjugated antigen used for immunization [108].

The question now arises whether each of these groups of antibody is homogeneous, or whether they in turn can each be further fractionated. To resolve this question for the group of antibodies reacting with hapten, they were precipitated with hapten analogs conjugated with another supporting protein. It was found that the nearer the structure of the analog approached that of hapten, the larger was the quantity of precipitate formed. Consequently, antibody molecules differ in their degree of affinity to hapten. On this basis, Landsteiner stated than an immune serum "contains not one antibody, but antibody fractions, which differ in their reactivity with heteroantigen" [146].

The antibody–hapten reaction has been quantitatively studied by Pauling et al. [180]. These investigators found that the experimental results did not always agree with theoretical calculations. Thus, in a study of inhibition of precipitation by hapten, the amount of precipitate decreased linearly with increase in the amount of hapten only at small hapten concentrations. At large amounts of hapten, a deviation from linearity was observed. This phenomenon can be explained by assuming that the serum contained antibodies with different capacities to combine with hapten. Pauling et al.

quantitatively characterized this microheterogeneity of antibodies in describing it by a distribution function, which is the error function in the free energy of the antibody—hapten interaction in competition with the precipitating antigen (Gaussian function).

Accurate quantitative data can be obtained from a study of the reaction of antihapten antibodies with haptens by means of equilibrium dialysis (cf. Chapter I). Eisen and Karush [71] used purified antibodies in such a study, and from the results constructed a curve on which 1/r was plotted against 1/c, where r is the moles bound hapten per mole antibody, and c is the concentration of free hapten. No linear relationship between 1/r and 1/c was found. This indicates that the antibodies studied were heterogeneous with respect to their antigen-combining capacity since on the basis of the low of mass action, the 1/r vs. 1/c curve should be linear if they were homogeneous.

Karush [132-134] later obtained similar data in a study of the reaction of haptens with purified antibodies. Figure 4 shows results of experiments on the reaction of Ip hapten dye (for formula see p.180) with anti-Ip-hapten antibody by equilibrium dialysis. Because of the heterogeneity of the antibody combining affinity for hapten, r/c was not a linear function of r. Karush, who with Pauling et al. assumed that this heterogeneity can be described by a Gaussian function, calculated the theoretical values (curve in Fig. 4) of these variables, and found that they coincided well with experimental points. Nisonoff and Pressman [163, 164] carried out an analogous set of experiments with antibodies to the p-azobenzoate group.

As in the case with protein polysaccharide antigens, the heterogeneity of antibodies to haptens can be shown by chromatography. Lerman [150] in such a study used a column filled with cellulose conjugated with hapten (p-aminophenylazo-p-phenylarsonic acid), through which he passed a rabbit serum containing antibody to p-phenylarsanilate. Some of the antibody was retained in the column by the fixed hapten. A series of fractions was subsequently obtained by elution with solutions of sodium arsanilate in successively increasing concentrations. The first and second halves of the total number of fractions were collected separately and after dialysis again adsorbed on the fixed hapten. In a repeat elution each of the groups of fractions kept its position on the chromatogram. This indicates differences in the strength of the bond with haptens among the antibody molecules. Pressman et al. [134a, 141] also fractionated antihapten antibodies with the aid of immunoadsorbents.

Thus, there is an appreciable body of evidence for the antigen-conjugating or binding-site heterogeneity of antibodies and this heterogeneity should be kept in mind in examining various problems in the biochemistry of antibodies.

The principal reasons for binding-site heterogeneity can be divided into two categories – those dependent on the antigen and those dependent on the organism of the animal producing the antibody.

In the first case, the immunological heterogeneity of antigens is intended. Antibodies formed as a result of immunization have binding sites. Their structure is complementary to sites on the surface of antigen molecules, the so-called antigenic determinants. If an antigen has many types of determinants, one can obviously expect reciprocal formation of many types of antibodies.

In a discussion of the multiplicity of antigenic determinants the heterogeneity of antigens should above all be kept in mind. In some instances it is possible that the antigens employed do not consist of like molecules. There are certain acknowledged difficulties involved in the purification of the most frequently employed antigens [19]. For example, six-time recrystallized ovalbumin can contain a conalbumin impurity [59].

Even such a widely used antigen as serum albumin, which is relatively simple to purify, has a considerable degree of microheterogeneity [90].

It can be imagined that even a completely homogeneous protein antigen has several different determinants on the surface of its molecules. This topic has been studied in detail by French investigators [148, 233, 234]. Lapresle and Webb found that proteolytic enzymes split the albumin molecule into two or three components. By electrophoresis it was shown that each of these molecular fragments had its own characteristic precipitation band with immune serum to native albumin. Thus, serum contained several groups of antibodies, each of which was capable of reacting with one of the fragments of the antigen molecule.

Later, these same investigators by successive elution of antibodies from an immunoadsorbent by solutions with different acidities obtained three fractions, each of which contained antibodies of different specificities. By binding an albumin fragment with a molecular weight of about 15,000 to cellulose, they succeeded in isolating antibodies which had apparently reacted with only two antigenic determinants. It is interesting that these antibodies could not be precipitated by adding antigen.

It is of prime importance to ascertain how the organism responds to the same determinant group, i.e., whether by the synthesis of one antibody or of a group of antibodies. The answer to this question might well be obtained in experiments in which antibodies to a definite chemical group (hapten) are studied. As was evident from the preceding discussion of such experiments, antihapten antibodies are apparently no less heterogeneous than antibodies to other antigens. However, it would be very premature to conclude that many different antibody molecules are formed for one determinant group. In conjugated proteins used to stimulate antihapten antibody formation, the hapten itself may be only a part of the antigen group, which in addition to it may contain the adjacent areas of the surface of the carrier protein. A diazo bond is most frequently used for linking haptens to proteins, and the diazo group, as already mentioned (cf. Chapter II) is capable of reacting with several amino acids. Consequently, hapten molecules may be bound with different amino acid residues, which would thus explain the formation of antibodies with differing binding sites.

Some haptens are known to combine with only one amino acid residue. Thus, for example, 2,4-dinitrophenol (DNP) combines with the epsilon-amino group of lysine. However, antibodies to this hapten are very heterogeneous, as a measurement of their equilibrium constants has shown [65, 158, 232]. This can be explained by the fact that the DNP-group, like other haptens, is only a part of various antigenic determinants on the surface of the carrier protein. According to Singer [206], noncovalent bonds of the hydrophobic type, which are responsible for the mutual attraction of hydrophobic residues in aqueous solutions [137, 194], play a considerable role in the formation of these hybrid determinants. For example, the DNP-group is hydrophobic, and may be attached by the attraction of any other hydrophobic protein residue to the surface of the molecules of the carrier protein. Thus, even when a hapten is bound to an amino acid residue of one type, many other antigenic determinants containing only hapten will appear. There need not exist any sharp demarcations between the different determinants of this type, and hence antibody molecules with only slight dissimilarities can form. Since such hybrid determinants should be quite rigid, they should have strong immunogenic properties.

It is not certain whether the foregoing will provide an exhaustive explanation for the heterogeneity of anti-DNP antibodies in immunization of animals with ribonuclease, whose molecule is linked to only one hapten molecule [72], or in immunization with a DNP-poly-L-lysine complex [73]. In any event, it is obvious that almost all the antigens in use possess a quite extensive variety of dissimilar antigenic determinants which when administered to an animal stimulate synthesis of heterogeneous antibody molecules.

The second set of reasons responsible for binding-site heterogeneity of antibodies is related to the organism itself, i.e., the immunized animal.

First of all the numerous references to the change in antibody properties during an immune response should be mentioned. Thus, Heidelberger et al. [112, 114] in a study of antibodies to ovalbumin observed that antibodies in serum obtained in later bleedings were capable of combining with antigen to form a precipitate with a high antibody–antigen nitrogen ratio as compared with antibodies in serums obtained in early bleedings. In the opinion of these authors, this phenomenon is explained by the fact that in the later stages of immunization antibodies which react with a greater number of antigenic determinants are formed. This explanation accords with data on the spreading of cross-reactions, which are frequently observed during prolonged immunization. Thus, Hooker and Boyd [122] discovered that in the earlier stages of immunization with p-aminobenzoic acid conjugated with hemocyanin the specific antibody–antigen reaction is not suppressed by a related heterologous hapten, in this case cyclohexane carboxylic acid. If however, immunization is continued, antibodies whose reaction with antigen is inhibited by a heterohapten appear in the serum.

Kosyakov and Reznikova [9] also point out that a primary immunization yields specific serums much more often than a prolonged repeated immunization (similar observations have also been made by other investigators [29]). The observations of Porter [185], who studied the properties of fragments of molecules of bovine serum albumin, are instructive on this point. After a brief period of immunization, an albumin molecule fragment with a molecular weight of 1200 reacted only with one-third of the antibodies in the serum which were specific for native albumin. In a longer immunization, this fragment was able to react with a larger number of antibodies, and sometimes with all.

An accurate characterization of the change in the combining affinity of antibody for antigen can be obtained by measuring the association constant (K_0) of hapten with antibody during immunization. Experiments of this type were performed with a DNP group and specific rabbit antibodies [73]. It is evident from the data in Table 17 that at low antigen doses K_0 can vary by many times the starting value with increase in the time after immunization. Karush [94b] has also shown that the association constant may increase appreciably during immunization.

During immunization, the antibody distribution among the globulin fractions obtained by salting-out with sodium sulfate [45] or by partition chromatography [123] undergoes variations. The precipitation properties of the antibodies also change [4]. Some results can be explained by assuming that soon after immunization γM-antibodies are synthesized, followed by formation of γG-antibodies (cf. Chapter V). Both types of antibodies can have not only different physicochemical, but also different immunological properties [195]. It is possible that during immunization the antibody distribution among subclasses of immunoglobulins also varies [135].

If, indeed, antibody properties are altered by immunization, then at any given moment the serum will contain antibodies of different populations: "earlier," i.e., those which have not had time to disappear from the blood stream, and "later," i.e., those which were formed not long before the blood samples were taken. These "age" differences are evidently one of the factors determining antibody heterogeneity.

There is much less information available on the other cause of antibody heterogeneity. It is quite likely that different cells synthesize slightly differing antibodies. At the present time, however,

TABLE 17. Binding of ε-DNP-L-lysine by Antibodies Isolated at Different Times after Adding Beef DNP-Globulin [73]

Amount of antigen added in incomplete adjuvant, mg	Rabbit	Association constant K_0 (in $1/mole \cdot 10^{-6}$) at different times after the first antigen injection		
		two weeks	five weeks	eight weeks
5	1	0.6	32.0	
	2	1.6	27.0	
	3	0.32	1.6	20.0
	4	1.0	5.9	250.0
	5	0.78	1.5	80.0
50	6		0.21	0.55
	7	0.21	0.36	0.97
	8	0.26	0.20	1.4
	9	0.78	2.7	32.0
	10	0.28	0.20	
100	11	0.21	0.29	
	12	0.37	0.59	0.89
	13	0.17	0.13	0.23
	14	0.87	1.0	0.55
	15	0.26	0.19	0.37
250	16	0.14	0.13	0.10
	17	0.36	0.23	0.38
	18	0.13	0.12	0.16
	19	0.26	0.19	0.11

it is difficult practically to verify this hypothesis. But there is some evidence that different organs produce slightly differing γ-globulins. Askonas, Humphrey, and Porter [36] studied the incorporation of ^{14}C-glycine into γ-globulins by slices of various organs taken from internally immunized and nonimmunized rabbits (Figs. 19 and 20). Spleen sections incorporated more of the label into the middle γ-globulin fractions, while bone marrow and lymph nodes incorporated more into the early emerging fractions in partition chromatography. These differences were probably due to the physicochemical properties of antibodies, but it is a reasonable assumption that the binding sites of antibodies synthesized by different lymphoid organs do also vary.

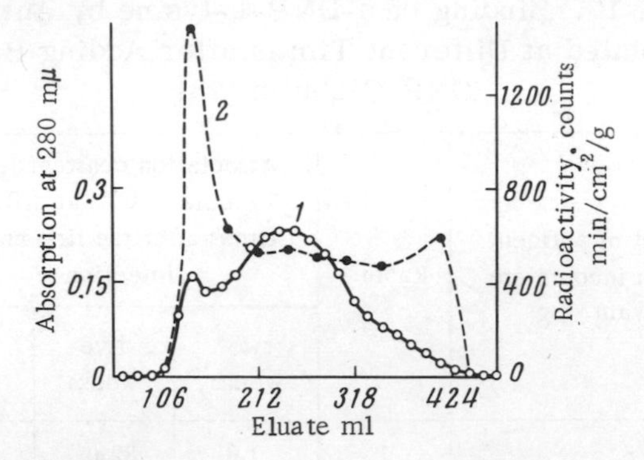

Fig. 19. Partition chromatography of γ-globulins synthesized by slices of cervical lymph nodes of an immune rabbit in vitro. 1) Absorption at 280 mμ; 2) radioactivity [36].

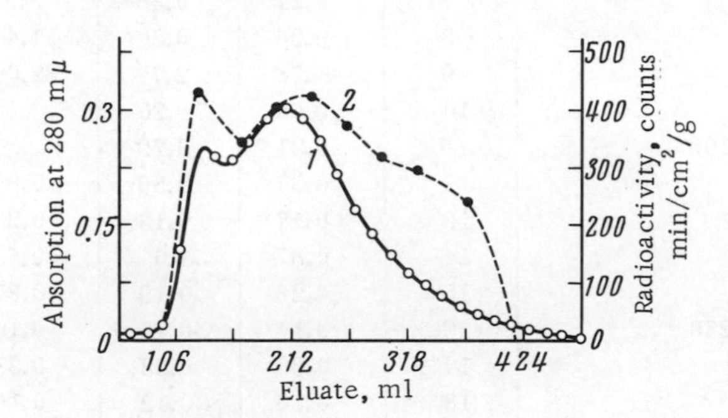

Fig. 20. Partition chromatography of γ-globulins synthesized by spleen slices of an immune rabbit in vitro. 1) Absorption at 280 mμ; 2) radioactivity [36].

The following observations also support the possibility of synthesis of slightly differing antibodies by lymph cells of different anatomical origin. Intraperitoneal injection of antigen into guinea pigs resulted in the synthesis of antibodies differing in electrophoretic mobility from antibodies which appeared after injection of an antibody in an adjuvant [40]. The relative participation of the different organs in antibody formation is dependent on the site of injection of the antigen (cf. Chapter IV). It is therefore evident that the different organs synthesize antibodies with somewhat differing properties, in particular, in their Fd fragments, which govern electrophoretic mobility.

Thus, the antibody molecules contained in any given preparation are not identical, but exhibit certain disparities. These differences may be sharply defined (such as the molecular weight difference between γM and γG antibodies), or graded (for example, when serum contains a whole range of antibodies aptly describable by a Gaussian distribution function). An example of the latter is the antigen-binding capacity of antibodies. Among the causes of the second type of heterogeneity, differences in the antigen used deserve prime mention. Although it is as yet uncertain whether a variety of antibodies to one determinant can be synthesized, it is nevertheless clear that the animal organism also contributes to antibody heterogeneity.

In any event, the fact that an entire population of antibodies is always found in serum should be taken into account in any biochemical investigation of antibodies. It is especially important to bear in mind the heterogeneity of antibodies in the study of their structure, where the minimal heterogeneity is desirable.

LITERATURE CITED

1. L. V. Abaturov, R. S. Nezlin, and J. M. Warshavsky, "Structure of rabbit γG-globulin and its subunits as revealed by H–D exchange and infrared spectroscopy" (1967), in preparation.
2. S. E. Bresler, Introduction to Molecular Biology [in Russian], Academy of Sciences of the USSR, Moscow-Leningrad (1963).
3. A. E. Gurvich and N. G. Karsaevskaya, "Study of serum proteins in ontogenesis by an electrophoresis–precipitation method," Biokhimiya 21(6):746 (1956).
4. A. E. Gurvich, R. B. Kapner, and R. S. Nezlin, "Isolation of pure antibodies with the aid of antigens linked to cellulose and a study of their properties," Biokhimiya 24(1):144 (1959).
5. A. E. Gurvich and R. S. Nezlin, "Nomenclature for human immunoglobulins," Biokhimiya 30(2):443 (1965).
6. A. E. Gurvich and A. M. Olovnikov, "Comparison of antigenic properties of pure antibodies and nonspecific γ-globulins," Biokhimiya 25(4):646 (1960).
7. L. A. Zil'ber, A. M. Gardash'yan, and Z. A. Avenirova, "On the characteristics of the antigenic structure of immune globulins," J. Hyg. Epidemiol. Microbiol, Immunol. 4(1):26 (1960).
8. B. Jirgensons, Natural Organic Macromolecules, Pergamon Press, New York (1962).

9. P. N. Kosyakov and M. N. Reznikova, "On the problem of immunological principals of antibody formation," Usp. Sovrem. Biol. 40(3):320 (1955).

10. G. N. Kryzhanovskii, L. N. Fontalin, and L. A. Pevnitskii, "On the problem of anti-antibody formation," Vest. Akad. Med. Nauk SSSR 15(10): 18 (1960).

11. A. Ya. Kul'berg, "Submolecular structure of antibodies," in: Virology and Immunology [in Russian], Nauka, Moscow (1964), p. 208.

12. A. Ya. Kul'berg and L. M. Bartova, "Isolation of low-molecular weight fragments of antibodies from the urine of immune rabbits," Vopr. Med. Khim. 9(5):514 (1963).

13. A. Ya. Kul'berg, I. A. Tarkhanova, and N. I. Khramkova, "Antigenic structure of rabbit γ-globulin," Folia Biol. 7:213 (1961).

14. N. V. Nartsissov, "On incomplete antibodies," in: Current Problems of Immunology [in Russian], Meditsina, Moscow (1964), p. 254.

15. R. S. Nezlin, "Certain problems of isolation and fractionation of pure antibodies," Biokhimiya 24(3):521 (1959).

16. R. S. Nezlin, "Isolation of pure antibodies and study of their properties," Dissertation, Acad. Med. Nauk, Moscow (1961).

16a. R. S. Nezlin and L. M. Kulpina, "Cellulose-fixed immunoglobulin subunits," Immunochemistry 4(4):269 (1967).

17. J. Northrop, M. Kunitz, and R. Herriot, Crystalline Enzymes, Columbia University Press, New York (1948).

18. V. I. Okulov and G. B. Troitskii, "Thermal electrophoretic homogenization of blood serum proteins," Biokhimiya 28(2):277 (1963).

18a. V. I. Okulov, G. V. Troitskii, and Yu. N. Gordeev, "Conformation of γ-globulin molecule as revealed by kinetic study of its destruction," Biokhimiya 31(4):768 (1966).

19. V. N. Orekhovich, "Contemporary state of protein problems," in: Current Problems of Contemporary Biochemistry, Vol. 1, Biochemistry of Proteins [in Russian], Akad. Med. Nauk SSSR, Moscow (1959), p. 5.

20. E. L. Rozenfel'd and O. M. Kostyukovskaya, "Fucose and other carbohydrate components of different γ-globulins," Vopr. Med. Khim. 7(6):620 (1961).

21. I. A. Tarakhanova and A. Ya. Kul'berg, "Participation of tryptophan in the structure of antibody binding sites," Vopr. Med. Khim. 8(2):163 (1962).

22. Yu. S. Tatarinov, "On the specificity of antigenic components of γ-globulin," revealed after spontaneous decomposition of the molecule," Byul. Eksperim. Biol. i Med. 57(2):84 (1964).

23. Yu. S. Tatarinov, Immunochemical Studies of Human Serum Proteins [in Russian], Doctoral dissertation, Central Institute for the Advanced Training of Physicians, Moscow (1965).

24. G. V. Troitskii, "Study of the conformation of blood plasma proteins," in: Proteins in Medicine and Industry [in Russian], Naukova Dumka, Kiev (1965), p. 5.

25. G. V. Troitskii and V. I. Okulov, "Comparison of the spectropolarimetric characteristics of denaturation of beef serum γ-globulin with other signs of denaturation," Biokhimiya 29(4):615 (1964).

26. G. V. Troitskii, V. I. Okulov, and I. F. Kiryukhin, "Disulfide skeleton and conformation of γ-globulin," Biokhimiya 30(2):268 (1965).

27. G. V. Troitskii, V. I. Okulov, and A. D. Sorkina, "On the possibility of transformation of albumin and γ-globulin of blood plasma into α- and β-globulin," Biokhimiya 26(1):44 (1961).

28. F. Franek and R. S. Nezlin, "Study of the role of different peptide chains of antibodies in the antigen–antibody reaction," Biokhimiya 28(2):193 (1963).

29. M. E. Adair and J. Hamilton, "The specificity of antisera against crystalline serum albumin," J. Hyg. 39(2):170 (1939).

30. R. A. Alberty, "A study of the variation of the average isoelectric points of several plasma proteins with ionic strength," J. Phys. Chem. 53(1):114 (1949).

31. J. Almeida, B. Cinader, and A. Howatson, "The structure of antigen–antibody complexes. A study in electron microscopy," Immunochemistry 2(2):169 (1965).

32. J. Almeida, B. Cinader, and D. Nylor, "Univalent fragments of antibody: a study by electron microscopy," Immunochemistry 2(2):169 (1965).

33. B. G. Arnason, C. de Vaux St. Cyr, and J. B. Shafferner, "A comparison of immunoglobulins and antibody production of the normal and thymectomized mouse," J. Immunol. 93(6):915 (1964).

34. B. G. Arnason, C. de Vaux St. Cyr, and E. H. Relyveld, "Role of the thymus in immune reactions in rats. IV. Immunoglobulins and antibody formation," Intern. Arch. Allergy Appl. Immunol. 25(4):206 (1964).

35. B. A. Askonas, "Heterogeneity of globulins produced by plasma cells," in: Immunochemical Approaches to Problems in Microbiology, Rutgers University Press, New Brunswick (1961), p. 343.

36. B. A. Askonas, J. H. Humphrey, and R. R. Porter, "On the origin of the multiple forms of rabbit γ-globulin," Biochem. J. 63(3):412 (1959).

37. R. E. Ballieux, G. M. Bernier, K. Tominaga, and F. W. Putnam, "Gamma-globulin antigenic types defined by heavy chain determinants," Science 145(3628):167 (1964).

38. E. W. Bassett, S. W. Tanenbaum, K. Pryzwansky, S. M. Beiser, and E. A. Kabat, "Studies on human antibodies. II. Amino acid composition of four antibodies from one individual," J. Exp. Med. 122(2):251 (1965).

39. J. S. Baumstark, R. J. Laffin, and W. A. Bardawil, "A preparative method for the separation of 7 S gamma globulin from human serum," Arch. Biochem. Biophys. 108(3):514 (1964).

39a. S. M. Beiser, E. W. Bassett, and S. W. Tanenbaum, "Studies in immunoglobulin chemistry. I. Reinvestigation of the C-terminal amino acids of 6.5 S rabbit and human γ-globulins," Biochemistry 5(2):652 (1966).

40. B. Benacerraf, Z. Ovary, K. J. Bloch, and E. C. Franklin, "Properties of guinea pig 7 S antibodies. I. Electrophoretic separation of two types of guinea pig 7 S antibodies," J. Exp. Med. 117(6):937 (1963).

40a. J. Berggård and H. Bennich, "Fc fragment of immunoglobulin in normal human plasma and urine," Nature 214(5089):697 (1967).

41. J. Berggård and G. M. Edelman, "Normal counterparts to Bence–Jones proteins: free L polypeptide chains of human γ-globulin," Proc. Nat. Acad. Sci. USA 49(3):330 (1963).

42. R. A. Binaghi, B. Benacerraf, K. J. Bloch, and Fr. M. Kourilsky, "Properties of rat anaphylactic antibody," J. Immunol. 92(6):927 (1964).

43. J. D. Bowman and F. Aladjem, "A method for the determination of heterogeneity of antibodies," J. Theoret. Biol. 4(3):242 (1963).

44. W. C. Boyd, Fundamentals of Immunology, Wiley, New York (1956).

45. W. C. Boyd and H. Bernard, "Quantitative changes in antibodies and globulin fractions in sera of rabbits injected with several antigens," J. Immunol. 33(2):111 (1937).

46. F. W. R. Brambell, W. A. Hemmings, C. L. Oakley, and R. R. Porter, "The relative transmission of the fractions of papain hydrolyzed homologous γ-globulin from the uterine cavity to the fetal circulation in the rabbit," Proc. Roy. Soc. B151(945):478 (1960).

47. P. Callaghan and N. H. Martin, "Optical rotatory dispersion and the conformation of human γ-globulin," Biochem. J. 87(2):225 (1963).

48. P. Callaghan and N. H. Martin, "The optical rotatory properties of 19 S γ-globulin," Biochim. Biophys. Acta 79(3):539 (1964).

49. K. A. Commack, "Molecular weight of rabbit γ-globulin," Nature 194(4830):745 (1962).

50. D. H. Campbell, R. H. Blacker, and A. B. Pardee, "The purification and properties of antibody against p-azophenylarsonic acid and molecular weight studies from light scattering data," J. Am. Chem. Soc. 70(7):2496 (1948).

51. J. R. Cann, D. H. Campbell, R. A. Brown, and J. G. Kirkwood, "Fractionation of rabbit antiserum (antiphenylarsonic-azo bovine globulins) by electrophoresis convection," J. Am. Chem. Soc. 73(10):4611 (1951).

52. M. E. Carsten and H. N. Eisen, "The specific interaction of some dinitrobenzenes with rabbit antibody to dinitrophenyl bovine γ-globulin," J. Am. Chem. Soc. 77(5):1273 (1955).

52a. R. Cecil and G. T. Stevenson, "The disulfide bonds of human and rabbit γ-globulins," Biochem. J. 97(2):569 (1965).

52b. J. J. Cebra and P. A. Small, "Polypeptide chain structure of rabbit immunoglobulins. III. Secretory γA-immunoglobulin from colostrum," Biochemistry 6(2):503 (1967).

53. H. Chaplin, S. Cohen, and E. M. Press, "Preparation and properties of the peptide chains of normal human 19 S γ-globulin (IgM)," Biochem. J. 95(1):256 (1965).

54. P. A. Charlwood, "Ultracentrifugal examination of digestion products from rabbit γ-globulin," Biochem. J. 73(1):126 (1959).

55. F. H. Chowdbury and P. Johnson, "Physico-chemical studies on bovine γ-globulin," Biochim. Biophys. Acta 66(2):218 (1963).

56. H. N. Claman and D. Merrill, "Quantitative measurement of human γ-2, β-2A, and β-2M serum immunoglobulins," J. Lab. Clin. Med. 64(4):685 (1964).

57. J. R. Clamp and F. W. Putnam, "The carbohydrate prosthetic group of human γ-globulin," J. Biol. Chem. 239(10):3233 (1964).

58. S. Cohen and R. R. Porter, "Structure and biological activity of immunoglobulins," Advan. Immunol. 4:287 (1964).

59. M. Cohn, L. R. Wetter, and H. F. Deutsch, "Immunological studies on egg white proteins. I. Precipitation of chicken-ovalbumin and conalbumin by rabbit and horse antisera," J. Immunol. 61(4):283 (1949).
60. G. Colacicco, "N-terminals of porcine γ-globulin," Nature 198(4882):784 (1963).
61. A. G. Cole, "Preparation and precipitin reactions of some crystallized ovalbumins," Arch. Pathol. 26(1):96 (1938).
62. P. Cornillot, R. Bourillon, J. Michon, and R. Got, "Isolation and characterization of urinary γ-globulins," Biochim. Biophys. Acta 71(1):89 (1963).
63. M. J. Crumpton and J. M. Wilkinson, "Amino acid compositions of human and rabbit γ-globulins and of the fragments produced by reduction," Biochem. J. 88(2):228 (1963).
64. J. E. Cushing and D. H. Campbell, Principals of Immunology, McGraw-Hill, New York (1957).
65. L. A. Day, J. M. Sturtevant, and S. J. Singer, "The kinetics of the reaction between antibodies to the 2,4-dinitrophenyl group and specific haptens," Ann. N. Y. Acad. Sci. 103(2): 611 (1963)
66. S. Dray, "Three γ-globulins in normal human serum revealed by monkey precipitins," Science 132(3436):1313 (1960).
67. G. Dreesman, C. Larson, R. N. Pinckard, R. M. Groyon, and A. A. Benedict, "Antibody activity in different chicken globulins," Proc. Soc. Exp. Biol. Med. 118(1):292 (1965).
68. J. T. Edsall and J. E. Foster, "Studies on double refraction of flow. IV. Human serum γ-globulin and crystallized bovine serum albumin," J. Am. Chem. Soc. 70(5):1860 (1948).
69. H. N. Eisen, "Ultraviolet absorption spectroscopy on immune precipitates," J. Immunol. 60(1):77 (1948).
70. H. N. Eisen, M. E. Carsten, and S. Belman, "Studies of hypersensitivity to low molecular weight substances. III. The 2,4-dinitrophenyl group as a determinant in the precipitin reaction," J. Immunol. 73(5):296 (1954).
71. H. N. Eisen and F. Karush, "The interaction of purified antibody with homologous hapten. Antibody valence and binding constant," J. Am. Chem. Soc. 71(1):363 (1949).
72. H. N. Eisen, E. S. Simms, J. R. Little, and L. A. Steiner, "Affinities of anti-2,4-dinitrophenyl (DNP) antibodies induced by ε-41-mono-DNP-ribonuclease," Federation. Proc. 23(2):559 (1964).
73. H. N. Eisen and G. W. Siskind, "Variations in affinities of antibodies during the immune response," Biochemistry 3(7):996 (1964).
74. S. J. Epstein, P. Doty, and W. C. Boyd, "A thermodynamic study of hapten—antibody association," J. Am. Chem. Soc. 78(14):3306 (1956).
75. J. L. Fahey, "Heterogeneity of γ-globulins," Advan. Immunol. 2:41 (1962).
76. J. L. Fahey, "Structural basis for the differences between type I and type II human γ-globulin molecules," Immunol. 91(4):448 (1963).
77. J. L. Fahey, "Contribution of γ-globulin subunits to electrophoretic heterogeneity: identification of a distinctive group of 6.6 S γ-myeloma proteins," Immunochemistry 1(2):121 (1964).

78. J. L. Fahey and H. Goodman, "Antibody activity in six classes of human immunoglobulins," Science 143(3606):588 (1964).

79. J. L. Fahey and A. P. Horbett, "Human γ-globulin fractionation on anion exchange cellulose columns," J. Biol. Chem. 234(10):2645 (1959).

80. J. L. Fahey and C. McLaughlin, "Preparation of antisera specific for 6.6 S γ-globulins, β_2A-globulins, γ_1-macroglobulins, and for type I and II common γ-globulin determinants," J. Immunol. 91(4):484 (1963).

81. J. L. Fahey and E. M. McKelvey, "Quantitative determination of serum immunoglobulins in antibody-agar plates," J. Immunol. 94(1):84 (1965).

82. J. L. Fahey, J. Wunderlich, and R. Mishell, "The immunoglobulins of mice. I. Four major classes of immunoglobulins: 7 S γ_2-, 7 S γ_1-, γ_1A (β_2A)- and 18 S γ_1M-globulins," J. Exp. Med. 120(2):223 (1964).

83. J. L. Fahey, J. Wunderlich, and R. Mishell, "The immunoglobulin of mice. II. Two subclasses of mouse 7 S γ_2-globulins: γ_2A- and γ_2B-globulins," J. Exp. Med. 120(2):223 (1964).

84. A. Feinstein, "The nature of the electrophoretic microheterogeneity of rabbit γ-globulin," Biochem. J. 85(2):16 (1962).

85. A. Feinstein and A. Rowe, "Molecular mechanism of formation of an antigen–antibody complex," Nature 205(4967):147 (1965).

86. J. D. Feldman, "Ultrastructure of immunologic processes," Advan. Immunol. Vol. 4, p. 175 (1964).

87. P. Fireman, E. Hershgold, F. Cordoba, K. Schmid, and D. Gitlin, "Low molecular weight γ-globulins of urine and plasma and their relation to 7 S γ_2-globulin," Nature 203(4940):78 (1964).

88. P. Fireman, W. E. Vannier, and H. Goodman, "The association of skin-sensitizing antibody with the β_2A-globulins in sera from ragweed-sensitive patients," J. Exp. Med. 117(4):603 (1963).

89. S. Fleisher, R. L. Hardin, J. Horowitz, M. Zimmerman, E. Grasham, J. E. Turner, J. P. Burnett, Z. Stary, and F. Haurowitz, "Composition of antibodies against acidic and basic azoprotein," Arch. Biochem. Biophys. 92(2):329 (1961).

90. J. F. Foster, M. Sogami, H. A. Petersen, and W. J. Leonard, "The microheterogeneity of plasma albumins. I. Critical evidence for and description of the microheterogeneity model," J. Biol. Chem. 240(6):2495 (1965).

91. F. Franek, I. Riha, and J. Sterzl, "Characteristics of γ-globulin lacking antibody properties in newborn pigs," Nature 189(4769):1020 (1961).

92. F. Franek and I. Riha, "Purification and structural characterization of 5 S γ-globulin in newborn pigs," Immunochemistry 1(1):49 (1964).

93. B. Frangione and E. C. Franklin, "Structural studies of human immunoglobulins. Differences in the Fd fragments of the heavy chains of G myeloma proteins," J. Exp. Med. 122(1):1 (1965).

94. E. C. Franklin, "Physicochemical and immunological studies of γ-globulins of normal human urine," J. Clin. Invest. 38:2159 (1959).

94a. M. H. Freedman and G. E. Connel, "Monomer and dimer forms of γ-globulin polypeptides in normal urine," Can. J. Biochem. 42(12):1815 (1964).

94b. H. Fujio and F. Karush, "Antibody affinity. II. Effect of immunization interval on antihapten antibody in the rabbit," Biochemistry 5(6):1856 (1966).
95. D. Givol, S. Fuchs, and M. Sela, "Isolation of antibodies to antigen of low molecular weight," Biochim. Biophys. Acta 63:222 (1962).
96. D. Givol, S. Fuchs, and M. Sela, "Isolation and fragmentation of antibodies to polytyrosyl gelatin," Biochemistry 3(3):444 (1965).
97. H. C. Goodman, "Antigen- and antibody-combining properties and their influence on the immune reaction of red cells," Nature 194(4832):934 (1962).
98. H. C. Goodman, "Chromatographic fractionation and characterization of antibody based on the thermal dissociation of the specific antigen–antibody complex," Nature 193(4813):350 (1962).
99. H. C. Goodman, "Chromatography of the antibody globulins on anion-exchange cellulose columns," in: Immunological Methods, Blackwell, Oxford (1962), p. 143.
100. H. J. Gould, T. J. Gill, and P. Doty, "The conformation and hydrogen ion equilibrium of normal rabbit γ-globulin," J. Chem. 239(9):2842 (1964).
101. R. S. Gordon, "The preparation and properties of cold hemagglutinin," J. Immunol. 71(4):220 (1953).
102. H. M. Grey, "Production of mercaptoethanol-sensitive slowly sedimenting antibody in the duck," Proc. Soc. Exp. Biol. Med. 113(4):963 (1963).
103. H. M. Grey, "Studies on changes in the quality of rabbit–bovine serum albumin antibody following immunization," Immunology 7(1):82 (1964).
104. H. M. Grey, M. Mannik, and H. G. Kunkel, "Individual antigenic specificity of myeloma proteins. Characteristics and localization to subunits," J. Exp. Med. 121(4):561 (1965).
105. H. M. Grey and H. G. Kunkel, "H-Chain subgroups of myeloma proteins and normal 7 S γ-globulin," J. Exp. Med. 120(2):253 (1964).
106. C. E. Hall, A. Nisonoff, and H. S. Slayter, "Electron microscopic observations of rabbit antibodies," J. Biophys. Biochem. Cytol. 6(3):407 (1959).
107. L. A. Hanson and E. M. Tan, "Characterization of antibodies in human urine," J. Clin. Invest. 44(5):703 (1965).
108. F. Haurowitz, "Separation and determination of multiple antibodies," J. Immunol. 43(4):331 (1942).
109. F. Haurowitz and P. Schwerin, "The specificity of antibodies to antigens containing two different determinant groups," J. Immunol. 47(2):111 (1943).
110. M. Heidelberger, E. A. Kabat, and M. Mayer, "A further study of the cross reaction between the specific polysaccharides of types III and VIII pneumococci in horse antisera," J. Exp. Med. 75(1):35 (1942).
111. M. Heidelberger and F. E. Kendall, "The precipitin reaction between type III pneumococcus polysaccharide and homologous antibody. III. A quantitative study and a theory of the reaction mechanism," J. Exp. Med. 61(4):563 (1935).
112. M. Heidelberger and F. E. Kendall, "A quantitative theory of the precipitin reaction. III. The reaction between crystalline egg albumin and its homologous antibody," J. Exp. Med. 62(5):697 (1935).
113. M. Heidelberger and K. O. Pedersen, "The molecular weight of antibodies," J. Exp. Med. 65(3):393 (1937).

114. M. Heidelberger, H. P. Theffers, and M. Mayer, "A quantitative theory of the precipitin reaction. VII. The egg albumin—antibody reaction in antisera from rabbit and horse," J. Exp. Med. 71(2):271 (1948).
115. R. Heimer, E. K. Schwart, R. L. Engle, and K. R. Woods, "The relationship of structure to the thermal solubility characteristics of a Bence—Jones protein," Biochemistry 2(7):1380 (1963).
116. R. Heimer, K. R. Woods, and R. L. Engle, "Amino acid analysis of rheumatoid factors and normal γ-globulins," Proc. Soc. Exp. Biol. Med. 110(3):496 (1962).
117. W. A. Hemmings and R. E. Jones, "The occurrence of macroglobulin antibodies in maternal and foetal sera of rabbits as determined by gradient centrifugation," Proc. Roy. Soc. B157(966):27 (1962).
118. J. F. Heremans, Serum Globulins of the γ-System, Masson, Paris (1960).
119. J. Heremans and J. P. Vaerman, "β_2A-globulin as possible carrier of allergic reaginic activity," Nature 193(4820):1901 (1967).
120. J. F. Heremans, J. P. Vaerman, and C. Vaerman, "Studies on immune globulins of human serum. II. A study of the distribution of anti-brucella and anti-diphtheria antibody activities among γ_{ss}-, γ_1m-, and γ_1A-globulin fractions," J. Immunol. 91(1):11 (1963).
121. A. Holasek, O. Kratky, P. Mittelbach, and H. Wawra, "Small angle scattering of Bence—Jones protein," Biochim. Biophys. Acta 79(1):76 (1964).
122. S. B. Hooker and W. C. Boyd, "Widened reactivity of antibody produced by prolonged immunization," Proc. Soc. Exp. Biol. Med. 47(1):187 (1941).
123. J. H. Humphrey and R. R. Porter, "An investigation of rabbit antibodies by the use of partition chromatography," Biochem. J. 62(1):93 (1956).
124. K. Imahori, "β-conformation in γ-globulin," Biopolymers 1(6):563 (1953).
125. K. Imahori and H. Momoi, "Globulin and myeloma: the structural characterization of the γ-globulin and myeloma protein," Arch. Biochem. Biophys. 97(2):236 (1962).
126. T. Ikenaka, D. Gittin, and K. Schmid,, "Preparation and characterization of low molecular weight human plasma 3 S γ_1-globulins," J. Biol. Chem. 240(7):2868 (1965).
126a. K. Ishizaka, T. Ishizaka, and M. M. Hornbrook, "Physico-chemical properties of reaginic antibody. V. Correlation of reaginic activity with γE-globulin antibody," J. Immunol. 97(6):840 (1966).
127. K. Ishizaka, T. Ishizaka, and E. M. Hathorn, "Blocking of Prausnitz—Küstner sensitization with reagin by 'A-chain' of human γ_1A-globulin," Immunochemistry 1(3):197 (1964).
127a. T. Iwasaki and K. Schmid, "Purification and characterization of the 2 S γ_2-globulin in normal human plasma," J. Biol. Chem. 242(10):2356 (1967).
128. H. Cacot—Guillarmod and H. Isliker, "Cleavage and reassociation of iso-agglutinins: preparation of antibodies (mixed)," Chimia 15(7):405 (1961).
129. B. Jirgensons, "Optical rotatory dispersion and conformation of various globular proteins," J. Biol. Chem. 238(8):2716 (1963).
130. B. Jirgensons, "The Cotton effects in the optical rotatory dispersion of proteins as new criteria of conformation," J. Biol. Chem. 240(3):1064 (1965).
131. E. A. Kabat, "The molecular weight of antibodies," J. Exp. Med. 69(1):103 (1939).

132. F. Karush, "The interaction of purified antibody with optically isomeric haptens," J. Am. Chem. Soc. 78(21):5519 (1956).

133. F. Karush, "Quantitative measurement of heterogeneity among antibodies," in: Mechanisms of Hypersensitivity, Little, Brown, Boston (1959), p. 19.

134. F. Karush, "Immunologic specificity and molecular structure," Advan. Immunol. 2:1 (1962).

134a. M. Kitagawa, Y. Yagi, and D. Pressman, "The heterogeneity of combining sites of antibodies as determined by specific immunoadsorbents. I and II," J. Immunol. 95(3):446, 455 (1965).

135. N. R. Kliman, J. H. Rockey, and F. Karush, "Equine antihapten antibody. II. The γG (7 S γ) components and their specific interaction," Immunochemistry 2(1):51 (1965).

136. N. R. Kliman, J. H. Rockey, and F. Karush, "Valence and affinity of equine nonprecipitating antibody to a haptenic group," Science 146(3642):401 (1964).

137. I. M. Klotz, "Noncovalent bonds in protein structure," in: Protein Structure and Function (Sympos. in Biology, No. 13, Brookhaven), p. 25 (1960).

138. M. Koshland and F. M. Engelberger, "Differences in the amino acid composition of two purified antibodies from the same rabbit," Proc. Nat. Acad. Sci. USA 50(1):61 (1963).

139. M. E. Koshland, F. M. Engelberger, and R. Shapenka, "Differences in the amino acid composition of a third rabbit antibody," Science 143(3612):1330 (1964).

139a. M. E. Koshland, F. M. Engelberger, and R. Shapanka, "Location of amino acid differences in the subunits of three rabbit antibodies," Biochemistry 5(2):641 (1966).

140. O. Kratky, I. Pilz, P. J. Schmitz, and R. Oberdofer, "Determination of the shape and size of the γ-globulin molecule by a small angle X-ray method," Z. Naturforsch. 18b(3):180 (1963).

141. V. P. Kreiter and D. Pressman, "Fractionation of anti-p-azobenzearsonate antibody by means of immunoadsorbents," Immunochemistry 1(2):91 (1964).

142. H. G. Kunkel, "Macroglobulins and high molecular weight antibodies," Plasma Proteins, No. 1, p. 279 (1960).

143. H. G. Kunkel, M. Mannick, and P. C. Williams, "Individual antigenic specificity of isolated antibodies," Science 140(3572):1218 (1963).

144. H. G. Kunkel and J. H. Rockey, "β_2A and other immunoglobulins in isolated anti-A antibodies," Proc. Exp. Biol. Med. 113(2):278 (1963).

144a. M. E. Lamm and P. A. Small, "Polypeptide chain structure of rabbit immunoglobulins. II. γM-immunoglobulin," Biochemistry 5(1):267 (1966).

145. K. Landsteiner, The Specificity of Serological Reactions, Dover, New York (1946).

146. K. Landsteiner and J. van der Scheer, "On cross reactions of immune sera to azoproteins," J. Exp. Med. 63(3):325 (1936).

147. K. Landsteiner and J. van der Scheer, "On a cross reaction of egg albumin sera," J. Exp. Med. 71(4):445 (1940).

148. C. Lapresle, M. Kaminski, and C. E. Tanner, "Immunochemical study of the enzymatic degradation of human serum albumin: an analysis of the antigenic structure of a protein molecule," J. Immunol. 82(2):94 (1959).

149. W.-P. Lay and W. J. Polglass, "Terminal amino acids of human and bovine gamma-globulin," Can. J. Biochem. Physiol. 35(1):39 (1957).
150. L. S. Lerman, "Antibody chromatography on an immunologically specific adsorbent," Nature 172(4379):635 (1953).
151. W. J. Mandy, M. K. Stambaugh, and A. Nisonoff, "Amino acid composition of univalent fragments of rabbit antibody," Science 140(3569):901 (1963).
152. E. Marler, C. A. Nelson, and C. Tanford, "The polypeptide chains of rabbit globulin and its papain-cleaved fragments," Biochemistry 3(2):279 (1964).
153. M. Mannick and H. G. Kunkel, "Classification of myeloma proteins, Bence–Jones proteins and macroglobulins into two groups on the basis of common antigenic characters," J. Exp. Med. 116(6):859 (1962).
154. M. Mannick and H. G. Kunkel, "Two major types of normal 7 S γ-globulin," J. Exp. Med. 117(2):213 (1963).
155. M. Mannick and H. G. Kunkel, "Localization of antibodies in group I and group II γ-globulins," J. Exp. Med. 118:817 (1963).
156. M. L. McFadden and E. L. Smith, "The free amino acid groups of γ-globulins of different species," J. Am. Chem. Soc. 75(10):2784 (1953).
157. M. L. McFadden and E. L. Smith, "Free amino acid groups and N-terminal sequence of rabbit antibodies," J. Biol. Chem. 214(1):185 (1955).
157a. M. L. McFadden and E. L. Smith, "Free amino acid groups of equine γ-globulins and a specific antibody," J. Biol. Chem. 216(2):621 (1955).
158. H. Metzger, L. Wofsy, and S. J. Singer, "Affinity labeling of the active sites of antibodies to the 2,4-dinitrophenyl hapten," Biochemistry 2(5):979 (1963).
158a. F. Miller and H. Metzger, "Characterization of a human macroglobulin. I. The molecular weight of its subunit," J. Biol. Chem. 240(8):3325 (1965).
158b. E. Mozes and M. Sela, "Distribution of antibody activities after simultaneous immunization with two antigens," Biochem. Biophys. Acta 130(1):254 (1966).
159. H. J. Müller-Eberhard and H. G. Kunkel, "The carbohydrate of γ-globulin and myeloma protein," J. Exp. Med. 104(2):253 (1956).
160. H. J. Müller-Eberhard, H. G. Kunkel, and E. O. Franklin, "Two types of γ-globulin differing in carbohydrate content," Proc. Soc. Exp. Biol. Med. 93(1): 146 (1956).
161. H. J. Müller-Eberhard and H. G. Kunkel, "Ultracentrifugal characteristics and carbohydrate content of macromolecular γ-globulins," Clin. Chim. Acta 4(2):252 (1959).
162. G. R. E. Naylor and M. E. Adair, "Studies of anti-horse crystalbumin sera. III. The heterologous reactions with human serum albumin and bovine serum albumin," J. Immunol. 78(3):185 (1957).
163. A. Nisonoff and D. Pressman, "Heterogeneity and average combining constants of antibodies from individual rabbits," J. Immunol. 80(6):417 (1958).
164. A. Nisonoff and D. Pressman, "Heterogeneity of antibody in their relative combining affinity for structurally related haptens," J. Immunol. 81(2):126 (1958).
165. A. Nisonoff, M. H. Winkler, and D. Pressman, "The similar specificity of the combining sites of an individual antibody molecule," J. Immunol. 82(3):201 (1959).

166. A. Nisonoff, F. C. Wissler, and D. L. Woernley, "Properties of univalent fragments of rabbit antibody isolated by specific adsorption," Arch. Biochem. Biophys. 82(2):241 (1966).

167. M. E. Noelken, C. A. Nelson, C. E. Buckley, III, and C. Tanford, "Gross conformation of rabbit 7 S γ-immunoglobulin and its papain-cleaved fragments," J. Biol. Chem. 240(1):218 (1965).

168. R. Norberg, "Carbohydrate content of normal 19 S γ-globulin," Clin. Chim. Acta 9(1):89 (1964).

169. "Nomenclature for human immunoglobulins," Bull. World Health Organ. 30:447 (1964).

170. R. S. Nussenzweig, C. Merryman, and B. Benacerraf, "Electrophoretic separation and properties of mouse antihapten antibodies involved in passive cutaneous anaphylaxis and passive hemolysis," J. Exp. Med. 120(2):315 (1964).

171. V. Nussenzweig and B. Benacerraf, "Differences in the electrophoretic mobilities of guinea pig 7 S antibodies of different specificities," J. Exp. Med. 119(3):409 (1964).

172. J. L. Oncley, G. Scatchard, and A. Brown, "Physical chemical characteristics of certain of the proteins of normal human plasma," J. Phys. Colloid Chem. 51(1):184 (1947).

172a. K. Onoue, Y. Yagi, A. L. Grossberg, and D. Pressman, "Number of binding sites of rabbit macroglobulin antibody and its subunits," Immunochemistry 2(4):401 (1965).

173. K. Onoue, Y. Yagi, and D. Pressman, "Multiplicity of antibody proteins in rabbit anti-p-azobenzenearsonate sera," Immunol. 92(2):173 (1964).

173a. K. Onoue, Y. Yagi, and D. Pressman, "Isolation of rabbit IgA antihapten antibody and demonstration of skin-sensitizing activity in homologous skin," J. Exp. Med. 123(1):173 (1966).

174. C. K. Osterland, M. Harboe, and H. G. Kunkel, "Anti-γ-globulin factors in human sera revealed by enzymatic splitting of anti-Rh antibodies," Vox Sanguinis 8(2):133 (1963).

175. J. Oudin and M. Michel, "A new allotypical form of rabbit serum γ-globulin, apparently associated with the function and specificity of antibody," C. R. Acad. Sci. 257:805 (1963).

176. Z. Ovary, "Passive cutaneous anaphylaxis," in: Immunological Methods, Blackwell, Oxford (1963), p. 259.

177. R. H. Pain, "The molecular weight of the peptide chains of γ-globulin," Biochem. J. 88(2):234 (1963).

178. R. H. Pain, "The homogeneity of horse 7 S γ-globulin," Biochem. Biophys. Acta 94(1):183 (1965).

179. A. M. Pappenheimer, "Anti-egg albumin antibody in the horse," J. Exp. Med. 71(2):263 (1940).

180. L. Pauling, D. Pressman, and A. L. Grossberg, "The serological properties of simple substances. VII. A quantitative theory of the inhibition by haptens of the precipitation of heterogeneous antisera with antigens and comparison with experimental results for polyhaptenic simple substance and for azoproteins," J. Am. Chem. Soc. 66(5):784 (1944).

181. B. Pirofsky and M. S. Cordova, "The nature of incomplete erythrocyte antibodies," Vox Sanguinis 9(1):17 (1964).
182. R. R. Porter, "A chemical study of rabbit antiovalbumin," Biochem. J. 46(4):473 (1950).
183. R. R. Porter, "The fractionation of rabbit γ-globulin by partition chromatography," Biochem. J. 59(3):405 (1955).
184. R. R. Porter, "The complexity of γ-globulin and antibodies," Folia Biol. 4(5):310 (1958).
185. R. R. Porter, "The isolation of an immunologically active fragment of bovine serum albumin," in: Symposium on Protein Structure, Wiley, New York (1958), p. 290.
186. R. R. Porter, "γ-Globulin and antibodies," Plasma Proteins 1:241 (1960).
187. R. R. Porter and E. M. Press, "The fractionation of bovine γ-globulin by partition chromatography," Biochem. J. 66(4):600 (1957).
188. J. W. Rosevear and E. L. Smith, "Glycopeptides. I. Isolation and properties of glycopeptides from a fraction of human γ-globulin," J. Biol. Chem. 236(2):425 (1961).
189. M. D. Prager and J. Bearden, "Blood group antibody activity among γ_1A-globulins," J. Immunol. 93(3):481 (1964).
190. E. M. Press and R. R. Porter, "N-terminal amino acids of bovine antibody," Nature 187(4731):59 (1960).
191. F. W. Putnam, "N-terminal groups of normal human gamma-globulin and of myeloma proteins," J. Am. Chem. Soc. 75(10):2785 (1953).
191a. F. W. Putnam, "Structure and function of the plasma proteins," in: The Proteins, Vol. 3, 2nd Ed., Academic Press, New York (1965), p. 153.
192. A. J. Rawson and N. M. Abelson, "Studies of blood group antibodies, VI. The blood group isoantibody activity of γ_1A-globulin," J. Immunol. 93(2):192 (1964).
193. J. S. Remington, E. Merler, A. M. Lerner, D. Gitlin, and M. Finland, "Antibodies of low molecular weight in normal human urine," Nature 194(4826):407 (1962).
194. F. M. Richards, "Structure of Proteins," Ann. Rev. Biochem. 32:269 (1963).
195. I. Riha, "The formation of specific 7 S and macroglobulin type antibodies in chickens," in: Molecular and Cellular Basis of Antibody Formation, Academic Press, New York (1965), p. 253.
196. J. H. Rockey, N. R. Kliman, and F. Karush, "Equine antihapten antibody. I. 7 S β_2A and 10 S γ_1 globulin components of purified anti-β-lactoside antibody," J. Exp. Med. 120(4):589 (1964).
197. J. H. Rockey and H. G. Kunkel, "Unusual sedimentation and sulfhydryl sensitivity of certain iso-agglutinins and skin-sensitizing antibody," Proc. Soc. Exp. Biol. Med. 110(1):101 (1962).
198. D. S. Rowe, "Human γ-globulin as antigen and antibody," Biochem. J. 88:20 (1963).
199. D. S. Rowe and J. L. Fahey, "A new class of human immunoglobulins. I. A unique myeloma protein. II. Normal serum IgD," J. Exp. Med. 121(1):171, 185 (1965).

200. G. Sandor, S. Korsch, and P. Mattern, "7 S globulin immunologically identical to 19 S γ-I (beta-2)-M-globulin, a new protein of horse serum" Nature 204(4960):795 (1964).

201. S. F. Schlossman and E. A. Kabat, "Specific fractionation of a population of antidextran molecules with combining sites of various sizes," J. Exp. Med. 116(4):535 (1962).

202. H. E. Schultze, H. Haupt, K. Heide, G. Moschlin, R. Schmidtberger, and G. Schwick, "Investigations of human serum γ-macroglobulins," Z. Naturforsch. 17b(5):313 (1962).

203. A. H. Sehon, "Heterogeneity of antibodies in allergic sera," in: Molecular and Cellular Basis of Antibody Formation, Academic Press, New York (1965), p. 227.

203a. M. Sela, "Chemical studies of the combining sites of antibodies," Proc. Roy. Soc., Ser. B, 166(1003):188 (1966).

204. M. Sela, D. Givol, and E. Mozes, "Resolution of rabbit γ-globulin into two fractions by chromatography on diethylaminoethyl-Sephadex," Biochim. Biophys. Acta 78(4):649 (1963).

204a. M. Sela and E. Mozes, "Dependence of the chemical nature of antibodies on the net electrical charge of antigens," Proc. Nat. Acad. Sci. USA 55(2):445 (1966).

205. H. I. Silman, J. J. Cebra, and D. Givol, "The carboxyl terminal amino acids of rabbit γ-globulin," J. Biol. Chem. 237(7):2196 (1962).

206. S. J. Singer, "On the heterogeneity of anti-hapten antibodies," Immunochemistry 1(1):15 (1964).

207. S. J. Singer and D. H. Campbell, "Physical chemical studies of soluble antigen—antibody complexes. I. The valence of precipitating rabbit antibody," J. Am. Chem. Soc. 74(7):1794 (1962).

208. P. A. Small, J. E. Kehn, and M. E. Lamm, "Polypeptide chains of rabbit γ-globulin," Science 142(3590):393 (1963).

209. J. D. Smiley and H. Horton, "Isolation and study of the function of human γ_2-globulin glycopeptide," Immunochemistry 2(1):61 (1965).

210. E. L. Smith and D. M. Brown, "The sedimentation behavior of bovine and equine immune proteins," J. Biol. Chem. 183(1):241 (1950).

211. E. L. Smith and N. H. Coy, "The absorption spectra of immune proteins," J. Biol. Chem. 164(1):367 (1946).

212. E. L. Smith and T. D. Gerlough, "The isolation and properties of the proteins associated with tetanus antitoxic activity in equine plasma," J. Biol. Chem. 167(3):679 (1947).

213. E. L. Smith, M. McFadden, A. Stockell, and V. Buettner-Janush, "Amino acid composition of four rabbit antibodies," J. Biol. Chem. 214(1):197 (1955).

214. O. Smithies, "Zone electrophoresis in starch gel: group variations in the serum proteins of normal human adults," Biochem. J. 61(4):629 (1955).

215. S. Stein, R. L. Nachman, and R. L. Engle, "Individual and subgroup antigenic specificity of Bence–Jones protein," Nature 200(4912):1180 (1963).

216. R. F. Steiner and H. Edelhoch, "Structural transitions in antibody and normal γ-globulins. II. Fluorescence polarization studies," J. Am. Chem. Soc. 84(11):2139 (1962).

216a. L. A. Steiner and S. Lowey, "Optical rotatory dispersion studies of rabbit γG-immunoglobulin and its papain fragments," J. Biol. Chem. 241(1):231 (1966).
217. P. Stelos and W. H. Taliaferro, "Separation of antibodies by starch zone electrophoresis," Anal. Chem. 31(5):845 (1959).
218. G. T. Stevenson, "Further studies of the gamma-related proteins of normal urine," J. Clin. Invest. 41(5):1190 (1962).
219. G. Strejan and I. Fletchner, "An antibody with α-I globulin mobility," Proc. Soc. Exp. Biol. Med. 115(2):352 (1964).
220. S. Takahashi and K. Schmid, "A low-molecular weight γA-globulin derived from normal human plasma," Biochim. Biophys. Acta 63(2):343 (1962).
221. K. Takatsuki and E. F. Osserman, "Demonstration of two types of low molecular weight γ-globulins in normal human urine," J. Immunol. 92(1):100 (1964).
222. S. Tekman and A. Ugur, "Paper electrophoresis of a purified specific antibody," Nature 175(4457):594 (1955).
223. A. I. Terr and J. D. Bentz, "Density-gradient sedimentation of skin-sensitizing antibody and β_2A-globulin in serum of allergic individuals," Proc. Soc. Exp. Biol. Med. 115(3):721 (1964).
224. G. J. Thorbecke, B. Benacerraf, and Z. Ovary, "Antigenic relationship between two types of 7 S guinea pig γ-globulin," J. Immunol. 91:670 (1963).
225. A. Tiselius and E. A. Kabat, "An electrophoretic study of immune sera and purified antibody preparations," J. Exp. Med. 69(1):119 (1939).
226. T. B. Tomasi, Jr., "Human γ-globulin," Blood 25(3):382 (1965).
226a. T. B. Tomasi, E. M. Tan, A. Soloman, and R. A. Pendergast, "Characteristics of an immune system common to certain external secretions," J. Exp. Med. 121(1):101 (1965).
227. B. T. Tozer, K. A. Cammack, and H. Smith, "Dissociation of serological complexes of ovalbumin and hemoglobin using aqueous carbon dioxide," Nature 182(4636):668 (1958).
228. H. P. Treffers, "Some contributions of immunology to the study of proteins," Advan. Protein Chem. 1:69 (1944).
229. H. P. Treffers and M. Heidelberger, "Quantitative experiments with antibodies to a specific precipitate. I," J. Exp. Med. 73(1):125 (1941).
230. H. P. Treffers, D. H. Moore, and M. Heidelberger, "Quantitative Experiments with antibodies to a specific precipitate. III. Antigenic properties of horse serum fractions isolated by electrophoresis and by ultracentrifugation," J. Exp. Med. 75(2):135 (1942).
231. P. Urnes and P. Doty, "Optical rotation and the conformation of polypeptides and proteins," Advan. Protein Chem. 16:401 (1961).
231a. J. P. Vaerman, H. H. Fudenberg, C. Vaerman, and W. J. Mandy, "On the significance of the heterogeneity in molecular size of human γA-globulins," Immunochemistry 2(3):263 (1965).
231b. R. C. Valentine and N. M. Green, "Electron microscopy of an antibody—hapten complex," J. Mol. Biol. 27(3):615 (1965).
232. S. Velick, C. Parker, and H. N. Eisen, "Excitation energy transfer and the quantitative study of the antibody—hapten reaction," Proc. Nat. Acad. Sci. USA 46(11):1470 (1960).

232a. G. Wallenius, R. Trautman, H. G. Kunkel, and E. C. Franklin, "Ultracentrifugal studies of serum," J. Biol. Chem. 225(1):253 (1957).

233. T. Webb and C. Lapresle, "Study of the adsorption on and desorption from polystyrene—human serum albumin conjugates of rabbit antihuman serum albumin antibodies having different specificities," J. Exp. Med. 114(1):43 (1961).

234. T. Webb and C. Lapresle, "Isolation and study of rabbit antibodies specific for certain of the antigen groups of human serum albumin," Biochem. J. 91(1):24 (1964).

235. T. Webb, B. Rose, and A. H. Sehon, "Biocolloids in normal human urine. I. Amount and electrophoretic characteristics," Can. J. Biochem. Physiol. 36(11):1159 (1958).

236. T. Webb, B. Rose, and A. H. Sehon, "Biocolloids in normal human urine. II. Physicochemical and immunochemical characteristics," Can. J. Biochem. 36(11):1167 (1958).

236a. R. C. Weir and R. R. Porter, "Comparison of the structure of the immunoglobulins from horse serum," Biochem. J. 100(1):63 (1966).

237. J. K. Weltman and M. Sela, "Hydrogen ion titration of rabbit γ-globulin and some of its subunits," Biochim. Biophys. Acta 93(3):553 (1964).

238. O. Wetter, C. Hartenstein, K. Jehnke, and R. Merten, "Ultracentrifugal investigations of the sedimentation behavior of isolated γ-globulins of healthy subjects and cancer patients," Z. Naturforsch. 19b(1):60 (1964).

239. M. Winkler and P. Doty, "Some observations on the configuration and precipitating activity of antibodies," Biochim. Biophys. Acta 54(3):448 (1961).

240. Y. Yagi, P. Maier, Y. Pressman, C. E. Arbesman, R. E. Reisman, and A. R. Lenzner, "Multiplicity of insulin-binding antibodies in human sera. Presence of antibody activity in γ-, β_2A-, and β_2M-globulins," J. Immunol. 90(5):760 (1963).

241. Y. Yagi, P. Maier, D. Pressman, C. Arbesman, and R. E. Reisman, "The presence of the ragweed-binding antibodies in the β_2A-, β_2M-, and γ-globulins of the sensitive individuals," J. Immunol. 91(1):83 (1963).

242. Y. A. Zagyansky, R. S. Nezlin, and L. A. Tuberman, "Flexibility of immunoglobulin G molecules as established by fluorescent polarisation measurements," Immunochemistry 6(6):787 (1969).

Chapter IV

The Structure of Antibodies

Considerable difficulties are involved in the study of immunoglobulins. Data on N-terminal amino acids, which have been useful for an understanding of the structure of many proteins, have not been of any essential help here. As mentioned earlier (cf. Chapter III), only one N-terminal amino acid — alanine (one mole alanine per mole protein) — plus a small quantity of aspartic acid and serine, has been found in rabbit γG-globulins. Three to five N-terminal amino acids have been found in equine and bovine γ-globulins, but their sum was less than one mole per mole protein. The γ-globulins of pigs and man contain two principal N-terminal amino acids. After reduction or denaturation, no new N-terminal amino acid residues appear. Hence, a reliable evaluation of the number of peptide chains in the immunoglobulin molecule cannot be obtained from these investigations.

Another difficulty derives from the rather considerable size of the immunoglobulin molecule; the molecular weight of γG-globulins is about 150,000. Thus, it can be reasonably concluded that a study of structure requires the splitting of the molecule into smaller subunits, or fragments. This can be accomplished either with various proteolytic enzymes or by reduction of some of the disulfide bonds [168].

FRAGMENTS OBTAINED BY PROTEOLYTIC CLEAVAGE OF IMMUNOGLOBULINS

The proteolytic effect of various enzymes on immunoglobulins does not disturb the specific activity of the latter under certain conditions. This fact was first established by Parfentjev [160], who treated antitoxic horse serum with pepsin at pH 4. As a result, only the antigen activity of the serum decreased, while the neutralizing activity remained unchanged. Later, this observation was confirmed by other investigators [164]. Petermann [162] showed that papain splits horse antitoxin into fragments, one of which has the capacity to flocculate the toxin.

Papain hydrolysis of human γ-globulin yielded fragments with a molecular weight of around 47,000, and only a very small quantity of nonprotein nitrogen [163]. Porter [164a] has shown that unpurified papain hydrolyzes immune rabbit globulin into fragments, having a molecular weight of about 44,000, which are not precipitated by antigen (ovalbumin) but which are able to inhibit the reaction of antigen with unrefined antibody. However, a more detailed study of these fragments was difficult because of the considerable amount of contamination by enzyme protein.

It was only after a highly purified crystalline papain preparation became generally available that a detailed study of hydrolysis was possible, since the small quantity of enzyme used in hydrolysis did not interfere. Below are discussed results of a study of proteolytic fragments of antibodies and γ-globulins. The data obtained by various investigators on this subject have been of considerable value in understanding the molecular structure of immunoglobulins.

Papain-Cleaved Fragments

For splitting γ-globulins, one milligram crystalline mercury papain per 100 mg substrate (at pH 7) is sufficient. Since the sulfhydryl groups of papain must be free if the latter is to exert its effect, cysteine (up to 0.01*M*) is added for activation. After incubation for 16 hr at 37°C, the bulk (90%) of the protein will not pass through a dialysis membrane. Consequently, under these proteolytic conditions, very little low molecular weight fragments are formed. Ultracentrifugation usually discloses one peak with a sedimentation constant of about 3.5 S, which corresponds to a molecular weight of approximately 50,000 [118, 166]. Since the molecular weight of γG-globulin is about 150,000, papain evidently splits γ-globulin into three approximately equal fragments.

Porter [166] was able to separate rabbit γ-globulin fragments with the aid of carboxymethylcellulose. Under the conditions employed, the hydrolyzate was divided into three main fractions, de-

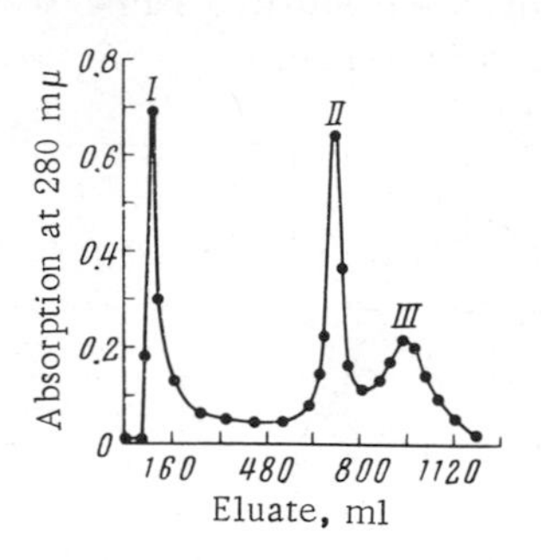

Fig. 21. Chromatography of papain-cleaved rabbit γ-globulin on carboxymethylcellulose (150 mg protein applied to a 30 × 2.4 cm column [166]. I, II, and III) The main fractions of the eluate. Gradient of sodium acetate solutions (pH 5.5) from 0.1 to 0.9*M*, beginning with 200 ml eluate.

signated as fragments I, II, and III (cf. Fig. 21), with a ratio of 1:0.8:0.9, respectively. Fragments I and II had quite similar properties, but differed sharply from fragment III. Absorption at 280 mμ (1 mg/ml in H_2O) was 1.4 for fragments I and II, and 1.0 for fragment III. Only fragment III could be crystallized, forming thin laminar crystals upon dialysis against a phosphate buffer at neutral pH.

Differences in the amino acid compositions of fragments I and II (Table 18) were relatively slight, but in this respect also these fragments differed considerably from fragment III.

Table 19 presents data on the carbohydrate content of γG-globulin and papain-cleaved fragments; it may be seen that the bulk of carbohydrates is associated with fragment III.

Although papain was able to hydrolyze precipitins, none of the resultant fragments were able to precipitate the specific antigen. However, fragments I and II were able to inhibit the reaction of the antigen with untreated antibody. In all probability, these fragments contained only one binding site, i.e., they are univalent [166]. When the antibody–antigen precipitate is subjected to hydrolysis, the antigen apparently remains bound only with fragments I and II, judging from results of electrophoresis or sedimentation analysis, or from the radioactivity count when a labeled antibody is used [23, 206].

Papain hydrolysis produces no marked change in the affinity for hapten. Thus, by means of equilibrium dialysis it was shown that fragments I and II of rabbit antibody to p-azobenzoate bound 90% of the amount of iodobenzoate-^{131}I bound by unpurified antibody [144]. Similar results were obtained by Karush, who used p-azophenyl-β-lactoside as hapten [105]. The average number of hapten-binding sites per fragment molecule, as determined by hapten-

TABLE 18. Amino Acid Composition of γ-Globulin Fragments (in % of total protein) from the Serum of a Nonimmune Rabbit [166]

Amino acid	Fragment			Amino acid	Fragment		
	I	II	III		I	II	III
Aspartic acid	5.66	6.01	6.18	Methionine	0.64	0.75	1.48
Threonine	10.42	11.56	5.61	Isoleucine	2.20	2.26	3.43
Serine	8.77	9.57	7.34	Leucine	5.17	5.34	5.19
Glutamic acid	5.99	6.18	7.83	Tyrosine	3.44	3.58	2.33
Proline	5.08	5.08	6.68	Phenylalanine	2.55	2.65	2.76
Glycine	7.57	6.93	3.18	Histidine	1.55	1.87	5.23
Alanine	5.65	6.12	2.82	Lysine	6.00	6.84	9.95
Valine	7.38	7.58	5.96	Arginine	5.99	6.22	13.93
				Cystine (1/2)	2.88	2.22	1.18

TABLE 19. Carbohydrate Content (in %) of Rabbit γ-Globulin and Its Papain-Cleaved Fragments [166]

Substance	Hexosamine	Hexoses	Total
γ-Globulin	0.6	0.5	1.1
Fragment I	0.5	0.6	1.1
Fragment II	0.1	0.5	0.25
Fragment III	1.3	0.9	2.2

binding measurements, is 0.9-1.0 [143]. The fragments differ appreciably in their antigenic properties: fragments I and II have similar antigen determinants which differ sharply from the determinants of fragment III. The site of action of papain is localized in the region of the disulfide bond (or bonds) which links two heavy chains together. At relatively low cysteine concentrations in the medium and a relatively short hydrolysis time, peptide bonds split nearer to the N-terminus from the disulfide bridge; the resultant fragment III contains an interchain disulfide bridge after oxidation. With increase in the cysteine concentration and the hydrolysis time, proteolysis is more complete, and a disulfide bond can no longer be obtained in fragment III [90a]. Insoluble papain (linked to a diazotized copolymer of leucine and p-amino-phenylalanine) splits the peptide chain on the other side of the disulfide bond, since univalent fragments can be obtained only after reduction [20, 22, 24].

The hydrolytic effect of papain results in cleavage of a relatively small number of peptide bonds [24]. Thus, if the hydrolysis time at pH 6.2 is brief (splitting 2-3 molecules in 15 minutes), only one new N-terminal amino acid – leucine or isoleucine – is generally obtained [180]. Papain initially splits only one of the univalent fragments from the molecule; the rest of the molecule is split into the two other fragments by the continued action of the enzyme [130].

At pH 7 papain does not further digest the fragments, but at acid pH values fragment III is hydrolyzable [172a]. Hydrolysis in 6*M* urea leads to decomposition of the fragments into small peptides. According to Kul'berg and Tarkhanova [3, 4] these peptides have a specific capacity to inhibit the antigen–antibody reaction. Goodman obtained antigen-active fragments with a molecular weight on the order of 2000-10,000 by peptic hydrolysis of fragment III [75].

The specific optical rotation of a dialyzed papain-cleaved hydrolyzate of γ-globulins is approximately the same as that of untreated rabbit γ-globulin. From this it would appear that the fragments are compact units, i.e., they are not unfolded or denatured. The decrease in viscosity after hydrolysis is an indication that the molecule is split perpendicularly to its long axis [143].

In his first report [166] Porter suggested that each γ-globulin molecule contains both fragments I and II. However, it was subsequently found that separation into fragments I and II takes place only with the chromatographic method used by Porter. In theory, an even greater number of such fractions can be obtained. It was discovered that the separation of a fraction into univalent fragments is due to the heterogeneity of the γ-globulin molecules themselves, and that fragments I and II are contained in diverse γ-globulin fractions. Thus, if rabbit γ-globulin is fractionated on carboxymethyl cellulose with the aid of sodium acetate solutions in 0.06, 0.1, and 0.5*M* concentrations, then each of the resultant fractions hydrolyzed by papain, and finally chromatography carried out again under the same conditions, the hydrolyzate of the first fraction will contain mainly fragment I and only a small quantity of fragment II, while the third fraction will contain fragment II and only a small quantity of fragment I [81, 156, 197, 207].

It has also been shown that the differences in charged amino acids in fractions of univalent fragments obtained by chromatography on carboxymethylcellulose coincide with those theoretically expected from the position of the fragment on the chromatogram [117]. Fragments I and II, isolated by Porter's method from pure antibodies specific for the p-azophenylarsonate group, differ in their capacity to react with iodine. Iodination of fragment II was 25-30% greater than for fragment I [207]. This also supports the hypothesis that these fragments are from different molecules. Finally, in immunization of the same rabbit with two different haptens conjugated with the same protein, antibodies were formed, all of whose univalent fragments were recovered, by fractionation on carboxymethylcellulose, in either the fraction corresponding to fragment I, or in the fraction corresponding to fragment II [82].

γM-Globulins have proved to be more susceptible to papain hydrolysis. Attempts to isolate the fragment corresponding to fragment II from the hydrolyzate of these proteins have been unsuccessful [215b]. Results obtained from papain hydrolysis of γ-globulin from human serum are, on the whole, similar. Approximately three-fourths of the human globulin is cleaved quickly by papain, while the remaining fourth requires prolonged hydrolysis with the enzyme (about 30 hr). About 10% of the protein was not hydrolyzed. During proteolysis, free amino acids or specific polypeptides did not appear. In addition, no products with sedimentation constants between 6.6 and 3.5 S were formed. Poulik, however, showed that prolonged hydrolysis in the presence of cysteine yielded a relatively small fragment, apparently a component of fragment III [169]. At the completion of proteolysis, only leucine (or isoleucine) appeared as the new N-terminal amino acid [89]. If proteolysis is carried out in an 8*M* urea solution, all the products will pass through a dialysis membrane.

A two-step decomposition can also be demonstrated for human γG-globulin as has been done for rabbit γG-globulin. Treatment with 1% mercuripapain without a reducing agent caused almost no change in the molecular weight of the protein, and the antibodies were precipitable. However, addition of mercaptophenol caused separating of the protein into 3.5 S fragments [34].

Papain-cleaved fragments of human γ-globulin are not readily separated on carboxymethyl cellulose by Porter's method, but by chromatography of DEAE-cellulose they can be separated into two peaks. In the first is found a substance with a sedimentation constant of 3.8 S, while the sedimentation constant of the second substance is 3.9 S. The protein in these peaks had different antigen properties: the first substance moved slowly in electrophoresis (S fraction), while the second (F fraction) moved rapidly. It was observed that each molecule contained both fractions. Thus, if γ-globulin is separated into slow and fast electrophoretic fractions, each of them will yield S and F fractions on proteolysis [41]. It was later found that the S fraction bears an antibody binding site and apparently corresponds to fragments I and II of rabbit γ-globulin, while the F fraction corresponds to the rabbit fraction III.

Under somewhat different conditions, Franklin separated a papain hydrolyzate on DEAE-cellulose into three fractions – A, B, and C [65], of which A and C resembled fragments I and II of rabbit γ-globulin, while B was similar to fragment III. Fractions A and C came from different molecules [209]. Gel filtration is a convenient method for separating the fragments from the whole molecule [212].

It should be noted that serum always contains a certain number of z-type molecules in which the inactive papain-cleaved fragment is split by papain [211]. Some of the molecules are totally nonsusceptible to hydrolysis by this enzyme [76]. Sehon studied the properties of antibody fragments found in the blood of allergic patients [71].

Papain hydrolysis of the γ-globulins of other animals, e.g., horse, cow, and pig [5, 8, 89, 181], mouse [48], and guinea pig [147] has also been studied. As a rule, the results agreed quite well with the above-described results for rabbit and human γ-globulins. As is evident from Table 20, rabbit γ-globulin and human myeloma proteins are most completely and rapidly hydrolyzed, while pig γ-globulin least readily submits to hydrolysis. Stefani, Kul'berg, and Shakhanina [9] discovered that antitoxic horse γ-globulin is still able to flocculate with antigen after papain hydrolysis, which indicates that the fragments obtained are bivalent. According to the data in Table 21, the new N-terminal amino acid is most often leucine (or isoleucine).

TABLE 20. Papain Hydrolysis of γ-Globulins from Different Animals (in %) [89]

γ-Globulin	Hydrolysis after 0 hr			Hydrolysis after 4 hr*		
	3.5S	6.6S	9S	3.5S	6.5S	5.5S
Man	74	26	0	88	0	12
Rabbit	67	33	0	100	0	0
Horse	5	95	0	65	0	35
Cow	7	85	8	72	0	28
Pig	0	100	0	42	22	36
Myeloma 6.6 S proteins:						
Wi, Ag {With same electrophoretic mobility as γG-globulins	100	0	0	—	—	—
	90	10	0	100	0	0
Jo {With electrophoretic mobility between γG- and β-globulins	41	59	0	100	0	0

*Hours are the incubation time before placing the cuvette in an ultracentrifuge.

Some biological and physicochemical properties of papain-cleaved fragments of antibodies and γG-globulins are given in Table 22.

Fragments Obtained by Hydrolysis with Pepsin and Other Proteases

As has been mentioned, pepsin was the first proteolytic enzyme used to digest antibodies [160]. Later, Bridgman [15] showed that prolonged peptic hydrolysis of human γ-globulin will yield 5.8 and 3.1 S fragments. However, phenomena taking place during pepsin proteolysis first became intelligible after publication of a series of carefully conducted experiments by Nisonoff et al. [140-142] on rabbit antibodies.

When rabbit γ-globulins were subjected to hydrolysis with pepsin (~2%) at pH 4, the sedimentation constant of 90% of the substrate protein fell from 7 to 5.25 S [141, 142]. Taking into account the diffusion constant ($4.7 \cdot 10^{-7}$ cm^2/sec) and the partial specific volume, the molecular weight of the resultant fragment was calculated at 106,000 [140]. In other determinations

TABLE 21. Formation of N-Terminal Groups in Papain Hydrolysis of γ-Globulins from Different Animals [184]

γ-Globulin source	γ-Globulin preparation	Composition with sedimentation const. of 3.5 S, %	Components with other sedimentation const., %	N-terminal amino acids, μmole per μmole γ-globulin							
				aspartic acid	glutamic acid	leucine (isoleucine)	valine	alanine	glycine	serine	threonine
Man	Initial	—	—	0.78	1.26	—	—	—	—		—
	Digested	87	—	1.00	1.03	0.93	—	—	—	—	
Rabbit	Initial	—	—	0.31	—	0.16	—	0.79	—	—	—
	Digested at										
	pH 6	100	—	0.33		0.69		0.91	0.25	0.15	0.25
	pH 7.5	76	24 (5,7S)	0.23	—	0.28	0.11	0.86	—	0.16	—
Pig	Initial	—	—	—	1.57	—	—	0.65	—	—	—
	Digested	M*	M*		2.02	0.18		0.80		0.13	—
Cow	Initial	—	—	0.15	0.09	—		0.1	—	0.12	—
	Digested	H+	H+	0.20		0.43				0.12	
Horse	Initial			0.17	0.08	—	—	—	—	0.1	—
	Digested	45	55 (5.9 S)	0.20	0.36	0.45	—			0.17	

* Mixture of 3.3 S and 4.1 S components.

† Heterogeneous mixture of 6.5 S and low-molecular-weight components.

TABLE 22. Properties of Papain-Cleaved γG-Globulin Fragments [47]

Properties	Rabbit			Human, mouse	
	I (Fab)	II (Fab)	III (Fc)	S (Fab)	F (Fc)
Physicochemical:					
Total charge (relative mobility toward anode at pH 6)	Small	Moderate	Slow	Slow	Fast
Molecular weight	45,000	45,000	55,000	–	–
Immunological:					
Antibody activity	Yes	Yes	No	Yes	No
Penetration through placenta	No	No	Yes	–	–
Skin fixation	"	"	"	–	–
Reaction with rheumatoid factor	"	"	"	No	Yes
Allotypical determinants:					
Rabbit (a, b)	Yes	Yes	No	–	–
Human Gm	–	–	–	No	Yes
Inv	–	–	–	Yes	No
Mouse Iga	–	–	–	No	Yes

this value was somewhat lower: 87,000 [22] and 92,000 [216]. Fragments with a sedimentation constant of 5 S were precipitable with antigen, whereby about 70% of the original antibody activity was retained. These fragments were presumably bivalent. In the presence of reducing agents (0.01*M* cysteine, 0.01*M* mercaptoethanolamine) the 5 S fragment was reduced to 3.5 S fragments (diffusion constant $6.1 \cdot 10^{-7}$ cm^2/sec, molecular weight 56,000), which could not be precipitated with antigen, but which inhibited precipitation of antigen with unpurified antibody and were able to bind hapten. The inhibition of specific precipitation by 3.5 S fragments of rabbit antiovalbumin antibody, which were obtained by papain or pepsin hydrolysis in the presence of cysteine [141] are given below (in % of control):

Blocking activity

Added milligrams of fragments	0.2	0.4	0.7	1.4	2.1
Papain-cleaved fragments	101	93	84	23	0
Pepsin-cleaved fragments	94	88	75	0	0

Thus, 3.5 S pepsin-cleaved fragments are univalent and similar to the papain-cleaved fragments I and II. Univalent pepsin-cleaved fragments were larger than papain-cleaved fragments by a peptide chain section with a molecular weight of 4000-5000 [22, 216]. One of the carbohydrate components of γ-globulin is linked to this small section containing an SH group which participates in the interchain disulfide bond [216].

According to the data of Utsumi and Karush [216] pepsin first split the γ-globulin molecule into a 5 S fragment and a fragment which was similar in its antigen properties to the papain-cleaved fragment III and had a sedimentation constant of 3.3 S. This latter fragment was broken down into smaller units by continued hydrolysis with the enzyme. One of these had a molecular weight of about 27,000 (it was possibly a dimer consisting of two subunits with a molecular weight of 13,500) has some of the antigenic determinants of the papain-cleaved fragment III, and contained no carbohydrates nor interchain disulfide bonds. The other subfragment, which had no antigenic activity, had a molecular weight of 14,000 (possibly a dimer of two subunits with a molecular weight of 7000), and contained all the fucose and 70% of the hexoses and hexosamine of the entire molecule. In addition to these, other smaller fragments, having molecular weights not exceeding 5000 and containing a considerable share of the antigenic determinants of the papain-cleaved fragment III, were obtained. Complete cleavage of 5 S fragments resulted in 1.7-2.9 free SH groups per mole of protein with molecular weight of 106,000; i.e., 0.85-1.45 disulfide bonds per mole were broken. Although the 5 S fragment contained 10-12 disulfide bonds, apparently only one of these, the most labile, bound the two 3.5 S fragments [136].

After extracting the reducing agent, the sulfhydryl groups of this labile bond can be oxidized and reconstitution of the bivalent fragment takes place [133].

It was found that when a mixture contained 3.5 S fragments from antibodies with different specificities, hybrid 5 S fragments precipitable by adding both antigens were formed. For example, if a mixture contained 3.5 S fragments from antibodies specific for ovalbumin and p-azobenzoate, the newly resynthesized hybrid 5 S fragments were precipitated only when ovalbumin and a hapten conjugated with a protein support were added simultaneously. Precipitating hybrids were analogously obtained from univalent fragments of antibodies to ovalbumin and beef γ-globulin [69, 135, 139]. Such hybrid 5 S fragments were formed according to the principle of random combination.

Tryptic hydrolysis also yields 3.5 S fragments. Thus, digestion of γ-globulin by trypsin (1%) for 72 hr yielded 45-60% 3.7 S fragments in addition to undigested molecules. Undialyzable material comprised 90%. The fragments thus obtained did not differ from papain-cleaved fragments in their sedimentation properties, or in their behavior on fractionation with DEAE-cellulose [193]. One of the trypsin-cleaved fragments is crystallizable [72a].

γ-Globulins undergo splitting during prolonged storage. Thus, Skvaril [200-202] observed that the antitoxic activity of γ-globulin had not changed after storage for 3-6 years at 4°C, although when subjected to immunoelectrophoresis new fractions appeared which were similar to those obtained by hydrolysis with proteases (trypsin, plasmin, and papain). Oudin [154] also observed splitting of γ-globulins during storage. A similar phenomenon was recently observed by James et al. [95] who detected 5 S fragments after prolonged storage of γ-globulin preparations. All of these observations have an important practical significance.

In conclusion, it should be emphasized that the most diversely acting proteolytic enzymes (pepsin, trypsin, vegetable proteases such as papain and bromelain, blood proteases) have approximately the same cleavage effect on antibodies and γ-globulins from diverse animal species. This is evidence, albeit indirect, of first, that a definite structural similarity exists between γ-globulins of different animals, and secondly, that the sites of action of different proteases are similar if not completely identical [94]. The γ-globulin molecule apparently has some special part which is susceptible to proteolytic splitting. However, it is not clear whether the peculiar features of this site are due to its unique position, which itself is dependent on the secondary structure of the molecule. The distinguishing characteristics of this part are obviously determined by its primary structure. In particular, it has been shown that this part of the heavy chain contains many proline residues, of which in one place there are three in a row (a situation known to occur in protein chemistry) with another not far away [203b]. As is known, this amino acid hinders formation of α-helices in peptides.

Classification of Proteolytic Fragments

In a classification worked out in 1964 [2], the term "fragment" is understood to be that part of the γ-globulin molecule obtained as a result of splitting of the peptide bond. The following terminology is proposed for papain-cleaved fragments: for univalent, antigen-binding fragments, Fab fragments (previously designated A, C, S, I, and II); for inactive, crystallizing fragments, Fc fragment (previously B, F, III). In addition, 5 S pepsin-cleaved fragments and univalent 3.5 S pepsin-cleaved fragments are designated as $F(ab')_2$ and Fab', respectively, and finally the heavy chain in an Fab fragment is termed Fd fragment.

SUBUNITS AND PEPTIDE CHAINS OBTAINED BY CHEMICAL TREATMENT OF MOLECULES OF IMMUNOGLOBULINS AND ANTIBODIES

Proteolytic fragments are artificial segments of γ-globulins; in biosynthesis in cells, γ-globulin is not built up from them [50]. These parts of the molecule obtained by reduction of disulfide bonds, i.e., peptide chains [6], are considerably more important for understanding the structure of antibodies, for it is they which are the natural components of γ-globulins and antibodies; i.e., they are the initial structural units for biosynthesis in cells.

The γ-globulin molecule contains more than 20 disulfide bonds. As these bonds vary in their susceptibility to reducing agents, the molecule can be broken down in stages by reduction. Larger components are obtained by splitting the minimum number of disulfide bonds, and these may be designated provisionally as "subunits." If approximately one-fourth of all the disulfide bonds in the molecule are reduced, the molecule breaks down into the constituents most frequently designated as peptide chains. The products of γ-globulin reduction have been the most studied.

Splitting of Disulfide Bonds

Edelman [35] was the first to show that the γ-globulin mole-

TABLE 23. Effect of Mercaptoethanol on Rabbit Antibody to Ovine Serum Proteins [195]

Mercapto-ethanol concentra-tion	Sedimentation constant		Precipitat-ing ca-pacity	Comple-ment-fixing activity	Number of disulfide bonds
	%	S_{20w}	% of unreacted antibody		
0	100	6.4	100	100	22.5
0.01	100	6.4	97.5	39	20
0.1	100	6.4	98	13	15.5
0.3			96.5	8	—
0.85	75	6.4	85	8	—
	25	3.5			
1.45	—	—	64	6	10.3
1.65	70	6.4	34	—	—
	30	3.5	—	—	—
$>$1.65	—	—	0	—	—

cule consists of subunits mutually connected by disulfide bonds. This investigator split the disulfide bonds in human γ-globulin with a 0.1*M* solution of β-mercaptoethanol in 6*M* urea at room temperature, with subsequent acetylation by iodacetamide. This treatment reduced the sedimentation constant to 2.3 S (molecular weight 48,000). Use of either a 6*M* urea solution or 0.1*M* reducing agent solution alone with subsequent centrifugation in the same solution altered the sedimentation properties of the protein very little. Under these conditions, 7-9 disulfide bonds were split.

Franek [56] discovered that not only human γ-globulin, but animal γ-globulins as well break down into constituents having a molecular weight of about 43,000 when their disulfide bonds are split (with centrifugation in 0.1*M* phosphate buffer, pH 7, in 6*M* urea solution). In these experiments, the disulfide bonds were split by sulfitolysis (S-sulfonation) in 8*M* urea by the method of Pechére et al. [161]. Okulov used a similar method [7]. The products of this reaction were insoluble in water or in salt solutions, and soluble only in concentrated urea solutions, characteristics which complicate the study of their properties, in particular their biological activity.

Subsequently, less drastic milder methods for antibody reduction were worked out. Fleischman et al. [51] split about five disulfide bonds of rabbit γ-globulin in one hour by reduction with

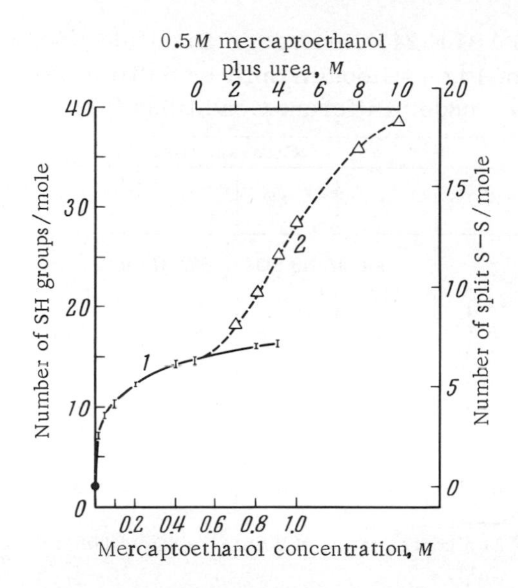

Fig. 22. Reduction of rabbit γ-globulin at different mercaptoethanol and urea concentrations [217]. 1) Without urea; 2) with urea.

0.75*M* β-mercaptoethanol at pH 8.2. This was sufficient to cleave all interchain bonds. Indeed complete reduction in a 10*M* urea solution did not diminish the molecular weight of the molecular subunits obtained. The same was true for reduction in 1*M* hydrazine solution (pH 10), in which the ether-like bonds in collagen were split [52].

Different numbers of disulfide bonds can be split by varying the mercaptoethanol concentration. This relationship is shown in Table 23.

Figure 22 presents data from a study of splitting of disulfide bonds of rabbit γ-globulin by mercaptoethanol with and without urea [217]. It is evident from the curve that even in 10*M* urea and 0.5*M* mercaptoethanol not all the disulfide bonds in the rabbit γ-globulin molecule were split. Maximum reduction was achieved with a 12*M* urea solution and 0.54*M* mercaptoethanol solution in 45 minutes at 40°C. Amperometric titration detected 43.7 sulfhydryl groups per mole, which corresponds to 21 disulfide bonds. Similar data were obtained by Pressman [119], who was able to

TABLE 24. Number of Disulfide Bonds of Pig γ-Globulin Split by Sulfitolysis under Different Conditions [63]

pH	Cleavage time, hr			
	1	6	24	48
	0.1 M Na_2SO_3 0.002 M Cu^{++}			
3.0	0.7	1.7	4.3	5.2
5.7	3.2	3.7	5.3	5.3
7.0	2.1	3.1	6.0	6.5
8.2	5.0	6.0	6.6	6.8
8.6	5.1	5.7	6.9	7.9
10.0	3.2	4.8	6.3	6.0
8.6*	6.1	7.4	8.0	8.0

*At a concentration of 0.25M Na_2SO_3, 0.005M Cu^{++}.

cleave a maximum of 21.5 disulfide bonds per mole with a 0.4M mercaptoethanolamine solution in 10M urea.

Franek [11, 63] used another method for splitting disulfide bonds. This method which was first used by Swan [210] is based on the following reactions:

$$RSSR + 2Cu^{++} + 2SO_3^{--} \rightarrow 2RSSO_3^- + 2Cu^+,$$
$$RSH + 2Cu^{++} + SO_3^{--} \rightarrow RSSO_3^- + 2Cu^+ + H^+.$$

A 0.15M solution of sodium sulfite (pH 8.6) in the presence of 0.005M Cu^{++} ions was used to obtain S-sulfonated horse γ-globulins and antibodies. All the sulfonated γ-globulins were readily soluble in a physiological solution, while S-sulfonated antitoxins retained approximately 50% of their original antigen-binding capacity (for a homologous toxin linked to cellulose). The number of disulfide bonds split may be varied by varying the conditions of sulfitolysis (Table 24). Jirginsons used a similar method [98].

Some investigators have split disulfide bonds by reduction in detergent solutions, in particular in a solution of sodium dodecylsulfate [96].

Thus there are now available methods which can cleave the most susceptible (i.e., interchain) disulfide bonds while still retaining a good percentage of the biological activity of the antibody.

Separation of Peptide Chains

The first attempt to separate polypeptide chains obtained by reduction of γ-globulins was made by Edelman and Poulik [44] who, by fractionation on carboxymethylcellulose, isolated two fractions. Although fractionation was not complete, the molecular weight of these fractions was determined as 17,000 and 103,000. The existence of two types of chains was confirmed by electrophoresis in a starch gel. Thus, it was established that the γ-globulin molecule consists of two polypeptide chains — a heavier one and a lighter one.

In view of the difference in size between the two types of chains, the method most suited for fractionation proved to be gel filtration with a Sephadex "molecular sieve." Fleischman et al. [51] obtained two peaks by gel filtration of reduced rabbit γ-globulin in acid through Sephadex G-75. The first contained heavy chains (A) and uncleaved fragments, while the second contained light chains (B) comprising about 25% of the molecular weight. Best results were obtained from gel filtration in 1*N* propionic acid. From a determination of N-terminal amino acids in these fractions these investigators concluded that the heavy chain fractions contained no more than 3% light chains. Although all N-terminal amino acids in rabbit γ-globulin are apparently attached to light chains, great care should be taken in using this as a criterion for assessing the purity of heavy-chain fractions. The rabbit γ-globulin molecule contains one mole N-terminal amino acids per mole protein, and since present evidence indicates that the molecule contains two light chains, a considerable quantity of light-chain impurity in heavy-chain preparations may easily be overlooked in relying on data from an analysis of N-terminal amino acids [20].

Better fractionation can be achieved on Sephadex G-100, for example by gel filtration in a 6*M* urea solution in 0.05*M* formic acid. Figure 23 gives results of fractionation under these conditions [11]. As is seen, the first peak is itself split into two smaller peaks. As was shown in a study of the peak proteins in a

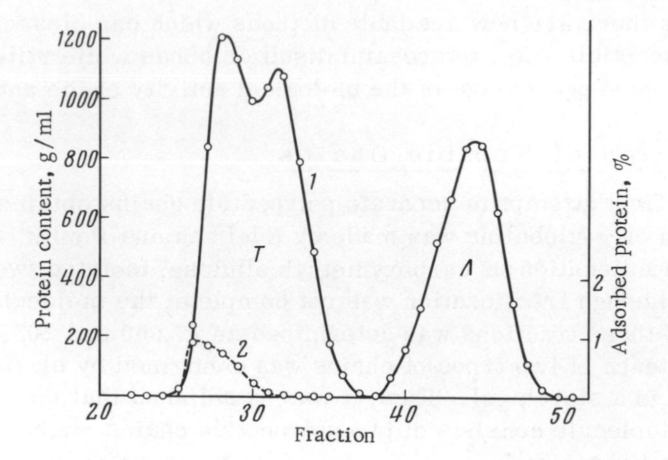

Fig. 23. Fractionation of diphtheria S-sulfonated antitoxin on a column containing Sephadex G-100 in 6*M* urea and 0.05*M* formic acid. 1) Protein content; 2) quantity of protein reacting with a fixed anatoxin (in % of total protein content in the fraction); H, L) peaks containing heavy and light chains, respectively [11].

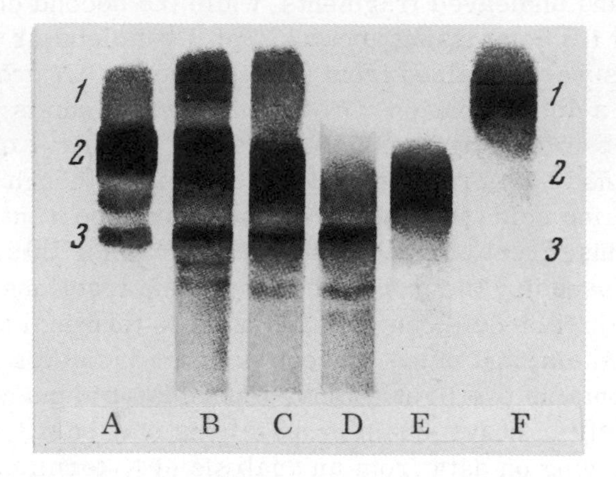

Fig. 24. Electrophoresis in a starch gel [11]. A) S-sulfonated horse globulin; B) S-sulfonated tetanus antitoxin; C) S-sulfonated diphtheria antitoxin; D, E, F) fractions from peaks H-1, H-2, and L, obtained by fractionation of S-sulfonated diphtheria antitoxin on a column containing Sephadex G-100. Bands: 1) Light chains; 2) heavy chains; 3) incompletely cleaved antitoxin fragments.

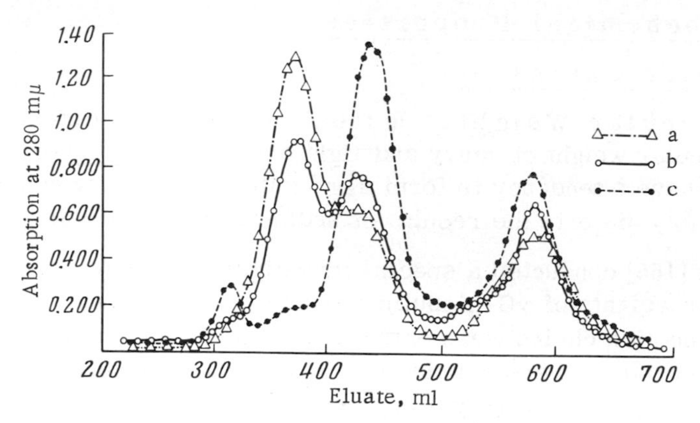

Fig. 25. Fractionation of reduced and alkylated rabbit γ-globulin on a column containing Sephadex G-200 in 0.05 *M* sodium decylsulfate and 0.02*M* Tris buffer at pH 8 [217]. The curves are compiled from the last peak (light chains). The number of split disulfide bonds are (a) 4, (b) 11, and (c) 18.

starch gel (Fig. 24), the first part of the first peak still contains partially cleaved γ-globulin fragments in addition to heavy chains. It is not impossible that dimers of heavy chains are also present. In gel filtration on Sephadex G-100 these two smaller peaks may be distinct from each other. According to data from electrophoresis in starch gel, the first peak contains none of the substances in the second, and vice versa.

Detergents are also used to separate chains by gel filtration. Thus, Utsumi and Karush [216, 217] employed sodium decylsulfate (0.05*M* solution in 0.02*M* Tris buffer, pH 8). Figure 25 gives results of their experiments on reduction of products obtained at different stages of reduction of rabbit γ-globulin by mercaptoethanol under these conditions.

It has recently been pointed out that the quantity of light chains in the molecule is slightly higher than 25%, apparently about 33% [63, 122, 203].

Thus, gel filtration in various solvent systems is the method most suited for fractionation of products of γ-globulin reduction. Investigations in this area have shed light on the properties of the subunits of these proteins presently designated as heavy and light polypeptide chains.

Physicochemical Properties of Peptide Chains

Molecular Weight. In the first attempts to determine the molecular weight of heavy and light chains it was discovered that they have a tendency to form aggregates, a property which considerably distorts the results of sedimentation analysis.

Pain [155] conducted a special investigation to determine the molecular weights of γG-globulin chains by a sedimentation method. The peptide chains were extracted by the Fleischman method. All measurements were carried out in an 8*M* urea solution at acid pH values to prevent aggregation. Results are given below [155].

Heavy chain of horse γ-globulin (pH 3.5)	50,300
Light chain of horse γ-globulin (pH 7.5)	19,400
Light chain of rabbit γ-globulin (pH 7.5)	20,000
Fragment I after papain hydrolysis of rabbit γG-globulin (pH 3.5)	40,700
Subunit of the heavy chain of fragment I of rabbit γG-globulin (pH 3.5)	21,600

In these same experiments it was shown that the molecular weight of intact horse γG-globulin is 151,000. It may hence be concluded that the γG-globulin molecule contains two light chains, each having a molecular weight of about 20,000, and two heavy chains, each with a molecular weight around 50,000.

According to Small [203, 203a] the molecular weight of the heavy and light chains of rabbit γG-globulin are 53,500 ± 4000 and 23,800 ± 1000, respectively, the weight of the γG-globulin molecule thus being approximately 140,000. According to Marler et al. [118] the molecular weights of the heavy and light chains of these proteins are 50,000 and 25,000, respectively.

The molecular weight of heavy chains of γM-globulins is possibly somewhat higher than that of γG-globulin heavy chains. Thus, the molecular weight of human and rabbit μ-chains is 65,000-70,000 [126, 112a].

The electrophoretic properties of peptide chains are distinctly brought out by electrophoresis in a starch gel with acid buffer solutions in 6-8*M* urea. Edelman and Poulik [44] were the first to study the electrophoretic mobility of antibody chains.

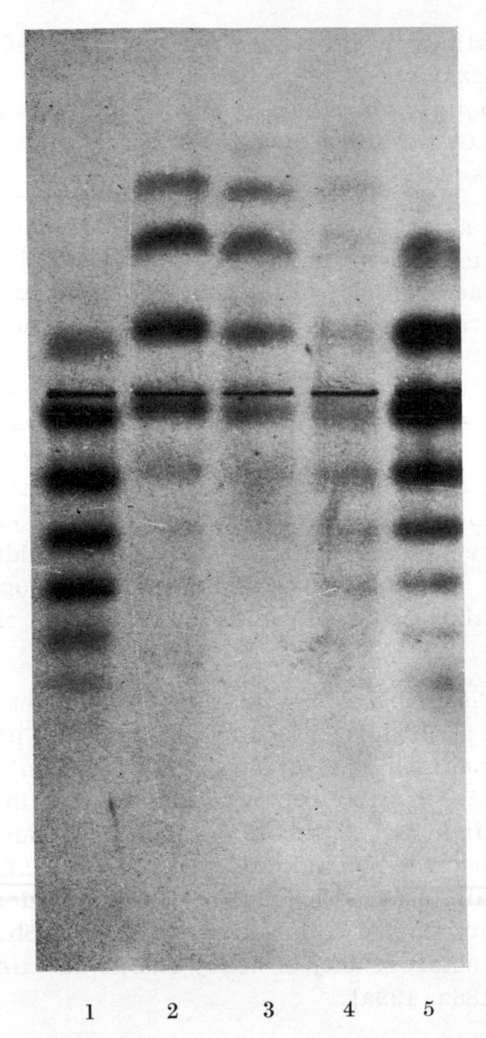

Fig. 25A. Electrophoresis in 8*M* urea glycine starch gel pH 7 of IgG light chains of guinea pig (1), cattle (2), horse (3), baboon (4), and human (5) [28].

Figure 24 gives typical data from a study of the electrophoretic mobility of S-sulfonated horse antitoxins and isolated chains of these antibodies [11]. As in other investigations, it was found that the light chains were clearly differentiated from the heavy chains by virtue of their greater electrophoretic mobility.

Certain qualitative differences can be discerned, by electrophoresis. For example, it is probable that the heavy chains of human γM-globulins, which have a lower electrophoretic mobility than chains of γG-globulins, also differ from the latter in several other properties [27].

Edelman et al. [37] found that the chains of guinea pig antibodies specific for haptens move in several bands instead of one band, and that each antibody has a characteristic number of bands. Similar data were obtained in an investigation of the electrophoretic properties of peptide chains of mouse antibodies [121]. Whereas light chains of reduced mouse γ-globulin and highly purified antibodies to ovalbumin moved in one diffusion band, antibodies to the p-iodo-phenylsulfonyl group formed 2-3 narrower and more distinct bands in this region. Certain differences in the electrophoregrams of light chains were also observed in a study of different antibodies from the same individual [42]. Grossberg et al. [83] noted variations in the electrophoregrams of reduced and alkylated papain-cleaved fragments of different antibodies.

However, Cohen and Porter [28] separated light chains of both γ-globulins and antibodies, including antibodies to protein antigens, into ten fractions by electrophoresis at pH values between 7 and 8 (Fig. 25A). Similar results were obtained in a study of different antibodies, and of antibodies and γG-globulins from different animal species. Each of the electrophoretic fractions was, in turn, chemically heterogeneous[61a]. Other evidence also exists for the heterogeneity of light chains [125a, 128b, 170]. The electrophoretic heterogeneity of heavy chains has also been described [184b, 186a, 199a].

Reynek et al. [184a] described the peptide chain of immunoglobulins of pig colostrum; the electrophoretic mobility of this chain was greater than that of the light chains of serum immunoglobulins.

Absorption at 280 mμ. Crumpton and Wilkinson [32] determined the extinction coefficient at 280 mμ for rabbit γ-globulin and individual peptide chains obtained by the Fleischman method. Results are given below (measurements were made in 0.01*M* HCl):

Preparation	$E^{1\%}_{1\,cm}$
γ-Globulin	13.5
Heavy chain	13.7
Light chain	11.8
Fd fragment	14.4

On the whole, absorption at 280 mμ coincides with the tryptophan and tyrosine contents (see below). However, although the contents of these amino acids were similar in the Fd fragment and the entire heavy chain, the coefficients of extinction differed markedly. The authors were unable to give a satisfactory explanation for this.

The values given above are somewhat lower than the results obtained by Utsumi and Karush [217], who measured the absorption of chains at 280 mμ in 0.04*M* sodium decylsulfate and 0.01*M* phosphate buffer at pH 7.2. $E^{1\%}_{1\,cm}$ was 14.5 and 13.2 for the heavy and light chains, respectively.

Solubility. Light chains have good solubility regardless of the method used to extract them. Heavy chains obtained by reduction of the molecule with mercaptoethanol and fractionation in acid solutions usually have poor solubility in neutral aqueous solutions. In such cases, addition of light chains will solubilize the heavy chains [151]. A modification of this method for reduction of immunoglobulins yields soluble fractions of heavy chains [31]. Heavy-chain fractions isolated by sulfitolysis without detergents and urea [11, 110a, 184a] and heavy chains isolated from poly-DL-alanyl γG-globulin [68a] were also readily soluble.

Chemical Composition of Peptide Chains of Immunoglobulins

The amino acid composition of individual peptide chains was first determined by Edelman and Poulik [44]. However, they were not able to separate completely the chains into distinct fractions. Porter's data [25, 29, 32] were somewhat more precise. This investigator studied peptide chains, isolated by the Fleischman method, in automatic amino acid analyzers. Data for peptide chains of human immunoglobulins are given in Table 26.

The greatest differences between the amino acid compositions of the two types of chains were found in the contents of histidine,

TABLE 25. N-Terminal Amino Acids of Rabbit γG-Globulin and Its Peptide Chains [51] (in moles/160,000 g)

Amino acid	γ G-globulin	Heavy chain	Light chain
Alanine.	1.1	0.3	3.3
Aspartic acid	0.3	0.15	0.6
Serine.	0.1	0	0

arginine, proline, alanine, methionine, tyrosine, and S-carboxymethylcysteine. It is evident from Table 26 that each light chain has one cysteine residue the SH group of which participates in the interchain disulfide bond, while each heavy chain contains 3 to 4 cysteine residues. This corresponds to 4-5 interchain disulfide bonds in a molecule of unreduced protein.

Fleischman et al. [51] studied N-terminal amino acids of rabbit γ-globulin and its peptide chains (Table 25) and found that all N-terminal amino acids reside in the light chains. The small quantity of N-terminal amino acids found in heavy chains is explained by the fact that the heavy-chain fractions contain light chain impurities. The N-terminal residue of heavy chains is pyrrolidonecarboxylic acid [168a].

The principal N-terminal amino acids of human γG-globulin are aspartic acid and glutamic acid, in concentrations of 1 mole acid per mole protein. After separating the chains, both of these N-terminal amino acids were found in both the heavy and the light chains, although glutamic acid was located chiefly in the heavy chain. The content of N-terminal amino acids in pathological globulins varies. In some cases, they cannot be detected at all [27]. The C-terminal amino acid of heavy chains of rabbit and human γG-globulins is glycine.

Franek and Keil [58] studied peptide chains of pig γG-globulins by a "fingerprint" method. The peptide maps thus obtained for heavy chains differed sharply from the light chain maps. Differences between the peptide maps of heavy chains of the γ, α, and μ types [64a] are most likely evidence of differences in the primary structure of these chains.

TABLE 26. Amino Acid Analysis of Heavy and Light Chains of Human γG- and γM-Globulins [25, 29] (Amino acid content given in moles/mole noncarbohydrate chain)

Amino acid	Heavy chain (molecular weight 50,000)			Light chain (molecular weight 20,000)		
	γG	γM	γM	γG	γM	γM
Lysine	29	20	21	9.7	9.6	9.5
Histidine	9.6	7.6	8.3	2.5	2.4	2.7
Arginine	12	17	18	5.8	5.3	4.6
Aspartic acid	33	33	35	13	13	14
Threonine	34	38	38	15	15	15
Serine	50	42	39	24	24	24
Glutamic acid	39	40	41	20	19	19
Proline	33	29	28	11	11	12
Glycine	28	28	29	11	12	13
Alanine	19	26	28	12	12	12
Valine	41	36	38	13	13	13
Methionine	3.9	5.3	5.0	0.6	0.7	0.3
Isoleucine	8.2	13	13	4.9	4.7	4.6
Leucine	31	30	32	13	13	12
Tyrosine	17	12	12	7.6	8.0	6.1
Phenylalanine	13	15	16	6.0	5.3	5.3
Cystine (1/2)	6.9	7.4	5.7	2.9	3.3	--
Carboxymethyl-cysteine	3.6	3.2	4.5	0.9	1.1	1.3
Tryptophan	7.5	7.6	7.8	2.3	2.7	2.8

The characteristic antigenic properties of the chief types of immunoglobulins are apparently dependent on the primary structure of heavy chains.

As has been mentioned, the rabbit γG-globulin molecule probably contains two carbohydrate components, both of which are in the heavy chains. By papain hydrolysis, the chief component is recovered in the Fc fragment; the lesser component occurs as a small glycopeptide, whereas in pepsin hydrolysis it is recovered in the 5 S fragments [52, 216]. This component is not found in all heavy chains. As has been recently shown, it is galactosamine linked to the threonine residue near the disulfide bridge binding the heavy chains [203b].

Pathological immunoglobulins of humans and mice contain a carbohydrate component attached to the light chains [2a, 25a].

TABLE 27. Carbohydrates of Immunoglobulin Peptide Chains in Man [25] (in g/100 g protein)

Carbohydrate	Heavy chain		Light chain	
	γG	γM	γG	γM
Hexose	1.42	7.0	0.26	–
Fucose	Present		–	–
Hexosamine	1.55	4.40	0.21	0.30
Sialic acid	–	1.0	–	–

The contents of carbohydrates in peptide chains of human immunoglobulins are given in Table 27.

Antigenic Properties of Peptide Chains

In the study of the antigenic properties of papain-cleaved fragments, two chief groups of antigenic determinants were found. The first of these – the S group – is associated with Fab fragment, while the second – the F group – is associated with Fc fragments [41, 48].

The very first studies of the antigenic properties of the two types of peptide chains showed sharp differences [150]. Light chains reacted with antiserum to Fab, but not with antiserum to Fc. In contrast, heavy chains reacted with anti-Fc, but not with anti-Fab antiserum. It was hence concluded, that the Fab fragment contains the light chain, while the Fc fragment consists of heavy chains. Anti-Fab serum was found to retain the capacity to react with the Fab fragment after depletion of light chains. Presumably, Fab fragments contain other determinants in addition to determinants peculiar to light chains. Although heavy chains did not react with anti-Fab antiserum, it cannot be ruled out that Fab fragments in some cases contain heavy chains, since the antigenic properties of the latter may undergo changes during reduction and alkylation of the γG-globulin molecule.

The foregoing characteristics are pertinently illustrated by photographs of Utsumi and Karush [217] (Fig. 26). Note should be made of the identity reaction between the heavy chain and Fc

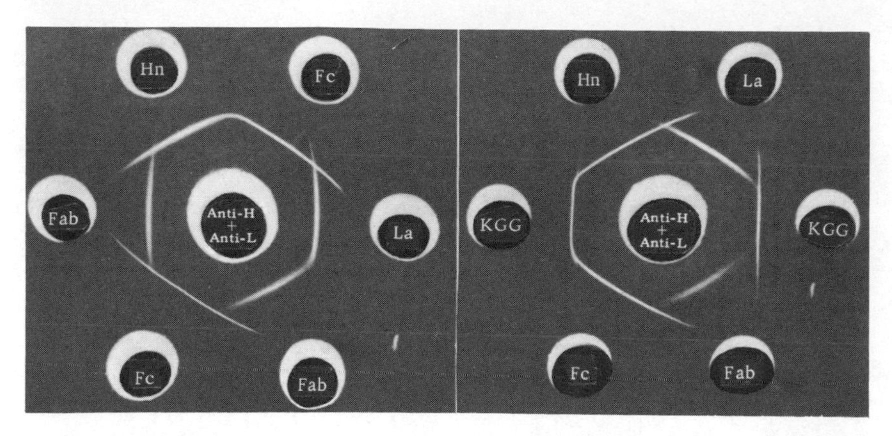

Fig. 26. Antigenic correlation between papain-cleaved fragments II (Fab) and III (Fc) and peptide chains [217]. Peptide chains were obtained as in Fig. 25; heavy chains (H), light chains (L), material from the first and second main peaks of Fig. 25, respectively. Anti-H and anti-L are antiserum to H and L; Hn is H from normal γ-globulin; La is from antibody to the β-lactoside group; KGG is normal γ-globulin.

fragment, which is an indication of the relatively weak antigenic properties of the heavy chain section (Fd fragment) not included in the Fc fragment.

The following results were obtained from a comparison of the antigenic properties of chains of different types of human immunoglobulins [26, 27].

Rabbit anti-γG-globulin antiserum, absorbed by light chains, did not react with γA- and γM-globulins. Anti-γM-globulin antiserum depleted by light chains from γG-globulins, reacted with normal γM-globulins. In other experiments, rabbit antiserum to light chains from human γG-globulins was specific for normal γG- and γM-globulins, while antiserum to heavy chains reacted only with γG-globulins. At the same time, light chains from pathological γG-, γM-, and γA-globulins were identical to light chains from normal γG-globulins in the reaction with antiserum to light chains. Antiserum to heavy chains from γG-globulins reacted only with heavy chains from pathological γG-globulins and not with heavy chains from pathological γA-globulins or γM-globulins. These studies provide convincing evidence that light chains from both normal and pathological globulins have antigenic properties common to all three chief types of immunoglobulins, while the specific determinants of each of these immunoglobulin types are associated with the heavy chains.

In recent years it has been established that both light and heavy chains can be divided into several subtypes. Such subtypes have been found for humans and several animal species [13, 112, 211, 213, 214]. Human γ-chains have the following subtypes [146a]:

Current	Occurrence as myeloma protein (%)	Proposed	Polypeptide heavy chain (γ-chain)
We or γ2b or C	70-80	IgG1 or γG1	γ_1
Ne or γ2a	13-18	IgG2 or γG2	γ_2
Vi or γ2c or Z	6-8	IgG3 or γG3	γ_3
Ge or γ2	3	IgG4 or γG4	γ_4

Two types of μ-chains [66a] and two types of α chains [111a] have also been found in humans. Human light chains have two principal types: K and L (I and II). Both of these light-chain types can be found in any of the chief types of immunoglobulins. They may be separated from each other by ion-exchange chromatography [202]. Similar light chain types are found in mice and probably also exist in mammals [63a, 224]. Some determinants of light chains are masked in the intact immunoglobulin molecule [129].

In the foregoing we have discussed isotypical antigenic determinants characterizing the antigen properties of γ-globulins of a given animal species. However, there are also allotypic antigenic determinants which can be detected by immunization of different individuals of this same species to γ-globulins. The following section will be devoted to the importance and properties of these determinants, which characterize the genetic subclasses of γ-globulins (cf. Chapter VI).

By ascertaining which fraction of antibodies from serum specific for immunoglobulin can absorb a fixed light or heavy chain, the relative immunogenicity of each chain can be determined. We conducted similar experiments with chains and proteolytic fragments of rabbit and rat γG-globulins and ass and rabbit antiserums, respectively [6a]. As is seen from Table 27A, a considerable portion (approximately one-fourth) of antigenic determinants of the whole molecule reside in the light chains. Fixed papain-cleaved fragments adsorb somewhat more antibody than the papain-cleaved fragments. This is fully explicable by the fact that

TABLE 27A. Immunogenicity of γG-Globulin Subunits

Fixed protein	Adsorbed antibodies from one ml antiserum on fixed			
	rabbit protein		rat protein	
	μg	%	μg	%
γG-globulin	1278	100	1522	100
Heavy chains	733	57	978	64
Light chains	334	26	321	21
$F(ab')_2$ fragments	968	76		
Fab fragments	939	73		

a portion of the heavy chain of the pepsin-cleaved fragment is slightly larger than the Fd fragment [217a]. The total quantity of antibody adsorbed with the aid of heavy and light chains is less than the quantity of antibodies which reacted with fixed γG-globulin. This is possibly attributable to the presence in the whole molecule of determinants built up of both heavy and light chains. Another explanation may be that part of the antigenic determinants are lost during isolation and fixation. It is evident from Table 27A that antigenic determinants are distributed relatively evenly among subunits of rabbit and rat γG-globulin.

Peptide Chains of Pathological Immunoglobulins

In certain diseases, immunoglobulins which are not found in normal serum appear in the blood. In particular, in multiple myelomatosis, so-called myeloma proteins appear in the serum. In contrast to normal immunoglobulins, these proteins are relatively homogeneous. In this same disease, pathological proteins, called Bence–Jones proteins after their discoverer, appear in the urine.

In his first experiments, Edelman [35] showed that reduction of the myeloma protein molecule is accompanied by a decrease in molecular weight. Later [44], it was found that the molecules of myeloma proteins are broken down by reduction into several subunits having different electrophoretic mobilities in a starch gel.

In some instances, one of these subunits was similar to heavy chains of normal γ-globulins in mobility, while others resembled light chains. However, other such subunits had no analogs with respect to mobility among chains of normal γ-globulins. Chains of myeloma proteins from different patients had their own characteristic electrophoretic picture.

Cohen [27] studied pathological immunoglobulins from several patients with multiple myelomatosis, macroglobulinemia, and cryoglobulinemia. In all cases, separation of the reduced proteins on Sephadex G-75 yielded the same results as separation of normal γG-globulins; the relative contents of the two types of chains were likewise the same. The heavy and light chains of pathological globulins were more homogeneous than chains of normal globulins.

Poulik and Edelman [171] studied the electrophoretic properties of reduced and alkylated Bence–Jones proteins and compared them with the properties of reduced myeloid proteins from the same patients. Some of the bands were characterized by the same electrophoretic mobility for the two types of proteins. Light chains of normal and myeloma globulins and Bence–Jones proteins exhibited antigenic similarities [38]. On the basis of these and other experiments indicating the chemical similarity between light chains of myeloma protein and Bence–Jones protein [196], Edelman advanced the hypothesis that Bence–Jones proteins are dimers of light chains of myeloma proteins. However, this is probably not always the case, since it has been shown that at least for K-type Bence–Jones proteins the molecule contains chains of different sizes [86].

Heavy chains of pathological as well as normal globulins are quite heterogeneous. In addition to the above-mentioned antigenic subclasses, differences in the susceptibility of Fc fragments to proteolysis should be noted [172, 211].

Role of Peptide Chains of Antibodies in the Antigen Reaction. Molecular Antibody Hybrids

As is evident from the foregoing, heavy and light chains differ sharply in their physicochemical, chemical, and antigenic properties. Hence, it is reasonable to inquire into the role of the different peptide chains in the specific properties of antibodies. The problem would seem to involve two distinct questions: which factors govern the antigen-binding capacity of antibody; and with what factors is the specificity of the reaction associated? Three theoretical possibilities can be distinguished: the antigen-binding capacity depends either on the heavy chain alone, or on the light chain alone, or on a complex of both chains. Lately, the third possibility, i.e., that both chains are involved, has been gaining in credence.

Franek and Nezlin were the first to demonstrate the correctness of this latter hypothesis in their research on the properties of polypeptide chains of horse antitoxins [11, 60]. The proteins were subjected to S-sulfonation and the peptide chains were separated on Sephadex G-100 (cf. Fig. 23). The fractions of heavy and light chains were diluted to reduce the urea concentration, and then the capacity of each fraction to react with its specific toxin linked to cellulose was tested. This method for determination of activity should also detect the presence of univalent fragments. It was found that neither light nor heavy chains were alone capable of reacting with a fixed anatoxin. However, when they were mixed together the antigen-binding capacity was restored to approximately the same level as that of a reduced antibody after treatment with acidified urea, i.e., under the same conditions obtaining when the chains were isolated (Table 28).

The low activity of the heavy chain fractions was most likely due to contamination by uncleaved molecules or subunits. In another series of experiments, the specific activity was restored by mixing chains isolated from different antibodies (Table 28). Such "molecular hybrids" reacted only with a fixed toxin having the same specificity as heavy chains. It follows that the specificity, and a good portion, if not all, of the binding site resides in the heavy chains and that both types of chains are able to form aggregates, thereby restoring activity, without covalent bonds.

The peptide chains of diphtheria antitoxin partially digested by pepsin have also been studied [61]. Although the amount of light fragments formed from partial enzymic digestion of heavy chains was larger, combining of chains still restored activity. It is important to note that not only the capacity to react with a fixed toxin, but also the neutralizing activity was restored.

TABLE 28. Restoration of Antitoxin Activity by Combining Different Peptide Chains Isolated from Diphtheria and Tetanus Antitoxin (H_d, H_s, L_d, L_s are the respective heavy and light chains from diphtheria and tetanus antitoxins [11, 60])

Chain type	Total protein in mixture, μg	Amount of protein adsorbed on the fixed anatoxin, μg	
		Diphtheria	Tetanus
H_d	3420	7	3
L_d	1860	1	2
H_s	2800	1	8
L_s	1560	0	2
$H_d + L_d$	5820	185	6
$H_s + L_s$	4360	1	25
$H_d + L_s$	4980	35	7
$H_s + L_d$	4660	5	19

All results of these investigations were confirmed by Franek, Kotynek, Simek, and Zikan [59, 110a] in experiments on pig and beef antibodies to the 2,4-dinitrophenyl group. After S-sulfonation and separation of the chains the hapten-binding capacity of the chains was studied by equilibrium dialysis.

As is seen in Table 29, none of the chains was by itself capable of binding hapten.

However, after mixing the chains isolated from antibodies, the activity was restored to 80% of the level obtaining after treatment of S-sulfonated antibodies with 6*M* urea and 0.05*M* formic acid ($r_{10^{-5}} = 0.86$), which in turn was 40% of the activity of the native antibody ($r_{10^{-5}} = 1.77$). Interspecies mixtures of pig and beef antibodies were inactive.

It is worth mentioning that addition of the light chain of nonspecific γ-globulin to the antibody heavy chain also restored activity, although not completely. This result was not obtained when the heavy chain from nonspecific γ-globulin was added to the antibody light chain. This confirms the decisive role of the heavy chain in determining specificity.

TABLE 29. Activity of Mixtures of Chains from Beef γG-Globulin and Antibody to the Dinitrophenyl Group [100a]

Heavy chain, mg/ml		Light chain, mg/ml		
anti-body	γG-globulin	anti-body	γG-globulin	$r_{10^{-5}}$
0.50	0	0	0	0.00
0	0	0.50	0	0.09
0.33	0	0.17	0	0.58
0.33	0	0	0.17	0.34
0.33	0	0	0.85	0.36
0	0.33	0.17	0	0.11
0	1.0	0.17	0	0.06

Note: $r_{10^{-5}}$ is the number of moles of hapten bound by 160,000 g protein in equilibrium with 10^{-5} molar free hapten at 4°C.

Thus, in the experiments discussed in the foregoing, it was found that peptide chains of antibodies are able to form aggregates with each other at neutral pH values by means of noncovalent bonds, thereby restoring biological activity. The results of these studies were fully borne out by experiments of Edelman et al. [43, 55], who showed that isolated chains of antibodies specific for bacteriophages *f*(1) and *f*(2) were practically incapable of neutralizing the bacteriophages. However, the neutralizing capacity was restored by mixing light and heavy chains. When chains from different antibodies were mixed, the specificity of the neutralizing reaction was determined by the heavy chain. Experiments conducted to determine the labeled-hapten-binding capacity of chains of antibodies to the dinitrophenyl group gave quite definite results: a mixture of the two chains had a much higher binding capacity. Edelman et al. offer two possible explanations: either a direct addition to the structure of the binding site takes place; or one of the chains assists in the formation of a stable binding site in the other chain. Roholt et al. [185, 186] also demonstrated the possibility of formation of a complex by the two chains with consequent restoration of activity. These investigators studied peptide chains of rabbit antibody to the hapten p-azobenzoate. Electrophoresis showed that although the two chains have different mobilities, after mixing they move as a single complex. On ultracentrifugation of a mixture of antibody chains, one main peak with a sedimentation constant of 6.1 S appeared, while for a complex of both chains of normal γG-globulin the sedimentation constant was 6.2 S. Reduced and alkylated γG-globulin had the same constant. A mixture of antibody chains had almost the same hapten-binding capacity as a reduced and alkylated antibody which had been dialyzed against 1*N* propionic acid. A complex of an antibody heavy chain, with a γG-globulin light chain also reacted with hapten, but to a markedly less extent.

It was subsequently discovered that addition of both heavy and light chains isolated from the same antibody type does not always restore activity [187, 188a]. For example, mixing of heavy chains of antibody to p-azobenzoate from one rabbit with the light chains of the same antibody but from a different rabbit yielded a low activity, although formation of a complex was still possible. Thus, to restore activity it is not sufficient that the two chains are combined. It is also necessary that the two chains possess a certain specificity if the antigen-binding capacity is to be restored. This also explains the low activity of hybrid antibody molecules (Table 28). Analogous data have been obtained from a study of recombination of Fd fragment with light chains [188c]. Heavy chains combine more readily with those light chains which will yield the highest activity. This has been demonstrated in experiments in which molecules of hybrid antibodies were mixed. One such hybrid consisted of heavy chains of anti-p-benzoate antibody isolated from one rabbit and the light chain of anti-p-benzenearsonate antibody isolated from another rabbit. A second hybrid was obtained with the opposite chain combination: a light chain from the former and a heavy chain from the latter antibody. Both hybrids had only a slight antigen-binding capacity. The activity underwent no changes during long storage in a medium with pH 8. However, if the mixture of hybrid molecules was placed in 1*N* propionic acid and the pH readjusted to 8, the antibody activity was almost completely restored since the chains did not recombine at random; i.e., the chains of one antibody combined with each other.

The experiments of Olins and Edelman confirmed the formation of a complex by mixing heavy and light chains [151]. They found that centrifugation of a mixture of heavy and light chains at a sucrose solution gradient sedimented this complex at the same rate as molecules with sedimentation constant of 7 S. If before mixing the heavy chains are labeled with ^{131}I and the light chains with ^{125}I, the reconstituted γG-globulin molecule will contain these isotopes in a proportion corresponding approximately to two light and two heavy chains per molecule. After papain hydrolysis the reconstituted γG-globulin molecules formed fragments similar to the Fab and Fc fragments of untreated protein. All ^{125}I was found in the Fab fragments, while both the Fab and Fc fragments contained ^{131}I. When chains from reduced, but unalkylated γ-globulin are mixed, the reconstituted molecules may contain new interchain bonds. These molecules require a second reduction with mercaptoethanol to fractionate them into peptide chains.

It is interesting that stable dimers of light chains were not capable of solubilizing heavy chains, and did not combine with them to form 7 S molecules [70]. This may mean that light chains are not in direct contact with each other in the intact molecule. Hybrid molecules, consisting of heavy chains of rabbit γG-globulin and light chains of human γ-globulin, and vice versa, were also obtained. Such intraspecies molecular hybrids were sedimented by sucrose gradient centrifugation similarly to proteins with a sedimentation constant of 7 S and had the same heavy-to-light chain proportions as γG-globulin from the same species [53].

If a mixture of heavy and light chains contains antigen, the latter may have an appreciable influence on the composition of the reconstituted molecules. Metzger and Mannick investigated this phenomenon [122]. First, light chains of antibody to the 2,4-dinitrophenyl group (AtL) were labeled with ^{125}I, while the light chains of nonspecific γ-globulins (gL) received a ^{131}I label. Then heavy chains from either nonspecific γ-globulin (gH) or from antibody (AtH) were added to a mixture of equal quantities of AtL and gL in the presence or absence of hapten, and the ^{125}I/^{131}I ratio in the 7 S complex, extracted by centrifugation in a sucrose density gradient, was measured. In the absence of hapten, AtH combined with gL and AtL to the same degree, but in the presence of hapten AtH combined preferentially with the antibody light chains (Table 30). The reconstituted molecules obtained in the presence of hapten were almost twice as active in a coprecipitation reaction as those formed without hapten.

It is still difficult to interpret these interesting results. A possible explanation is offered by Metzger and Mannick with the

TABLE 30. Recombination of Heavy and Light Antibody Chains in the Presence of Hapten (Dinitrophenylaminocapreate) [122]

Specimen*	Sucrose density gradient fraction	In the presence of hapten			Without hapten		
		I^{131}, %	I^{125}, %	I^{125}/I^{131},	I^{131}, %	I^{125}, %	I^{125}/I^{131}
gH + AbL^{125}I, γL^{131}I	> 7 *S*	5.2	3.5	1.5	4.3	3.2	1.9
	7 S peak	39.2	33.4	1.9	38.5	30.8	2.0
	Light-chain peak	55.6	63.1	2.6	57.2	66.0	2.9
Total†		—	—	2.3	—	—	2.5
AbL^{125}I, γL^{131}I	—	—	—	2.3	—	—	2.5
AbH + AbL^{125}I, γL^{131}I	> 7 *S*	2.8	4.3	3.5	4.8	5.5	1.5
	7 S peak	26.6	54.0	4.7	37.6	41.5	1.9
	Light-chain peak	70.6	41.7	1.4	57.6	53.0	2.6
Total†				2.3			2.3
AbL^{125}I, γL^{131}I	—	—	—	2.3	—	—	2.3

* AbH, AbL; γH, γL – heavy and light chains of antibody and nonspecific γ-globulin, respectively.

† ^{125}I counts 20,500-27,500, ^{131}I counts 9000-12,000.

assumption that when the complex is formed, the hapten alters the configuration of the antibody heavy chains in such a way as to make it easier for them to combine with the corresponding light chains. It should be noted that recombination took place even without hapten, and that the activity of the reconstituted molecules was governed by the heavy chain, inasmuch as a complex of the heavy chain of nonspecific γG-globulin with light chains was practically inactive. This is in agreement with results of other experiments on chain recombination.

It is interesting that chain recombination not only generates a specific antigen-binding capacity, but also gives rise to antigen properties not found in the individual chains. We are referring, namely, to the antigenic determinant Gm^3 (Gm^W), which is found in the Fab fragment of γG-globulin in certain groups of humans. Alone, neither the light nor the heavy chain of this protein is capable of reacting with anti-Gm^3 antibody. However, when the chains are mixed, the antigenic activity of Gm^3 is reconstituted. The Gm^3 determinant itself is probably localized in the heavy chains, since hybrids of heavy chains of γG-globulin Gm(3-) and light chains from γ-globulin Gm(3+) are totally inactive [205]. From studies of reconstitution of antibody activity and antigenic activity of Gm^3 by chain recombination, it can be concluded that when the complex is formed, the configuration of the heavy chain is changed and as a result, the specific antibody binding site and Gm^3 antigenic determinant regain their activities.

From the above discussion of chain recombination and formation of molecular hybrids of antibodies it follows that for reaction with antigen a complex of both heavy and light chains is necessary and that the heavy chain is the decisive factor in determining specificity. Data accumulated for some time past indicates that isolated chains still retain a slight antigen-binding capacity.

Utsumi and Karush [217] have shown that isolated heavy chains of antibody to p-amino-phenyl-β-lactoside have a slight hapten-binding capacity in equilibrium dialysis against hapten.

Weir and Porter [220a] have shown that a portion of heavy chains of diphtheria antitoxin reacts with antigen. It was found that after acidification the reduced antitoxin had 60% of the activity of the unacidified preparation. Separately, the heavy chain had 25%, and the light chain only 4%, of the activity of the reduced

TABLE 31. Properties of Peptide Chains of γG-Globulins

Property	Heavy chain	Light chain
Content in molecule	~70%	~30%
Molecular weight	~50,000	~20,000
Presence of carbohydrates	Yes	No
Antigen determinants common with γA- and γM-globulins	No	Yes
Antigenic determinants specific for γG-globulins	Yes	No
Allotypic determinants in humans:		
Gm	Yes	No
Inv	No	Yes
Allotypic determinants in rabbits:		
a	Yes	No
b	No	Yes
Specificity in the reaction with antigens linked to cellulose, with hapten, and in bacteriophage neutralization	Yes	No
Presence of Fab and Fc in a papain fragment:		
Fab	Yes	Yes
Fc	Yes	No

antibody. An equimolecular mixture of chains had 65% activity. These investigators maintain that a slight impurity of light chains is not able to produce such a high heavy-chain activity. According to them, there are four types of heavy chains: 1) capable of reacting with antigen without light chains (20-25%); 2) activated after addition of specific or nonspecific light chains (15-20%); 3) activated only by addition of specific light chains (20%); incapable of activation (40%).

Haber and Richards [85a] studied chains of antibodies to the dinitrophenyl and trinitrophenyl groups, and obtained results similar to those of Weir and Porter. Special attention was devoted to determination of impurities of light chains in heavy-chain preparations. The quantity of light-chain impurity in heavy-chain preparations was reduced to approximately 1% by repeated gel filtration. However, these purified heavy chains had about 23% of the activity of the reduced antibody. Addition of light chains raised the activity to 50-80%.

The role of light chains in antibody activity is still unclear. As indicated above, there is some evidence that light chains act as "activators" for heavy chains. According to other investigators, light chains are able to participate directly in the formation of an antibody binding site. In the first experiments on this possibility, Edelman et al. [37] found that the number and mobility of bands in starch gel electrophoresis of light chains of guinea pig antibodies are characteristic for every type of antibody. On the basis of this indirect evidence these investigators hypothesized that the binding site is located either completely or partially in light chains.

Roholt et al. [188] obtained the following experimental evidence of the possibility of participation of light chains in the structure of an antibody binding site. A "double tag" method was used. A portion of the purified antibody to p-azophenylarsonic acid was treated with hapten and then iodinated with ^{131}I. Another portion of the same antibody, but in free form, was iodinated with ^{125}I. Then both antibody preparations were reduced with mercaptoethanol, and the peptide chains were separated by the Fleischman method and subjected to peptic hydrolysis. The peptides thus obtained were fractionated by chromatography and high-voltage paper electrophoresis and the $^{125}I/^{131}I$ ratios in peptides isolated from free antibody chains and from chains blocked by hapten before iodination were compared. In almost all cases, this ratio was nearly one in peptides from heavy chains, whereas in three peptides from light chains it was appreciably higher than one, indicating that these peptides were blocked by hapten. It is not unreasonable to assume, then, that they are also associated with the binding site, a part of which would consequently reside in the light chains. However, as these investigators point out, the same results can be expected assuming the antibody undergoes changes in configuration in the reaction with hapten. Then the binding site could be localized in any of the chains.

The experiments of Singer et al. [123-125] also seem to support the hypothesis that light chains participate in the binding site. They studied the capacity of diazotized haptens to react with antibody (method of affinity labeling). These haptens reacted specifically with the binding site of the corresponding antibody, thereby forming a stable covalent bond with tyrosine residues localized in the region of the binding site. It was found that after addition of these haptens to antibody and fractionation of the reduced antibody into its chains, about 65-80% of the hapten was bound with the heavy chain, and 20-35% was bound with the light chain. The authors regard this as evidence that amino acid residues of both chains participate simultaneously in the construction of the binding site. However, the method used still does not offer direct proof of this view. It is also possible that hapten may be linked to areas near the binding site but not directly contained in it.

However, it has recently been demonstrated that light chains have a slight binding capacity. A study of antibody to the 4-azo-naphthalene-I-sulfonate group showed that the light chains of this antibody have a slight hapten-binding capacity. This was inferred from the increase in fluorescence after mixing light chains with hapten, i.e., the same effect as occurs when whole antibody is mixed with hapten [225]. Thus, it is probable that light chains are able directly to supplement the binding site structure with their own amino acids. In addition, combining of heavy chains with light chains is accompanied by a change in their conformation.

γG-Globulin Subunits Containing Both Types of Chains

Under mild reduction conditions, only some of the interchain disulfide bonds in γ-globulin are broken, and the resultant subunits can contain both types of chains.

Nisonoff et al. [134, 137, 158, 159] conducted a series of experiments on cleavage of rabbit γ-globulin into half-molecules consisting of a heavy and a light chain.

Reduction was carried out in 2-mercaptoethylamine for 75 minutes at 37° and pH 5. The samples were treated with different concentrations of mercaptoethylamine, placed in a solution with pH 2.4, and studied in an analytical ultracentrifuge. At neutral pH values, the reduced γ-globulin had the same sedimentation properties as the native protein. However, at pH 2.4 it was dissociated into half-molecules with a sedimentation constant of 2.8-2.9 S and molecular weight of 80,000 (Fig. 27). For dissociation of approximately 66% of all molecules, it was necessary to split only one very labile disulfide bond, as was evident from the quantity of sulfhydryl groups (Table 32), and from the content of S-

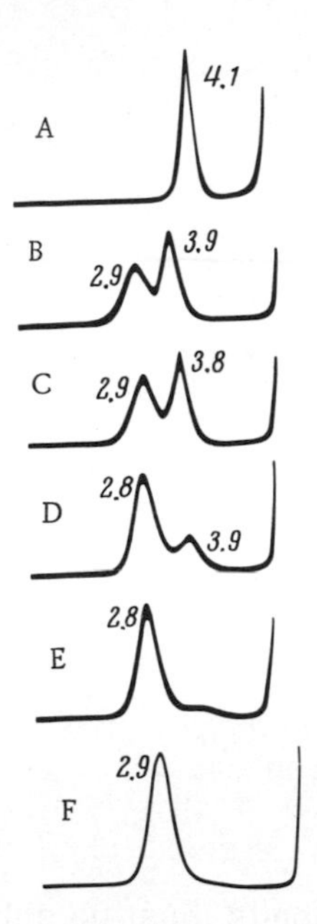

Fig. 27. Sedimentation of normal and reduced rabbit γ-globulin (128 min, 59,780 rpm, 20°) in 0.025*M* NaCl, pH 2.4 [158]. A) Unreduced γ-globulin; B, C, D, E, F) γ-globulin reduced by 0.01, 0.015, 0.02, 0.03, and 0.1*M* 2-mercaptoethylamine hydrochloride, respectively. The numbers are sedimentation constants.

carboxymethylcysteine in the alkylated protein (~2 residues per γ-globulin molecule). To dissociate all molecules into halves it was necessary to split 2-3 disulfide bridges. However, it is not certain whether in the remaining one-third of the molecules the halves are united by this number of bridges, or whether they are associated by only one disulfide bond, which, however, is more stable, and the splitting of which is accompanied by cleavage of other bonds of similar susceptibility.

The subunits obtained were apparently identical since they gave a symmetrical sedimentation peak (Fig. 27). In an amino acid analysis, the subunits were found to be identical with the native proteins. The quantity of light chains obtained after reduction under the mild conditions described by gel filtration in 1*N* propionic acid was negligible.

When the pH was adjusted to neutral values, the half-molecules recombined into a 7 S molecule, and the antibody regained almost all its activity [88, 159].

Under certain conditions, SH groups can be oxidized again to a disulfide bond, to form a 7 S molecule indistinguishable from the original [204].

Finally, it was found that the labile disulfide bond between the two half-molecules is identical with the bond formed between univalent 3.5 S fragments to constitute bivalent 5 S fragment by pepsin hydrolysis [157]. If 0.005*M* mercaptoethanolamine is used for reduction, 30% 5 S pepsin-cleaved fragments and 30% of γ-globulin

TABLE 32. Reduction of Rabbit γG-Globulin by 2-Mercaptoethylamine (2-MEA) [158]*

2-MEA concentration, M	No. of free SH group per molecule with a MW of 150,000	S_{20w} = 2.9 S, %	No. of free SH groups per dissociated molecule
0.010	0.8	35	2.3
0.015	1.2	46	2.6
0.020	1.6	66	2.4
0.030	3.5	80	4.4
0.050	4.2	88	4.8
0.10	8.4	96	8.8

*Before reduction the protein was treated with iodoacetamide. Centrifugation was done at pH 2.4.

are dissociated; in both cases, 2-3 disulfide bonds must be split for complete dissociation [134].

Isolation of subunits of γG-globulins has been described by several investigators. Thus, Franek and Zikan [63] found a relatively small quantity of subunits, designated as I and J, after S-sulfonation of pig γG-globulin. In electrophoresis in a starch gel in urea solutions J-subunits moved more slowly than heavy chains, while I-subunits were even slower, with a velocity only slightly different from that of γ-globulin. Both types of subunits were larger than heavy chains, since in gel filtration through Sephadex G-100 they emerged before the latter. The subunits apparently consisted of both types of chains, inasmuch as in prolonged sulfonation the percentage of heavy and light chains increased as the quantity of subunits decreased [57].

Cyanogen Bromide Fragments

Treatment with cyanogen bromide splits specifically the peptide chain at those places where methionine residues existed and the latter are converted to homoserine. The number of fragments obtained depends, of course, on the number of methionine residues and also on the reaction conditions. Under mild conditions (0.3*M*

HCl) in which about half of the methionine residues are converted, it is possible to isolate an active bivalent fragment – rabbit γG-antibody – similar in properties to a pepsin-cleaved fragment. A 5 S fragment of human γG-globulin and antibody was analogously obtained (the reaction was carried out in 0.05*M* HCl [112b]. Under harsher conditions (70% formic acid) all peptide bonds associated with methionine can be split. For example, treatment of the heavy chain of rabbit γ-globulin yielded six fragments. These were fractionated and their relative positions along the chain precisely determined [72b]. Cyanogen bromide-cleaved fragments have a definite composition, in contrast to proteolytic fragments and hence can be very useful for determining the primary structure of polypeptide chains [163a].

Peptide Chains and Subunits of γA- and γM-Globulins

Under mild reduction conditions, macroglobulins (γM-globulins) are broken down into subunits containing both light and heavy chains. Their molecular weight for humans is 185,000, which is somewhat higher than that of γG-globulins [126]. Since the molecular weight of macroglobulins is approximately 900,000, one molecule evidently contains five subunits.

Macroglobulin antibodies usually lose their activity after reduction. However, their subunits do not completely lose their antigen-binding capacity [91]. This has been demonstrated in experiments by Onoue et al. [152], who reduced rabbit antibodies (including 39% macroglobulins) to p-azophenylarsonic acid with a 0.1*M* mercaptoethanol solution. The resultant subunits of the macroglobulin antibodies were able to bind ^{131}I-labeled insulin conjugated with hapten, as was discovered by radioimmunoelectrophoresis. The binding site was apparently not destroyed by cleavage of γM-antibody into subunits.

Kunkel et al. [194] came to analogous conclusions in a study of 19 S anti-γ-globulins isolated from the serum of rheumatic patients. After reduction with ethylmercaptan or mercaptoethanol and alkylation, these proteins lost their ability to precipitate with aggregated human γ-globulin, but were still able to agglutinate erythrocytes covered with incomplete antibodies to γ-globulin.

Data from sedimentation analysis also testified to the ability of subunits to react with γ-globulins. Reassociation yielded molecules similar to macroglobulins.

Deeper reduction of macroglobulins split interchain bonds. Cohen [26] showed that gel filtration of products of human macroglobulin reduction gave the same results as an analogous study with γG-globulins.

Similarly, Carbonara and Heremans [19] found that reduction of human γA-globulins yielded subunits which behaved similarly to heavy and light chains of γG-globulins in starch-gel electrophoresis.

Products of reduction of several other proteins of the immunoglobulin class were also studied. Franek and Riga [62] conducted a series of experiments with S-sulfonated 5 S γ-globulin found in the blood of newborn pigs. Gel filtration on Sephadex G-100 yielded fractions resembling heavy and light chains of γG-globulins. However, distinct differences were revealed when the peptide maps of the light chains of γ-globulins of newborn pigs were compared with maps of the corresponding chains of adult pig γG-globulins. At the same time, the peptide maps of both heavy and light chains of γ-globulin of newborn pigs bore a partial resemblance to the maps of heavy chains of γG-globulins of adult pigs.

Classification of Peptide Chains

In the 1964 classification [2], the two chief groups of peptide chains of human immunoglobulins were designated as "heavy" and "light" (previously H- and L- or A- and B-chains). The heavy chains were designated by the lower case Greek letters corresponding to the Latin capitals in the immunoglobulin class nomenclature.

Immunoglobulin class	Heavy chain
γG	γ
γA	α
γM	μ
γD	δ

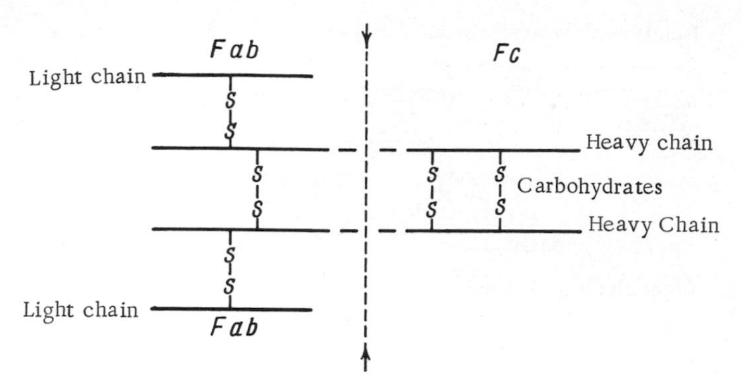

Fig. 28. Model for the structure of the rabbit γG-globulin molecule according to Porter [167]. Dashed line; site of papain cleavage.

For light chains of type K (I) and L (II) immunoglobulins the symbols kappa and lambda were proposed. The molecular formulas of the principal types of immunoglobulins will then take the following form:

Immunoglobulin class	Formula
γ GK	$\gamma_2\kappa_2$
γ GL	$\gamma_2\lambda_2$
γ AK	$\alpha_2\kappa_2$
γ AL	$\alpha_2\lambda_2$
γ MK	$(\mu_2\kappa_2)_5$
γ ML	$(\mu_2\lambda_2)_5$

Structural Model of γG-Globulin Molecules

It was possible to construct a general model for the structure of γG-globulins and antibodies on the basis of results of extensive research. Any proposed model should take into account the following. The γG-globulin molecule is symmetrical and splits into half-molecules on reduction of 1, 2, or 3 disulfide bonds. According to existing data, the molecule contains two light and two heavy peptide chains. Moreover, the light chain and a section of the heavy chain are recovered in the Fab fragment, while the other section of the heavy chain is recovered in the Fc fragment [52]. The light chains apparently are not linked in the molecule since dimers of

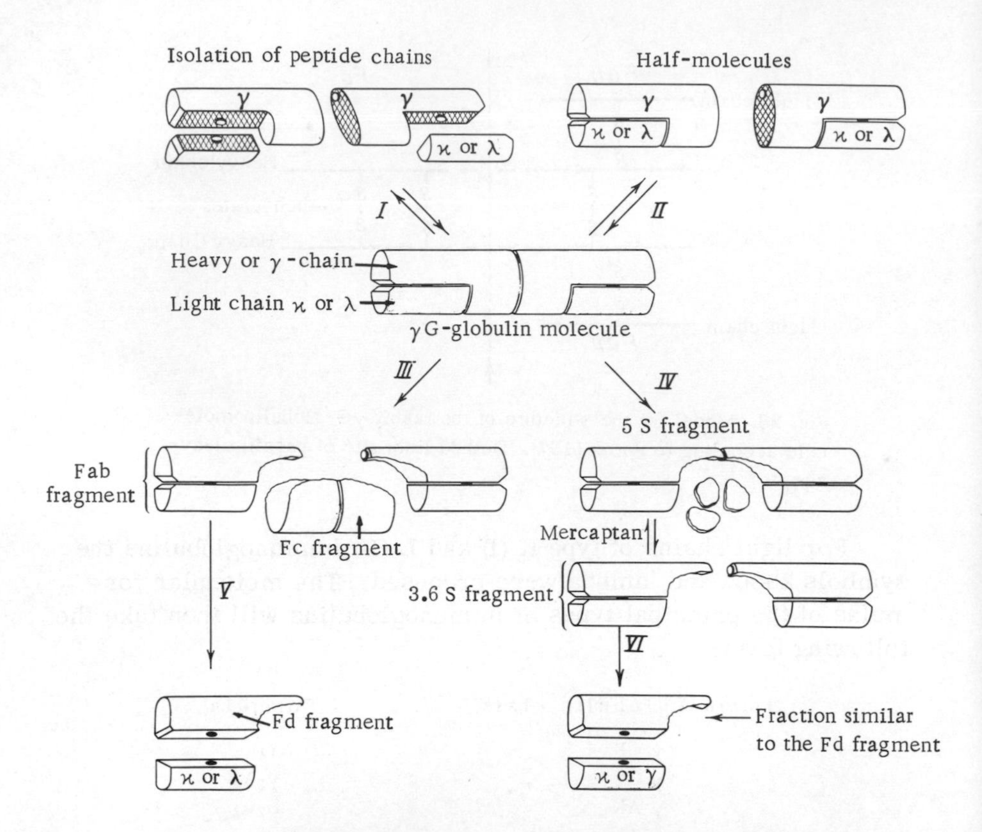

Fig. 29. Model for the structure of the γG-globulin molecule according to Edelman [54]. Disulfide bonds and half-cystine residues are represented by small black circles between chains. I) Reduction and alkylation, propionic acid + urea; II) mild reduction, dilute HCl; III) papain hydrolysis (with activated mercaptan); IV) peptic hydrolysis; V) reduction and alkylation of the Fab fragment (gel filtration through Sephadex G-100 in 6 *M* urea, 1*N* propionic acid); VI) reduction and alkylation of the 5 S fragment (gel filtration through Sephadex G-100 in 6*M* urea, 1*N* propionic acid).

light chains are not capable of forming aggregates with heavy chains. On the basis of these data, Porter [167] proposed his now classical model for the four-chain structure of rabbit γG-globulin (Fig. 28).

Most of the evidence available supports this model. In several recently proposed modifications of this model, attempts are made to illustrate the steric placement of the chains in the molecules (Figs. 29, 30, 31).

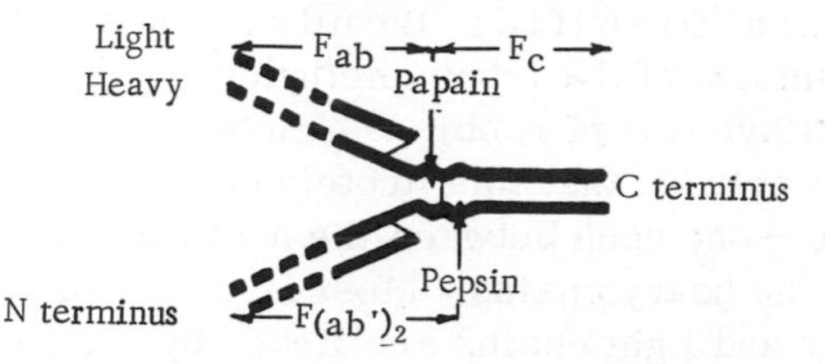

Fig. 30. Diagrammatic representation of IgG. Broken lines show stretches of peptide chains in which heterogeneity occurs (or is presumed). The undulating portion of the heavy chain represents the area susceptible to proteolytic digestion [27b].

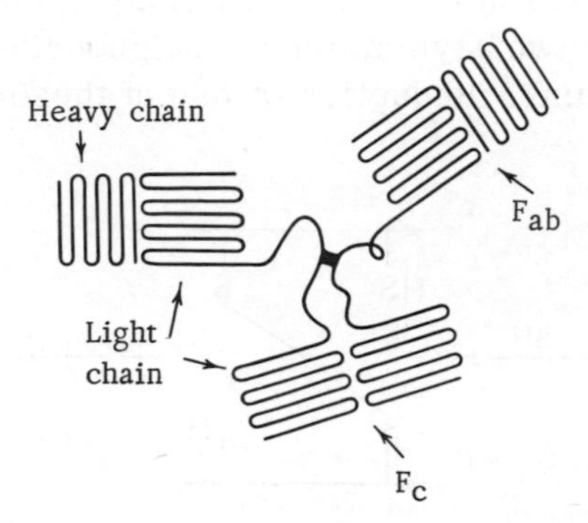

Fig. 31. Model for the structure of the γG-globulin molecule according to Tenford et al. [167, cf. Chapter III].

In the past few years it has been suggested that light and heavy polypeptide chains of immunoglobulins are in turn built up of two shorter chains. This hypothesis was based on the fact that the N-terminal portions of the chains are variable among different immunoglobulins of the same type, while the C-terminal parts are stable (see below). This is indirectly confirmed by the presence of a fragment, similar in properties to the Fc fragment of immunoglobulins, in the urine of many patients with the so-called "heavy-chain disease" [66, 68, 153]. However, a study of chain biosynthesis showed that both heavy and light chains are most likely formed as an integral whole at the translation level (cf. Chapter V). It has also been hypothesized that the γG-globulin molecule contains more than two light chains [63].

Interchain Disulfide Bonds. Porter et al. [52] determined the amount of S-carboxymethylcysteine in chains after reduction and alkylation of rabbit γG-globulin, and on the basis of their results concluded that this protein contained five interchain disulfide bonds – one each between the heavy and light chains, and three between the heavy chains. Most investigators accept the view that heavy and light chains are linked by such a bond, but there is disagreement on the number of bonds between heavy chains. Palmer and Nisonoff [157, 158] are of the opinion that two-thirds of rabbit G-globulin molecules have only one disulfide bond between heavy chains and that this bond is recovered on peptic hydrolysis in the 5 S fragment. Other data indicate a large number of bonds between heavy chains.

Karush [216] has advanced the hypothesis that the γG-globulin molecule contains two asymmetrical, neighboring disulfide bonds between heavy chains. Reduction of one of these is followed by an interchange:

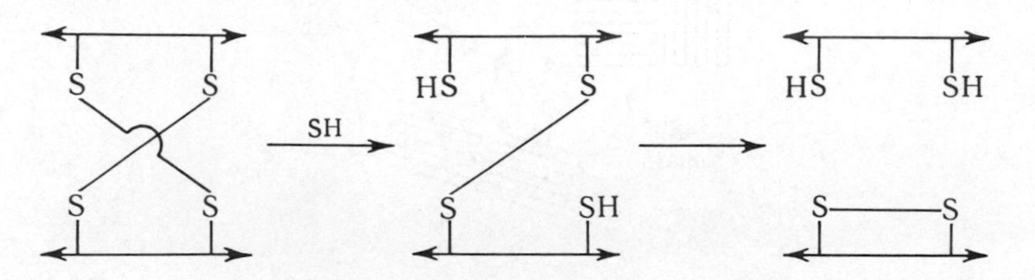

This gives the impression that there is only one bond between heavy chains. The following reactions are possible in hydrolysis with insoluble papain:

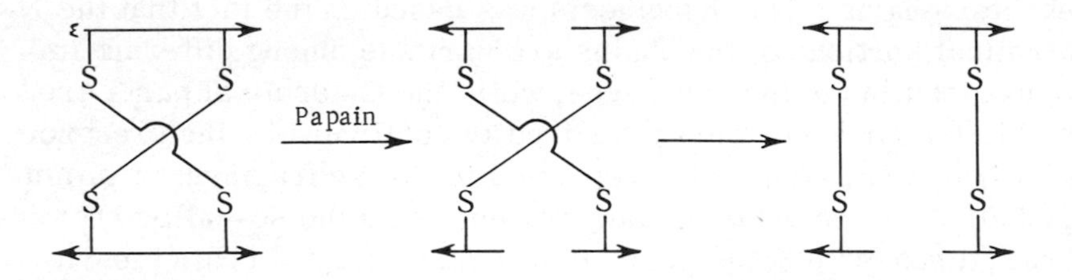

In papain hydrolysis, in the presence of sulfhydryl reagents, heavy chains are split at two points:

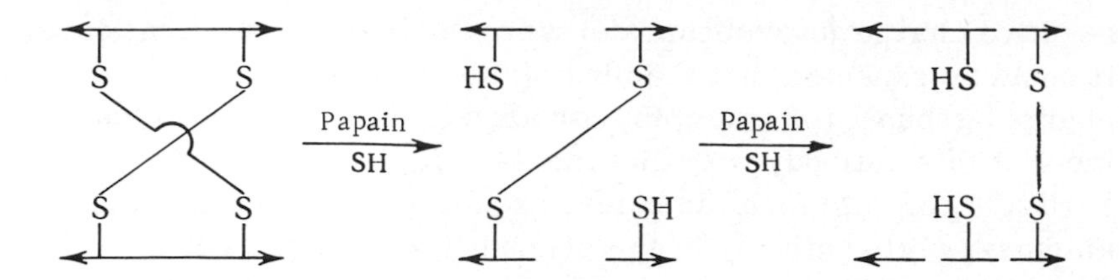

However, it is quite possible that not all molecules have the same number of disulfide bonds between heavy chains.

THE ANTIBODY BINDING SITE

The available data indicate that the antibody component responsible for antigen-binding specificity (binding site) is located principally in that section of the heavy chain which is recovered in the Fab fragment as a result of proteolytic digestion. This heavy chain section, or Fd fragment, has been little studied. It is known to have weak antigen properties in immunization with a whole γ-globulin molecule. There is some evidence that the Fd fragment exhibits sharp variations in comparison with other fragments of the molecule [64].

Chemical Structure

Numerous experiments have been carried out to identify certain chemical groups or amino acids in the binding site. Reagents which have a more or less specific reaction capacity for the given group or amino acid residue were used in these experiments. However, in evaluating the results the following must be borne in mind: (1) The majority of such reagents are not absolutely specific and will react with more than one amino acid; (2) the susceptibility to substitution can be altered by the interaction of side chains in the intact molecule; and (3) a decrease in antibody activity can be a secondary result of general changes in the configuration or charge, and not only of a structural modification in the binding site [45]. However, such experiments can be instructive if these potential limitations are taken into account in analyzing the results.

Pressman [174] has made intensive investigations of the chemical properties of the binding site. Antibodies to hapten with a positive charge (p-azophenyltrimethylammonium group) and a negative charge (p-azobenzoate) were studied. In the first case, it could be expected that the binding site would contain negatively charged groups, for example, carboxyl groups. Indeed, treatment of this antibody with diazoacetamide, which reacts with the carboxyl group to form an ester, results in inactivation of the antibody, while antibody of the second type does not change its properties. If a complex of antibody to the phenyltrimethylammonium group with hapten is esterified and then the hapten is removed, the binding site does not lose its hapten-binding capacity. The hapten evidently protects the carboxyl groups in the binding site, and it is precisely these groups which are responsible for loss of activity as a result of such treatment. The presence of negatively charged groups in the binding site of these antibodies is confirmed by the fact that a reaction of positively charged hapten with the antibodies is inhibited by inorganic cations but not by anions [79].

By analogy, it can be expected that the binding site of antibody to negatively charged hapten (p-azobenzoate) contains positively charged groups, since the reaction of this hapten with antibody is inhibited by univalent inorganic anions but not by cations [175]. The activity of this type of antibody was decreased by acetylation [138], but this effect was more likely associated with acetylation of the hydroxyl group of tyrosine than with blocking of the amino groups. The effect of acetylation was partially relieved at pH 10 [178] or by treatment with hydroxylamine [80] which caused hydrolysis of the acetyl group of the acetylated tyrosine residue. Acetylated amino groups are not decomposed by this treatment. Additional evidence of the presence of tyrosine in the binding site of antibenzoate antibody is the rapid decrease in the hapten-binding capacity at pH 9, i.e., under conditions in which the hydroxyl group of this amino acid is dissociated. Finally, the reduction in the activity of several antibodies as a result of iodination, i.e., a reaction in which the principal participant is the tyrosine residue, is also an indication of the presence of tyrosine in the binding site of the antibodies investigated. This reaction did not take place after the binding site was blocked by hapten [107, 108, 176, 179]. Thus, it has been shown that tyrosine may be present in binding sites of antibodies to uncharged (i.e., neutral), positively charged, and negatively charged haptens.

Results of Singer et al. [49, 124, 222], who employed a method of affinity labeling, also support the hypothesis that tyrosine participates in the binding site of certain antibodies. It has already been mentioned that in these experiments haptens capable of forming covalent bonds with certain amino acid residues were added to antibodies. Thus, in experiments with rabbit antibody to the 2,4-dinitrophenyl group, p-nitrophenyldiazonium fluoborate, which forms a stable bond with protein, was used. From the absorption spectrum, which was similar to the spectrum of a model compound – p-nitrophenylazo-(N-chloracetyl)-tyrosine – it was concluded that after being added to the antibody the hapten forms a stable bond with the tyrosine residues located somewhere in the binding site or near it. Similar results were obtained with rabbit antibody to phenylarsonic acid. One piece of evidence that some of the reacting tyrosyl residues are actually located in the binding site itself is the fact that the loss of antibody activity in the reaction was proportional to the number of azo-tyrosine groups formed [124].

This method, the "double tag" method with two different iodine isotopes developed by Pressman, as well as other methods [110], may be useful in attempts to isolate peptides from the binding site of antibodies, and perhaps also of enzymes.

However, it should be kept in mind that tyrosine is not necessarily present in the binding sites of all types of antibodies. For example, there is evidence that tyrosine is not present in the binding site of antibodies to the p-aminophenyltrimethylammonium group. Even after complete iodination of these antibodies in 10*M* urea, in which 140 iodine atoms are incorporated in one antibody molecule, about 14.6% of the activity remains. On the other hand, addition of only 60 atoms iodine to each molecule completely inactivates antibody to p-azophenylarsonic acid. The same amount of iodine atoms added to the first type of antibody results in only 40% inactivation, whether or not the binding site was blocked with hapten [109].

Experiments were conducted to determine the possible role of antibody amino groups in the antibody–antigen reaction. Activity was suppressed by acetylation of 40% of all free amino groups in the antibody. However, this effect was nonspecific, since guanidination of 75% of the amino groups did not change the antigen-binding capacity [84]. In other experiments by Singer, amidination of free amino groups of antibodies was studied. It was shown that imido esters cause almost complete amidination of free amino groups of lysine; however, the activity of antibodies including three antibodies to negatively charged haptens, decreased very slightly. Consequently, lysine, which contains a free amino group, is at least not the chief constituent of the binding site of the antibodies studied [223].

A spectrofluorometric method (cf. Chapter I) yielded data according to which the binding site of antibody to the 2,4-dinitrophenyl group contains tryptophane residues [220]. Formation of an antibody–antigen complex results in extinction of about 70% of the antibody fluorescence for which tryptophane residues are responsible. In the reaction with hapten, each molecule of bound hapten absorbs the excitation energy of 8-9 tryptophane residues. Since a quite close interaction of tryptophane with hapten is necessary for this effect, it may be inferred from the results obtained that tryptophane residues are constituents in the binding site of the antibody studied. New evidence for the presence of tryptophane in binding sites of antibodies to polynitrobenzene has recently been obtained [77a, 114a].

Another indication of the participation of tryptophane in the binding site of antibodies is given by experiments of Tarkhanova and Kul'berg [10] who studied inactivation of Fab fragments of antibody to serum albumin by N-bromosuccinimide.

Thus, numerous investigations have provided evidence of the possible participation of aromatic amino acids in the binding site of several antibodies. Methods and procedures used for investigation of enzyme properties will perhaps prove of use in the further study of antibody binding sites [17].

An important question is whether antibody binding sites are built up of amino acid residues situated in sequence in the peptide chain, i.e., comprising one section of this chain, or whether amino acids from different sections of the chain participate in the binding site. The latter possibility now seems more probable, being supported, for example, by results of experiments on the interaction of antibodies and their Fab fragments with the hapten dinitrophenyl group. On the basis of optical rotatory measurements in the far ultraviolet region, it was found that attachment of a hapten stabilizes the structure of antibodies [19a]. This may be regarded as proof that amino acid residues which are remote from each other in the peptide chain participate in hapten binding. Here it should be recalled that differences in primary structure between different light chains are not confined to certain sections but are quite diffusely scattered along the chain. Of course, it is still uncertain what relationship these differences bear to antibody activity, but it appears very probable that some such relationship exists. A comparison with binding sites of enzymes, e.g., lysozyme [14a], also supports the view that enzyme binding sites are constructed of amino acid groups scattered diffusely along the peptide chain.

There is at present too little experimental data to answer the question whether the binding sites of different types of antibody or antibodies with the same specificity but from different species are constructed of the same amino acids. Data obtained by an affinity labeling method [99b] provide indirect evidence that in the latter case binding sites may differ in structure. Thus, a comparison of horse and rabbit antibodies of β-lactoside showed two different spectra after reaction of the horse and rabbit antibodies with diazo-hapten [224]. Studies on this subject are of considerable interest, since they can give an answer to the question as to whether the same specific configuration in a protein is constructed of different amino acids.

Size of Antibody Binding Sites

An estimation of the size of that part of antibodies which is directly responsible for antigen binding, and consequently for antibody specificity, is of theoretical importance for an elucidation of the structure of antibodies as a whole, and for an understanding of the mechanism of biosynthesis of these proteins. The larger the binding site, the more specific information is necessary, obviously, for the synthesis of any antibody.

After the classic studies of Landsteiner [113], who demonstrated the possibility of formation of antibodies to relatively small chemical groups, it became clear that the antibody section to which the antigen was linked need not be very large. In some investigations, attempts were made to determine the size of the binding site from the size of a hapten capable of effectively inhibiting the antigen-binding reaction.

TABLE 33. Inhibition of Binding of a Hapten Dye (Ip) by Structurally Similar Molecules at 25° [103]*

Inhibitor	$K_I \cdot 10^{-4}$	$-\Delta F°$
D-$O_2NC_6H_4$-CO-NH—CH—C_6H_5 (COO⁻)	6.16	6.56
L-$O_2NC_6H_4$-CO-NH—CH—C_6H_5 (COO⁻)	0.0371	3.51
D-C_6H_5-CO-NH-CH-C_6H_5 (COO⁻)	2.80	6.08
CH_2-C_6H_5 (COO⁻)	0.184	4.47

*For antibody to D–Ip+D–Ip–dye: $K_0 = 8.5 \cdot 10^4$ · $-\Delta F^0$ = 6.74 kcal/mole.

Some interesting information can be derived from the studies of Karush [103, 104] on antibodies to two haptens, Ip and Lac groups: phenyl-(p-azobenzoylamino)-acetate and p-azophenyl-β-lactoside.

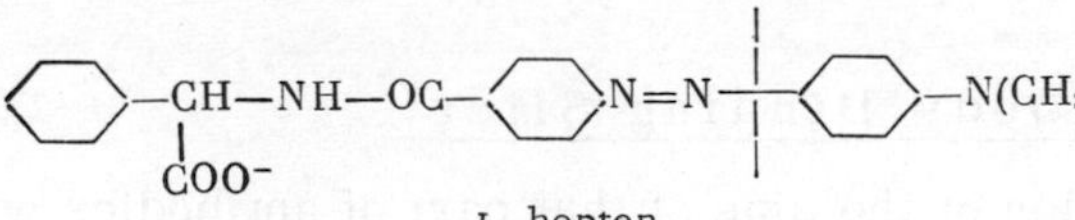

Ip-hepten

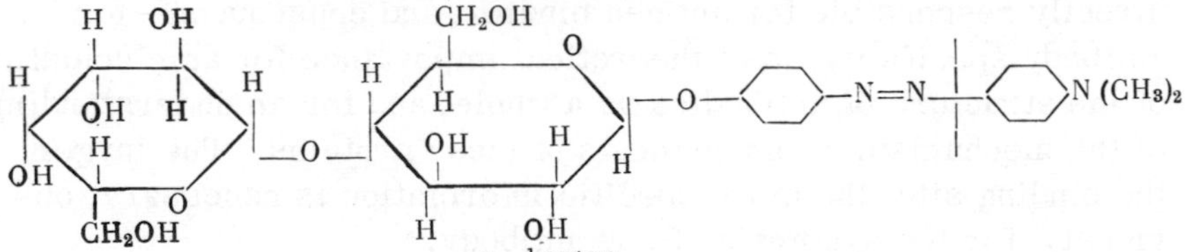

Lac-hepten

Karush used equilibrium dialysis to study the inhibitory effect exerted by colorless hapten analogs on binding of hapten dyes (haptens linked with dimethylaniline) by antibodies. In the experiments with Ip hapten it was found that substitution of the dimethylamino-p-phenylazo group by a much smaller nitro group diminished the binding energy ($-\Delta F^0$) by only 0.18 kcal (Table 33).

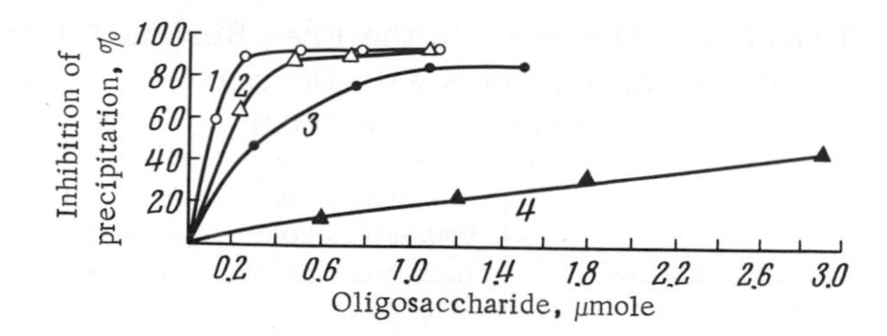

Fig. 32. Comparison of the inhibitory effect of oligosaccharides of different sizes on the reaction of dextran with human anti-dextran serum [102]. 1) Isomaltohexose; 2) isomaltopentose; 3) isomaltotetrose; 4) isomaltotriose.

It was therefore concluded that the energetically effective part of the region responsible for the reaction with antigen was not very much (if at all) larger than was required for contact with the Ip group. Analogous results were obtained in experiments with the Ip group. Analogous results were obtained in experiments with Lac hapten: a substance smaller than hapten and with the structures of p-nitro-phenyl-β-lactoside had a $-\Delta F^0$ value only 0.4 kcal less. Thus, the determination of the size of the binding site in these experiments was based on the size of the hapten, which was assumed to be the chief constituent of the antigen determinant. However, as was already pointed out in the discussion of antibody heterogeneity, the precise dimensions of the antigen determinant in the case of a hapten-conjugated protein are not known, since it is unclear how much of the region adjacent to the site of attachment of the hapten to the protein carrier is encompassed by the determinant. This problem was overcome in Karush's experiments by using a large hapten. However, this measure does not completely resolve the difficulty.

Consequently, it is especially interesting that Karush's results agreed with data obtained from a study of oligosaccharide inhibition of the reaction of polysaccharides with corresponding antibodies [77, 99-102, 116]. Isomaltohexose, which contains six hexose residues, had the strongest inhibitory effect among glucose polysaccharides on precipitation of dextran with human anti-dextran serums (Fig. 32). Isomaltopentose, which contains five maltose residues, was less effective. In other experiments, isomaltoheptose, with seven glucose residues, had a somewhat greater inhibitory effect.

TABLE 34. Decrease in the Free Binding Energy ($-\Delta F^0$) of Oligosaccharides with Decrease in the Degree of Polymerization [102]*

Oligosaccharide	Change in ΔF^0 (isomaltohexose minus other oligosaccharides), cal	$-\Delta F^0$ of oligosaccharides in comparison with $-\Delta F^0$ of isomaltohexose, %
Isomaltopentose	160	98
Isomaltotetrose	330	95
Isomaltotriose	730	90
Isomaltose	3000	60
Glucose	4600	39

*$-\Delta F^0$ for isomaltohexose = 7500 cal.

Similar results were obtained with horse and rabbit serums to pneumococcus II polysaccharides and with horse antiserums to S XX polysaccharides of the same origin, i.e., with serums giving a cross-reaction with dextran. Isomaltose oligosaccharides with 5, 6, and 7 glucose residues had approximately identical inhibitory effects. Oligolevans containing 5-7 residues had a greater inhibitory effect than oligolevans with 2-3 residues on the reaction of antilevan human serum with antigen. These results are presented in Table 34. It is seen from the table that the fifth and sixth glucose residues augment the total binding energy only slightly. Apparently, in the cases analyzed, the binding site of antidextran antibody is indeed comparable in size with isomaltohexose (34 × 12 × 7 Å).

In experiments with antibody to benzylpenicilloyl-hapten, which consists of two principal groups (phenylacetylamine and thiozolidinecarboxyl), Levine [114] demonstrated that the bulk of the antibody is specific for the whole hapten rather than for any of its individual parts. The antibody binding site must obviously be commensurable with the hapten, the length of which is 17 Å and even somewhat greater, since the lysine residue in the protein support also makes a small contribution to the total hapten–antibody binding energy. These data stand in contrast to results of Landsteiner and van der Scheer [113], who in immunization with the bifunctional hapten phthalyl-glycine-leucine obtained antibody to the individual amino acid groups in the hapten rather than to the hapten as a whole. However, in general, Levine's experiments are in good agreement with results of all other contemporary investigations, including those of Kreiter and Pressman [111], who in a control of the experiments of Landsteiner and van der Scheer showed that the antibody formed by immunization with phthalyl-glycyl-leucine did not have the same restricted specificity as had been originally supposed.

According to Cebra's data [21], rabbit antibody to silk fibroin is at least complementary to octapeptide or dodecapeptide, indicating that the binding site of this antibody is quite elongated.

Another system, consisting of the antibody to poly-*l*-alanine and peptides with different numbers of alanine residues, was studied by a method involving inhibition of the antibody–antigen reaction [191]. It was found that peptides with five alanine residues had the greatest inhibitory effect on the antibody–antigen reaction. It was therefore concluded that the binding site had the dimensions of penta-alanine, i.e., $\sim 25 \times 11 \times 6.5$ Å. This peptide does not contain α-spirals, so that the assumption that the binding site is small owing to folding of penta-hapten is in this case unsubstantiated. Schechter and Sela [192] and Arnon et al. [12] obtained analogous results.

It is interesting to compare the antibody binding site with enzyme binding sites with respect to size. Lysozyme is one of the most studied enzymes; not only its primary structure but also the steric configuration of the molecule is known [14a]. It has been found that a 6-unit oligosaccharide may be contained in the binding site cavity. A determination of the size of the papain binding site revealed a length corresponding to seven amino acid residues, i.e., approximately 25 Å [192a]. The binding sites of both types of specific proteins are very similar in size. It is possible that the size is no chance occurrence, but has been established during the course of evolution as the minimal necessary to retain two molecules in a stable complex by noncovalent bonds.

There is little experimental data on the shape of the binding site. It is hypothesized that the binding site is a more or less deep depression or slot on the surface of the molecule. Its shape is probably different for different antibodies, since it is supposed to correspond to different antigen determinants. Here an apt comparison can again be made with the lysozyme binding site which is formed by 24 amino acid residues [14a].

PRIMARY STRUCTURE OF ANTIBODIES AND IMMUNOGLOBULINS

Since the properties of antibodies and immunoglobulins depend on their primary structure, the study of the latter is important. For antibodies, in particular, it is important to know to what extent their specific antigen-binding capacity is associated with primary structure. An elucidation of this problem would shed considerable light on the as yet thoroughly obscure phenomenon of formation of a specific antibody as a response to addition of antigen. However, the study of the sequence of amino acids in polypeptide chains of antibodies is still in its initial stages, and we are totally ignorant as to the degree to which peptides of binding sites of different antibodies differ among themselves. Methodological difficulties encountered in experiments on this subject are associated first and foremost with the considerable size and heterogeneity of antibody molecules.

Recently, some light has been cast on the primary structure of sections of some immunoglobulins. In addition, enzymic hydrolyzates of antibodies have been studied with the aid of peptide maps.

Primary Structure of Certain Sections of Immunoglobulins. Amino Acid Sequence in Polypeptide Chains of Immunoglobulins

Over the past two years, information has been accumulating extremely rapidly on the sequence of amino acid residues in immunoglobulin polypeptide chains. These investigations, which have been indisputably the most important achievement of recent years, have already had a considerable influence on the evolution of hypotheses to explain the mechanism of antibody biosynthesis [27a, 167a].

As has been pointed out, the greatest obstacle to the chemical study of immunoglobulins is their heterogeneity. Hence, patholog-

ical myeloma proteins of humans and mice, which are usually homogeneous, have been used for research.

The principal result obtained from the analysis of the primary structure of several light chains was the discovery that these chains are composed of two approximately equal structural components [87, 87a, 127, 215a]. It has been found that almost all differences between chains of the same class are concentrated in the N-terminal half, i.e., the variable part (V), while the C-terminal half (C) is the same, except for one residue which determines the allotypic characteristics of the chain: in type K chains at positions 191 valine in Inv(b+) proteins and leucine in Inv(a+) proteins. Similar data were obtained in an analysis of human light chains: there is only one substitution in the C-half (position 190 either lysine or arginine) [128]. The fragment Fd of heavy chains is probably constructed in the same way.

Table 34A presents the amino acid sequence for N-terminal halves of several light chains of humans and mice, and for the heavy chain section in humans. The length of the variable part varies from 105 to 111 residues for the type K light chains studied, while the length of the C section is always constant at 107 residues. The length of the C part of type L chains is two residues less. Type K light chains have differences of 10-40 residues in the variable section. In other words, approximately 60% of the residues are identical in this part of the chain. As is evident from Fig. 32A, variability occurs in certain sections of the V part. The stability of the invariable sections is supported by the fact that the V part exhibits very few of the species-specific variations which appear in a comparison of type K chains of humans and mice [102a]. Analogously, the V part of three chains isolated from three different proteins have approximately 40 residues which are different [183].

If the genetic causes of the variability of amino acid residues in the V part are analyzed, it is seen that approximately 75% of all substitutions are attributable to variations in one triplet base, 25% to changes in two bases, and less than 1% to changes in three bases. The number of exclusions and insertions is relatively slight. Thus, variability is explained well by point mutations [113a].

The data obtained on the primary structure of polypeptide chains of immunoglobulins provide a basis for an understanding of

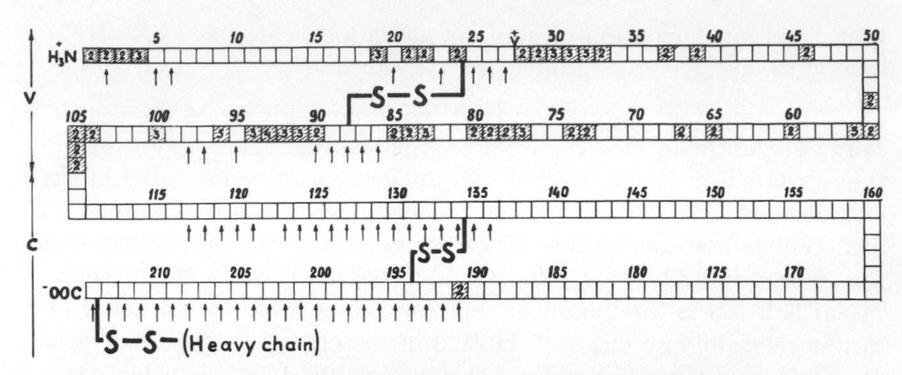

Fig. 32A. The number and position of variations observed in comparative sequence analyses of human kappa chains. Numbers outside boxes indicate the position of the residue. Empty boxes are those residues at which no variants have been observed. Shaded boxes represent variant residues; the enclosed numbers indicate the number of variants which have been found at each position. Arrows show positions at which a particular residue has been observed in at least five proteins. +) An insertion of four residues in a mouse kappa chain has been reported in this stretch [27b].

the variability of these proteins and of those evolutionary processes which have led to the appearance of many types and subtypes of immunoglobulins. There is another question of equal importance: whether the data on the variability of immunoglobulin peptide chains bear any relationship to the specific activity of antibodies. A definite positive answer to this question cannot now be given, mainly because the structure of antibody peptide chains has not yet been analyzed. Such a study will evidently soon be carried out since suitable objects for such investigations are now available, especially myeloma proteins which have antibody activity against proteins [121a] as well as hapten [44a]. The discovery of this type of antibody, which is supposed to be homogeneous, removed the main difficulty from the study of antibody structure, i.e., their heterogeneity. Nevertheless, it is still probable that the differences discussed in primary structure are responsible for the specific antigen-binding capacity of antibodies.

Attempts to ascertain the primary structure of heavy chains have been made in recent years. The structure of the crystalline papain-cleaved fragment of rabbit γG-globulin [72b, 86a] has been determined, and the structure of the heavy-chain sections to which the carbohydrate component is attached [146] has been established. The study of the structure of the N-terminal half of heavy chains has begun [173].

Peptide Maps of Antibodies

One of the most widely used methods for the study of protein structure is the peptide map or fingerprint method. First the native or denatured protein is subjected to hydrolysis by proteolytic enzyme. The mixture obtained is fractionated on a large sheet of chromatographic paper, first in one direction by electrophoresis, and then in the other direction by chromatography. The paper sheet is sprayed with ninhydrin or with specific reagents for given amino acids. The peptide maps thus obtained are compared with respect to the number and positions of peptide spots. Such maps disclose even very slight differences in the structures of various proteins. Thus, small structural differences were found among the different types of hemoglobins [90]. This method has, of course, attracted the attention of immunochemists, since differences among antibodies are apparently confined to very small portions of the molecule.

Several investigators have carried out such experiments, and in almost all cases differences in a very small number of peptides were found among maps of different antibodies or antibodies and γ-globulins.

Among the first published were the experiments of Gurvich, Gubernieva, and Myasoedova [1] who studied hydrolyzates of rabbit antibody to horse and human albumins isolated with the aid of antigens linked to cellulose. Two antibodies from the serum of the same rabbit and nonspecific γG-globulins were denatured with trichloracetic acid and hydrolyzed, first with trypsin, and then with chymotrypsin. Up to 40 peptide spots were then found on the peptide maps of the hydrolyzate. A comparison of the peptide maps showed that the hydrolyzate of γG-globulins contained two peptides which in antibody hydrolyzates were present in only very small quantities if at all.

Gitlin and Marler studied enzymic hydrolyzates of antibody to pneumococcal polysaccharides [72, 120]. Subtilisin hydrolysis of antibodies denatured by heating or oxidized with performic acid yielded 100-110 peptides, which was only half of the expected number. Tryptophan was found in only four peptides (there were a total of 20 tryptophan residues in the molecule), tyrosine was found in six (54 per molecule), and histidine was found in 12 peptides (15 per molecule). A study of the maps of hydrolyzates of four antibodies denatured by heating revealed differences in three peptides, but these were not found on the peptide maps of hydrolyzates of native antibodies. Consequently, it is suggested that the data obtained can be explained by small differences in the degree of denaturation of the different antibodies studied. Maps of hydrolyzates of antibodies oxidized by performic acid did not differ from each other. Tryptic or chymotryptic hydrolysis of antibodies oxidized with performic acid yielded, respectively, 60-70 and 50 peptides, which was only half of the number expected. No differences were found between the maps of these hydrolyzates.

In other experiments, hydrolyzates of papain-cleaved fragments of antipolysaccharide antibodies were studied. The fragments were first treated with performic acid and then hydrolyzed with subtilisin or chymotrypsin. In the first case, hydrolysis of Fab fragments yielded 70 peptides, while for Fc fragments 70-75 spots were found. In the second case, 34-40 and 50-55 spots were found for Fab and Fc fragments, respectively. Since subtilisin hydrolysis of the whole molecule disclosed 100-110 peptides, and the sum of peptides in the hydrolyzate of all fragments is 210-215 it can be concluded that the antibody molecule is built up of two identical components. These data were confirmed by Seijen and Gruber [199] and by Nelson et al. [131] for Fc fragments. An examination of peptide maps of fragments revealed no differences between hydrolyzates of fragments I and II; however, these maps differed sharply from maps of hydrolyzates of Fc fragments. Differences in two peptides were found between fragments of different antibodies hydrolyzed with subtilisin or chymotrypsin [72].

These experiments suggested a slight difference in the primary structure of the antibodies studied, but as the investigators themselves note, did not prove this difference. An interpretation of these and other experiments is complicated by the fact that small differences in the degree of denaturation of the proteins under comparison or differences due to the absence of enzyme specificity and to variations in the course of hydrolysis cannot be ruled out. The quite pronounced heterogeneity of antibodies should also be borne in mind, although Gitlin and Marler attempted to obviate this difficulty by using in each case serums from 2000-6000 rabbits which had been subjected to prolonged immunization.

Haurowitz [78] carried out two series of experiments on enzymic hydrolyzates of papain-cleaved fragments of rabbit antibodies to positively and negatively charged haptens conjugated with beef albumin. The Fab fragments of the purified antibodies were denatured by heating and hydrolyzed with trypsin. Differences in one peptide were observed on the peptide maps of two antibodies isolated from the serums of different rabbits. However, the very same differences were observed in hydrolyzates of antibodies to the protein support (beef albumin) which were isolated from the same rabbits. Hence, the observed difference was attributable to allotypic variations in γG-globulins of the rabbits. It is interesting that a small portion of the peptides (4-8%) remained at the point of application and were rich in sulfur-containing amino acids. Since the protein was not reduced before hydrolysis, it is possible that these peptides were situated near inter chain disulfide bonds.

However, in the next series of experiments [74] fully reliable differences between antibodies to these same haptens were found. Beef albumin to whose molecule both haptens had been attached simultaneously were used for immunization. It could be assumed that in this case both haptens entered the same cells, and to a certain extent this assumption helped to avoid the problem of antibody heterogeneity due to synthesis in different cells. The authors observed regular differences in one or two peptides between the maps of tryptic hydrolyzates of both antibodies. These discoveries are regarded as fundamental for a comparative understanding of the structures of the two proteins.

Givol and Sela [73] studied hydrolyzates of papain-cleaved fragments of rabbit antibodies to lysozyme and poly-*l*-tyrosylgelatin. The fragments were reduced with mercaptoethanol and treated with iodoacetate, then hydrolyzed with pepsin or bacterial proteinase. In the first instance, each of the papain-cleaved fragments gave 40-50 peptide spots on the maps; in the second case, 50-60 spots were obtained. The maps of Fc fragments differed from the Fab-fragment maps. Reproducible differences, which were more pronounced in peptic hydrolyzates, were found between the maps of hydrolyzates of papain-cleaved fragments I and II of γ-globulin. In particular, one positively charged peptide found in the fragment-II hydrolyzate was absent in the hydrolyzates of fragment I. It is quite possible that differences in adsorption of the two fragments on carboxymethylcellulose are attributable to this peptide. In a study of the hydrolyzate of the papain-cleaved fragment I (nagarase enzyme) three peptides were found in the hydrolyzate of antilysozyme antibody which were not present in hydrolyzates of fragment I of antibody to polytyrosylgelatin; moreover, two of these peptides were missing in the same fragment of γ-globulins. These three peptides were absent from the hydrolyzates of fragment II of all the protein studied, although other differences between fragments of different origin were found. Peptide hydrolyzates of fragments I and II and of γ-globulins also differed from each other. In the experiments under discussion, the proteins were isolated from the serum of several rabbits, and as the authors themselves point out, in interpreting the results it is necessary to take into account the possibility of allotypic variations. In addition, it should be recalled that a small quantity of bound antigen remains in antibody to polytyrosyl-gelatin after purification.

The purpose of the series of studies discussed was to discover the differences between peptide maps of hydrolyzates of different antibodies or of antibodies and γ-globulins in order to demonstrate the dissimilarity of the primary structure of these proteins. Indeed, in practically all experiments, with the exception of those of Nilson et al. [132], small but reliable differences were detected in one of the three peptides. An interesting conjecture is whether these differences are due to structural differences in the antibody binding sites. However, the correctness of this interpretation cannot yet be verified. It would be pertinent here to reiterate what has been already pointed out in the discussion of the significance of small differences in amino acid composition among antibodies of different specificities (cf. Chapter III). A solution of this question is complicated by the pronounced heterogeneity of γ-globulins and antibodies with regard to several properties. It is possible that amino acid variations are also due to differences in the primary molecular structure. In one case of heterogeneity of allotypic antigen determinants, the differences were indeed due to primary structure. In studies of the tryptic hydrolyzate of an Fc fragment of human γ-globulin with the allotypic characteristic Gm (a+), one peptide has always been found which was not in hydrolyzates of γ-globulin of the Gm(a-) type. For Gm(b+) γ-globulins another spot was found which was lacking for Gm(b-) protein. The differences were so distinct that it was possible to predict allotypic antigen properties of γ-globulin from the peptide map [67]. The following conclusions were drawn. In the first place, in spite of the pronounced heterogeneity of γ-globulins, small but consistent differences between the proteins being compared can be detected with the aid of peptide maps. The foregoing evidence of small differences in one or two peptides among different antibodies clearly deserves special consideration. Secondly, in all similar investigations it is necessary to take into account the genetic origin of the γ-globulin or antibody under study. In addition to the aforementioned study, Small et al. have also collected evidence of differences in the peptide maps of hydrolyzates of light chains from rabbit γ-globulins with different allotypic subclasses at the (b) site (cf. Chapter VI).

Dependence of the Secondary Structure of Antibodies on the Primary Structure

The question of the dependence of the secondary structure of protein on the amino acid sequence in the peptide chain, i.e., on the primary structure, is of theoretical importance for the chemistry of proteins. Although this question has not been definitely resolved, there are still grounds to assume that the amino acid sequence determines the secondary structure of protein. Thus, the voluminous research conducted by Anfinsen, as well as other investigators, has led to the conclusion that the foregoing assumption is correct for ribonuclease and lysozyme and possibly even for other proteins [46]. The assumed dependence of secondary structure on primary structure is supported by the correlation between amino acid composition of proteins and the percent helicity of the molecule [33].

As has already been mentioned, antibody activity is determined by higher structural orders; when inactive chains are mixed, activity is restored. Hence, data confirming that the secondary structure of antibodies is in its turn determined by the amino acid sequence in peptide chains would be of considerable value.

Some investigations have provided evidence indicating that this dependence is indeed possible. In the first place, carefully carried out analyses of amino acid composition have disclosed slight, but nevertheless quite consistent, differences among the amino acid compositions of three different rabbit antibodies (cf.

Chapter III). In the second place, as is evident from the preceding section, peptide-map studies of antibodies are able to reveal differences in the primary structure of different antibodies. In almost all experiments, differences have been found between antibodies and nonspecific γ-globulins or among different antibodies.

Important evidence of the existence of a definite dependence of secondary on primary structure is provided by experiments on reversible denaturation of antibodies. The principle of such experiments is as follows. If a univalent papain-cleaved fragment of antibody or γ-globulin is placed in a 6-7*M* solution of guanidine chloride, peptide chains are completely unfolded as a result of destruction of the entire secondary structure. This process can be followed with the aid of optical rotation. However, after gradual removal of the denaturant, the protein regains its original tertiary structure as a result of prolonged dialysis, as may be seen from data on sedimentation and optical rotation. Using such a method, Buckley et al. [16] showed that a guanidine-chloride-denatured fragment I from rabbit antibody to beef serum albumin regained approximately 70-80% of its original specific activity after dialysis, judging from the capacity to form soluble complexes with antigen and from the capacity to bind antigen linked to polystyrene. In other experiments [145], antibody to the 2,4-dinitrophenyl group was studied. Analogously, after removal of guanidine chloride, fragment I of these antibodies was able to conjugate with hapten, with 70% restoration of activity. The kinetics of unfolding of fragment I of the antibody were the same as for fragment I from nonspecific γ-globulin, as was evident from a study of denaturation and from the final degree of unfolding at all concentrations of guanidine chloride. It may be concluded that differences between antibodies and nonspecific γ-globulins are confined to a small section of the molecule.

The experiments performed, however, have one essential shortcoming: namely, the fragments which were subjected to denaturation and reconstitution retained their disulfide bonds intact. Hence, it could not be ruled out that it was precisely these bonds which were responsible for formation of a distinct secondary structure. This reservation was avoided in the experiments of Haber [85] who studied this process with fragments of antibodies to ribonucleases, in which the disulfide bonds had been split. A study of optical rotation in the ultraviolet region showed that the

Fab fragment had completely lost its secondary structure after reduction in 6*M* guanidine chloride. However, prolonged dialysis against mercaptoethanol and then against a buffer with pH 8 lead to oxidation of the fragment and restoration of 20-27% of its original antigen-binding capacity. The bulk of the reconstituted fragment had a sedimentation constant of ~3.2 S, i.e., the same as the untreated fragment. Reconstitution and reoxidation at pH 3.5 under conditions in which the disulfide bonds of reduced ribonuclease are formed at random did not restore activity. This fact is evidence against the assumption that in denaturation some small section of the binding site remains folded and determines the subsequent restoration of specific activity. The values obtained for restoration of activity were 400 times greater than the theoretical, which was calculated under the assumption that disulfide bonds are formed at random in the fragment during reconstitution. Similar data were obtained by Tanford [219]. It is interesting that in the presence of hapten restoration of activity was 1.5-2 times greater. After unfolding, not only the antibody binding sites, but also several other parts of the molecule, regain their structure. Freedman and Sela [68b, 68c] showed that after reconstitution of completely reduced antibody the antigen-binding capacity as well as the capacity to react with antibodies to γG-globulin are restored; i.e., the original antigenic determinants are regained.

These experiments are of considerable theoretical significance for the understanding of, first, antibody structure, and second, antibody synthesis. If, indeed, antibody specificity is governed by the primary structure, then by analogy with the synthesis of other proteins (for example, enzymes) it can be assumed that information necessary for formation of peptide chains of antibody resides in the sequence of DNA nucleotides. However, the differences in the primary structure are still not definite proof of this.

In conclusion, it should be stressed that research into antibody structure is advancing rapidly. Investigations on proteolytic fragments and peptide chains have made it possible to reconstruct a general model of the structure of the γG-globulin molecule. Experiments on the reassociation of peptide chains with accompanying restoration of activity and on the formation of molecular hybrids of antibodies have made it possible to determine more precisely the localization of the binding site. More knowledge is constantly being acquired on the primary structure of immunoglobulins.

However, the key point still remains to be elucidated, since the exact nature of the differences between binding sites of antibodies of different specificities remains obscure.

LITERATURE CITED

1. A. E. Gurvich, L. M. Gubernieva, and K. N. Myasoedova, "Comparison of enzymic hydrolyzates of nonspecific γ-globulins and antibodies," Biokhimiya 26(3):468 (1961).
2. A. E. Gurvich and R. S. Nezlin, "Nomenclature for human immunoglobulins," Biokhimiya 30(2):443 (1965).

2a. L. N. Kaigorodova, E. D. Kaversneva, "Chemical characterization of a cryogel-macroglobulin," Molec. Biology (USSR) 1(2):224 (1967).

3. A. Ya. Kul'berg and I. A. Tarkhanova, "On the problem of papain cleavage of immune γ-globulin," Vopr. Med. Khim. 7(5):520 (1961).
4. A. Ya. Kul'berg and I. A. Tarkhanova, "Role of disulfide bonds in the structure of the antibody binding site," Biokhimiya 29(2):246 (1964).
5. N. E. Kuchinskaya, A. Ya. Kul'berg, and V. S. Tsvetkov, "Immunochemical analysis of papain-cleavage products of bovine γ-globulin," Biokhimiya 30(5):1065 (1965).
6. R. S. Nezlin, "Peptide chains of γ-globulins and antibodies," Usp. Sovrem. Biol. 58(2):201 (1964).

6a. R. S. Nezlin and L. M. Kulpina, "Cellulose-fixed immunoglobulin subunits," Immunochemistry 4(4):296 (1967).

7. V. I. Okulov, "Effect of cleavage of disulfide bonds on various properties of bovine γ-globulin," in: Proteins in Medicine and National Economy [in Russian], Naukova Dumka, Kiev (1965), p. 53.
8. D. V. Stefani and A. Ya. Kul'berg, "Immunological analysis of products of papain proteolysis of horse antidiphtheria antitoxic γ-globulin," Vopr. Med. Khim. 10(3):279 (1964).
9. D. V. Stefani, A. Ya. Kul'berg, and K. L. Shakhanina, "Distinguishing features of the immunochemical structure of horse antitoxic β_2A-globulin," Vopr. Med. Khim. 11(4):34 (1965).
10. I. A. Tarkhanova and A. Ya. Kul'berg, "Participation of tryptophan in the structure of the antibody binding site," Vopr. Med. Khim. 8(2):163 (1962).
11. F. Franek and R. S. Nezlin, "Study of the role of different peptide chains of antibodies in the antigen–antibody reaction," Biokhimiya 28(2):193 (1963).
12. R. Arnon, M. Sela, A. Yaron, and H. A. Sober, "Polylysine-specific antibodies and their reaction with oligolysines," Biochemistry 4(5):948 (1965).
13. R. E. Balliex, G. M. Bernier, K. Tominage, and F. W. Putnam, "Gamma globulin antigenic types defined by heavy determinants," Science 145(3628):167 (1964).
14. B. Benacerraf, C. Merryman, and R. A. Binaghi, "Studies of the chains of guinea pig and rat anti-2,4-dinitrophenyl antibodies," J. Immunol. 93(4):618 (1964).

14a. C. C. F. Blake, D. F. Koenig, E. A. Mair, A. C. T. North, D. C. Phillips, V. R. Sarma, "Structure of hen egg-white lysozyme," Nature 206(4986):757 (1965).

15. W. B. Bridgman, "The peptic digestion of human gamma globulin," J. Am. Chem. Soc. 68(5):857 (1946).
16. C. E. Buckley, P. L. Whitney, and C. Tanford, "The unfolding and renaturation of a specific univalent antibody fragment," Proc. Nat. Acad. Sci. USA 50(5):827 (1963).
17. M. Burrand and D. E. Koshland, "Use of "reporter groups" in structure function studies of proteins," Proc. Nat. Acad. Sci. USA 52(4):1017 (1964).
18. H. J. Cahnmann, R. Arnon, M. Sela, "Isolation and characterization of active fragments obtained by cleavage of immunoglobulin G with cyanogen bromide," J. Biol. Chem. 241(14):3247 (1966).
19. A. O. Carbonara and J. F. Heremans, "Subunits of normal and pathological 1A-globulins (β_2A-globulins)," Arch. Biochem. Biophys. 102(1):137 (1963).
19a. R. E. Cathou and E. Haber, "Structure of the antibody combining site. I. Hapten stabilization of antibody conformation," Biochemistry 6(2):513 (1967).
20. J. J. Cebra, "The effect of sodium dodecylsulfate on intact and insoluble papain hydrolyzed immune globulin," J. Immunol. 92(6):977 (1964).
21. J. J. Cebra, "Studies on the combining sites of the protein antigen silk fibroin. III. Inhibition of the silk fibroin—antifibroin system by peptides derived from the antigen," J. Immunol. 86(2):205 (1961).
22. J. J. Cebra, "Some properties of reductively fragmented gamma-2 and gamma-1 M immune globulins," in: Molecular and Cellular Basis of Antibody Formation, Academic Press, New York (1965), p. 103.
23. J. J. Cebra, D. Givol, and E. Katchalski, "Soluble complexes of antigen and antibody complexes," J. Biol. Chem. 237(3):751 (1962).
24. J. J. Cebra, D. Givol, H. I. Silman, and E. Katchalski, "A two-stage cleavage of rabbit globulin by a water-insoluble papain preparation followed by cysteine," J. Biol. Chem. 236(6):1720 (1961).
25. H. Chaplin, S. Cohen, and E. M. Press, "Preparations and properties of the peptide chains of normal human 19 S-globulin (IgM)," Biochem. J. 95(1):256 (1965).
25a. T. J. Coleman, R. D. Marshall, and M. Potter, " Preparation of a glycopeptide from an immunoglobulin K polypeptide chain from the mouse. "Biochim. Biophys Acta. 147(2):396 (1967).
26. S. Cohen, "Properties of the separated chains of human gamma-globulin," Nature 197(4864):253 (1963).
27. S. Cohen, "Properties of the peptide chains of normal and pathological human-globulins," Biochem. J. 89(2):334 (1963).
27a. S. Cohen and S. Gerden, "Dissociation of n- and λ- chains from reduced human immunoglobulins," Biochem. J. 97(2):460 (1965).
27b. S. Cohen and C. Milstein, "Structure of antibody molecules," Nature 214(5087): 449 (1967).
28. S. Cohen and R. R. Porter, "Heterogeneity of the peptide chains of γ-globulin," Biochem. J. 90(2):278 (1964).
29. S. Cohen and R. R. Porter, "Structure and biological activity of immunoglobulins," Advan. Immunol. 4:287 (1964).
30. F. E. Cominga, "A third γ-globulin chain?," Lancet 2(7311):786 (1963).
31. R. S. Criddle, "Dissociation and separation of gamma globulin into subunits," Arch. Biochem. Biophys. 106(1-3):101 (1964).

32. J. Crumpton and J. M. Wilkinson, "Amino acid compositions of human and rabbit γ-globulins and the fragments produced by reduction," Biochem. J. 88(2):228 (1963).

33. D. R. Davies, "A correlation between amino acid composition and protein structure," J. Mol. Biol. 9(2):605 (1964).

34. H. F. Deutsch, E. R. Stiehm, and J. J. Mortone," Action of papain on human serum globulins," J. Biol. Chem. 236(8):2216 (1961).

34a. K. Y. Dorrington, M. H. Larlengo, and C. Tanford, "Conformational change and complementarity in the combination of H and L chains of immunoglobulin G. Proc. Nat. Acad. Sci. USA, 58(3996) (1967).

35. G. M. Edelman, "Dissociation of γ-globulins," J. Am. Chem. Soc. 81(12):3155 (1959).

36. G. M. Edelman and B. Benacerraf, "On structural and functional relations between antibodies and proteins of the gamma-system," Proc. Nat. Acad. Sci. USA 48(6):1035 (1962).

37. G. M. Edelman, B. Benacerraf, Z. Ovary, and M. D. Poulik, "Structural differences among antibodies of different specificities," Proc. Nat. Acad. Sci. USA 47(11):1751 (1961).

38. C. A. Edelman and J. A. Gally, "The nature of Bence–Jones proteins. Chemical similarities of polypeptide chains of myeloma globulins and normal γ-globulins," J. Exp. Med. 116(2):2207 (1962).

39. G. M. Edelman, "Formation of active 7 S antibody molecules by reassociation of polypeptide chains," in: Molecular and Cellular Basis of Antibody Formation, Academic Press, New York (1965), p. 113.

40. G. M. Edelman, J. F. Heremans, M. Heremans, and H. G. Kunkel, "Immunological studies of human globulin. Relation of precipitin line of whole globulin to those of the fragments produced by papain," J. Exp. Med. 112(1):203 (1960).

41. G. M. Edelman and J. A. Gally, "A model for the 7 S antibody molecule," Proc. Nat. Acad. Sci. USA 51(5):846 (1964).

42. G. M. Edelman and E. A. Kabat "Studies on human antibodies. I. Starch gel electrophoresis of the dissociated polypeptide chains," J. Exp. Med. 119(3):443 (1964

43. G. M. Edelman, D. E. Olins, J. A. Gally, and N. D. Zinder, "Reconstitution of immunologic activity by interaction of polypeptide chains of antibodies," Proc. Nat. Acad. Sci. USA 50(4):753 (1963).

44. G. M. Edelman and M. D. Poulik, "Studies on structural units of the gamma globulins," J. Exp. Med. 113(5):861 (1961).

44a. H. N. Eisen, J. R. Little, K. Osterland, and E. S. Simms, "A myeloma G-protein with antibody (anti-2, 4-dinitrophenyl) activity." Abstr. Cold Springs Harbor Symp., p. 13 (1967).

45. H. N. Eisen and J. H. Pearce, "The nature of antibodies and antigens," Ann. Rev. Microbiol. 16:101 (1962).

46. C. J. Epstein, R. F. Goldberger, and C. B. Anfinsen, "The genetic control of tertiary protein structure: studies with model system," Cold Spring Harbor Symp. Quart. Biol. 28:439 (1964).

47. J. L. Fahey, "Heterogeneity of γ-globulins," Advan. Immunol. 2(41) (1962).

48. J. L. Fahey and B. A. Askonas, "Enzymatically produced subunits of proteins formed by plasma cells in mice. I. γ-globulin and myeloma proteins," J. Exp. Med. 115(3):623 (1962).

49. J. W. Fenton and S. J. Singer, "Affinity labeling of antibodies to the p-azo-phenyltrimethylammonium hapten and a structural relationship among antibody active sites of different specificities," Biochem. Biophys. Res. Commun. 20(3):315 (1965).

50. J. B. Fleischman, "The synthesis of γ-globulin by rabbit lymph node cells," J. Immunol. 91(2):163 (1963).

51. J. B. Fleischman, R. R. Porter, and R. H. Pain, "Reduction of γ-globulins," Arch. Biochem. Biophys., Suppl. 1, p. 174 (1962).

52. J. B. Fleischman, R. R. Porter, and E. M. Press, "The arrangement of the peptide chains in γ-globulin," Biochem. J. 88(2):220 (1963).

53. M. Fougereau and G. M. Edelman, "Resemblance of the gross arrangement of polypeptide chains in reconstituted and native γ-globulins," Biochemistry 3(8):1120 (1964).

54. M. Fougereau and G. M. Edelman, "Corroboration of recent models of the γG-immunoglobulin molecule," J. Exp. Med. 121(3):373 (1965).

55. M. Fougereau, D. E. Olins, and G. M. Edelman, "Reconstitution of antiphage antibodies from L and H polypeptide chains and the formation of interspecies molecular hybrids," J. Biol. Chem. 120(3):349 (1964).

56. F. Franek, "Dissociation of animal 7 S γ-globulins by cleavage of disulfide bonds," Biochem. Biophys. Res. Commun. 4(1):28 (1961).

57. F. Franek, "Some properties of higher subunits of pig γG-globulin," Collection Czech. Chem. Commun. 30(6):1947 (1965).

58. F. Franek, and B. Keil, "Structural differences between γ-globulin chains," Collection Czech. Chem. Commun. 29(3):847 (1964).

59. F. Franek, O. Kotynek, L. Simek, and J. Zikan, "S-Sulfonated antidinitrophenyl antibodies. Some specific features of the interaction between isolated H and L subunits," in: Molecular and Cellular Basis of Antibody Formation, Academic Press, New York (1965), p. 125.

60. F. Franek and R. S. Nezlin, "Recovery of antibody combining activity by interaction of different peptide chains isolated from purified horse antitoxins," Folia Microbiol. 8(2):128 (1963).

61. F. Franek, R. S. Nezlin, and F. Skvaril, "Antibody binding capacity of different peptide chains isolated from digested and purified horse diphtheria antitoxin," Folia Microbiol. 8(4):197 (1963).

61a. F. Franek, V. Paces, B. Keil, and F. Sorn, "On proteins. CXI. Isolation of electrophoretically homogeneous components of normal pig γG-globulin light chains and demonstration of their structural heterogeneity," Collection Czech. Chem. Commun. 32(9):3242 (1967).

62. F. Franek and I. Riha, "Purification and structural characterization of 5 S γ-globulin in newborn pigs," Immunochemistry 1(1):49 (1964).

63. F. Franek and J. Zikan, "Limited cleavage of disulfide bonds of pig γ-globulin by S-sulfonation," Collection Czech. Chem. Commun. 29(6):1401 (1964).

63a. F. Franek and O. M. Zorina, "On proteins. CX. Isolation and characterization of lambda and pi chains representing structural types of pig γG-globulin light chains," Collection Czech. Chem. Commun. 32(9):3229 (1967).

63b. B. Frangione and E. C. Franklin, "Structural studies of human immunoglobulins. Differences in the Fd fragments of the heavy chains of G myeloma proteins," J. Exp. Med. 122(1):1 (1965).

64. B. Frangione, F. Prelli, and E. C. Franklin, "The structure of Fd fragments of myeloma proteins," Immunochemistry 4(2):95 (1967).

64a. B. Frangione and E. C. Franklin, "Structural studies of human immunoglobulins. Differences in primary structure of heavy chains of normal and pathologic G, A, and M immunoglobulins," Arch. Biochem. Biophys. 111(3):603 (1965).

65. E. C. Franklin, "Structural units of human 7 S γ-globulin," J. Clin. Invest. 39(12):1933 (1960).

66. E. C. Franklin, "Structural studies of human 7 S γ-globulin (G immunoglobulin). Further observations of a naturally occurring protein related to the crystallizable (fast) fragment," J. Exp. Med. 120(5):691 (1964).

66a. E. C. Franklin and B. Frangione, "Two serologically distinguishable subclasses of mu-chains of human macroglobulins," J. Immunol. 99(4):810 (1967).

67. E. C. Franklin, H. Fudenberg, B. Frangione, and M. Meltzer, "Structural basis for differences between γ-globulins of different genetic (Gm) types. Studies of normal γ-globulins and para-proteins," in: Molecular and Cellular Basis of Antibody Formation, Academic Press, New York (1965), p. 193.

68. E. C. Franklin, J. Lowenstein, B. Bigelow, and M. Meltzer, "Heavy chain disease – a new disorder of serum γ-globulins. Report of the first case," Am. J. Med. 37(3):327 (1964).

68a. S. Fuchs and M. Sela, "Preparation and characterization of poly-DL-alanyl rabbit γ-globulin," J. Biol. Chem. 240(3):327 (1964).

68b. M. H. Freedman and M. Sela, "Recovery of antigenic activity upon reoxidation of completely reduced polyalanyl rabbit immunoglobulin Gm," J. Biol. Chem. 241(10):2383 (1966).

68c. M. H. Freedman and M. Sela, "Recovery of specific activity upon reoxidation of completely reduced polyalanyl rabbit antibody," J. Biol. Chem. 241(22):5225 (1966).

69. H. H. Fudenberg, G. Drews, and A. Nisonoff, "Serological demonstration of dual specificity of rabbit bivalent hybrid antibody," J. Exp. Med. 119(1):151 (1964).

70. J. A. Gally and G. M. Edelman, "Protein–protein interactions among L-polypeptide chains of Bence–Jones proteins and human γ-globulins," J. Exp. Med. 119(5):817 (1964).

71. L. Gyenes, A. H. Sehon, S. O. Freedman, and Z. Ovary, "The properties of fragments of skin-sensitizing and blocking antibodies as revealed by the Prausnitz–Küstner, passive cutaneous anaphylaxis and hemagglutination reactions," Intern. Arch. Allergy and Appl. Immunol. 24:106 (1964).

72. D. Gitlin and E. Merler, "A comparison of the peptides released from related rabbit antibodies by enzymatic hydrolysis," J. Exp. Med. 114(2):317 (1961).

72a. D. Givol, "The cleavage of rabbit immunoglobulin G by trypsin after mild reduction and aminoethylation," Biochem. J. 104(3):390 (1967).

72b. D. Givol and R. R. Porter, "The chemical structure of the heavy chain of rabbit immunoglobulin G (IgG)," Israel J. Chem. 4:68 (1966).

73. D. Givol and M. Sela, "A comparison of fragments of rabbit antibodies and normal γ-globulin by the peptide-map technique," Biochemistry 3(3):451 (1964).

74. E. F. Gold, D. L. Knight, and F. Haurowitz, "Peptide maps of antibodies against an antigen containing two different determinant groups," Biochem. Biophys. Fes. Commun. 18(1):76 (1965).

74a. J. W. Goodman, "Antigenic determinants in fragments of gamma globulin from rabbit serum," Science 139(3561):1292 (1963).

75. J. W. Goodman, "Immunologically active fragments of rabbit γ-globulin," Biochemistry 3(6):857 (1964).

76. J. W. Goodman, "Heterogeneity of rabbit γ-globulin with respect to cleavage by papain," Biochemistry 4(11):2350 (1965).

77. J. W. Goodman and E. A. Kabat, "Immunochemical studies on cross = reactions of anti-pneumococcal sera," J. Immunol. 84(4):333 (1960).

77a. D. Griffin, D. K. Tachibana, B. Nelson, and L. T. Rosenberg, "Contribution of tryptophan to the biological properties of antidinitrophenyl antibody," Immunochemistry 4(1):23 (1967).

78. J. L. Groff and F. Haurowitz, "Comparison of the peptide maps of antibodies against an acidic and a basic determinant group," Immunochemistry 1(1):31 (1964).

78a. W. R. Gray, W. J. Dreyer, and L. Hood, "Mechanism of antibody synthesis: size differences between mouse kappa chains," Science 155(3761):465 (1967).

79. A. L. Grossberg, C. C. Chen, L. Rendina, and D. Pressman, "Specific cation effects with antibody to a hapten with a positive charge," J. Immunol. 88(5):600 (1962).

80. A. L. Grossberg and D. Pressman, "Effect of acetylation on active site of several antihapten antibodies: further evidence for the presence of tyrosine in each site," Biochemistry 2(1):90 (1963).

81. A. L. Grossberg and O. A. Roholt, "Further evidence that antibody fragments in fractions I and II are from different kinds of antibody molecules," Federation Proc. 21(2):28 (1962).

82. A. L. Grossberg, O. A. Roholt, and D. Pressman, "Different distribution of antibodies of two specificities among γ-globulins of an individual rabbit," Biochemistry 2(5):989 (1963).

83. A. L. Grossberg, P. Stelos, and D. Pressman, "Structure of fragments of antibody molecules as revealed by reduction of exposed disulfide bonds," Proc. Nat. Acad. Sci. USA 48(7):1203 (1962).

84. A. F. S. A. Habeeb, H. G. Cassidy, P. Stelos, and S. J. Singer, "Some studies of the chemical modification of antibodies," Biochem. Biophys. Acta 34(2):439 (1959).

85. E. Haber, "Recovery of antigenic specificity after denaturation and complete reduction of disulfides in a papain fragment of antibody," Proc. Nat. Acad. Sci. USA 52(4):1099 (1964).

85a. E. Haber and F. F. Richards, "The specificity of antigenic recognition of antibody heavy chain," Proc. Roy. Soc., Ser. B 166(1003):176 (1966).

86. R. Heimer, E. R. Schwartz, R. L. Engle, and K. R. Woods, "The relationship of structure to the thermal solubility characteristics of a Bence–Jones protein," Biochemistry 2(6):1380 (1963).

86a. R. L. Hill, R. Delaney, H. E. Lebovitz, and R. E. Fellows, "Studies on the amino acid sequence of heavy chains from rabbit immunoglobulin," Proc. Roy. Soc., Ser. B 166(1003):159 (1966).

87. N. Hilschmann and L. C. Craig, "Amino acid sequence studies with Bence–Jones proteins," Proc. Nat. Acad. Sci. USA 53(6):1403 (1965).

87a. N. Hilschmann, "The chemical structure of two Bence–Jones proteins (Roy and Cum) of the K-type," Hoppe–Seyler Z. Phys. Chem. 348(8):1077 (1967).

88. R. Hong, J. L. Palmer, and A. Nisonoff, "Univalence of half-molecules of rabbit antibodies," J. Immunol. 94(4):603 (1965).

88a. R. Hong and A. Nisonoff, "Relative labilities of the two types of interchain disulfied bond of rabbit γG-immunoglobulin," J. Biol. Chem. 240(10):3883 (1965).

89. Hsiao Shu-hsi and F. W. Putnam, "The cleavage of human γ-globulin by papain," J. Biol. Chem. 236(1):122 (1961).

90. V. M. Ingram, Hemoglobin and Its Abnormalities, Charles C. Thomas, Springfield, Ill. (1961).

90a. F. P. Inman and A. Nisonoff, "Reversible dissociation of fragment Fc of rabbit γG-immunoglobulin," J. Biol. Chem. 241(2):322 (1966).

91. H. Jacot-Guillarmond and H. Isliker, "Reversible cleavage of subunits," Vox Sanguinis 9(1):31 (1964).

92. H. Jaquet, B. Bloom, and J. J. Cebra, "Retention of antibody activity by reductively dissociated immune globulin," Federation Proc. 22(2):496 (1963).

93. H. Jaquet, B. Bloom, and J. J. Cebra, "The reductive dissociation of rabbit immune globulin in sodium dodecylsulfate," J. Immunol. 92(6):991 (1964).

94. H. Jaquet and J. J. Cebra, "Comparison of two precipitating derivatives of rabbit antibody: fragment L dimer and the product of pepsin digestion," Biochemistry 4(5):954 (1965).

95. K. James and D. R. Stanworth, "Structural changes occurring in 7 S globulins," Nature 202(4932):563 (1964).

96. B. Jirgensons, D. Yonezawa, and V. Gorguraki, "Structural studies on human serum gamma-globulins and myeloma proteins. I. Fragmentation and denaturation of the globulins with detergent and alkali," Makromol. Chem. 60:25 (1963).

97. B. Jirgensons, M. E. Adams-Mayne, V. Gorguraki, and P. J. Migliore, "Structure studies on human serum γ-globulins and reconstitution of the macromolecules," Arch. Biochem. Biophys. 111(2):283 (1965).

98. D. Yonezawa, P. J. Migliore, S. C. Capetillo, and B. Jirgensons, "Structural studies on human serum gamma-globulins and myeloma proteins. II. Cleavage of disulfide bonds of globulins with sulfite and recombination of the fragments," Makromol. Chem. 77:191 (1964).

99. E. A. Kabat, "Some configurational requirements and dimensions of the combining site on an antibody to a naturally occurring antigen," J. Am. Chem. Soc. 76(14):3709 (1954).

100. E. A. Kabat, "Heterogeneity in extent of the combining regions of human antidextran," J. Immunol. 77(6):377 (1956).

101. E. A. Kabat, "The upper limit for the size of the human antidextran combining site," J. Immunol. 84(1):82 (1960).

102. E. A. Kabat, Experimental Immunochemistry, Charles C. Thomas, Springfield, Ill. (1961).

102a. E. A. Kabat, "The paucity of species-specific amino acid residues in the variable regions of human and mouse Bence–Jones proteins and its evolutionary and genetic implications," Proc. Nat. Acad. Sci. USA 57(5):1345 (1967).

103. F. Karush, "The interaction of purified antibody with optically isomeric haptens," J. Am. Chem. Soc. 78(21):5519 (1956).

104. F. Karush, "The interaction of purified anti-β-lactoside antibody with haptens," J. Am. Chem. Soc. 79(13):3380 (1957).

105. F. Karush, "Properties of papain-digested purified anti-hapten antibody," Federation Proc. 18(1):1577 (1959).

106. F. Karush, "Immunologic specificity and molecular structure," Advan. Immunol. 2:1 (1962).

107. M. E. Koshland, F. M. Engelberger, and S. M. Gaddone, "The effect of iodination on the binding of haptene to arsonic acid," J. Immunol. 89(4):517 (1962).

108. M. E. Koshland, F. M. Engelberger, and S. M. Gaddone, "Identification of tyrosine at the active site of anti-p-azobenzenearsonic acid antibody," J. Biol. Chem. 238(4):1349 (1963).

109. M. E. Koshland, F. M. Engelberger, and S. M. Gaddone, "Evidence against the universality of a tyrosyl residue at antibody combining sites," Immunochemistry 2:115 (1965).

110. M. E. Koshland, F. Engelberger, and D. E. Koshland, "A general method for labeling of the active site of antibodies and enzymes," Proc. Nat. Acad. Sci. USA 45(10):1470 (1959).

110a. O. Kotynek and F. Franek, "Unequal importance of different polypeptide chains for the determination of antibody specificity in bovine antidinitrophenyl antibodies," Collection Czech. Chem. Commun. 30(9):3153 (1965).

111. V. P. Kreiter and D. Pressman, "Antibodies to a hapten with two determinant groups," Immunochemistry 1(3):151 (1964).

111a. H. G. Kunkel and R. A. Prendergast, "Supergroups of γA-immune globulins," Proc. Soc. Exp. Biol. Med. 122(3):910 (1966).

112. H. G. Kunkel, J. C. Allen, H. M. Grey, L. Martensson, and R. Grubb, "A relationship between H chain groups of 7 S γ-globulin and the Gm system," Nature 203(4943):413 (1964).

112a. M. E. Lamm and P. A. Small, "Polypeptide chain structure of rabbit immunoglobulins. II. γM-immunoglobulin," Biochemistry 5(1):267 (1966).

112b. M. Lahav, R. Arnon, and M. Sela, "Biological activity of the cleavage product of human immunoglobulin G with cyanogen bromide," J. Exp. Med. 125(5):787 (1967).

113. K. Landsteiner, The Specificity of Serological Reactions, 2nd Edition, Harvard University Press, Cambridge (1945).

113a. E. S. Lennox and M. Cohn, "Immunoglobulins," Ann. Rev. Biochem. 36, Pt. 1:365 (1967).

114. B. B. Levine, "Studies on the dimensions of rabbit antibenzylpenicilloyl antibody combining sites," J. Exp. Med. 117(1):161 (1963).

114a. J. R. Little and H. N. Eisen, "Evidence for tryptophan in the active sites of antibodies to polynitrobenzenes," Biochemistry 6(10):3119 (1967).

115. M. L. MacFadden and E. L. Smith, "Free amino acid groups and N-terminal sequence of rabbit antibodies," J. Biol. Chem. 214(1):185 (1955).

116. R. Mage and E. A. Kabat, "The combining regions of the type III pneumococcus polysaccharide and homologous antibody," Biochemistry 2(6):1278 (1963).

117. W. J. Mandy, M. K. Stambaugh, and A. Nisonoff, "Amino acid composition of univalent fragments of rabbit antibody," Science 140(3569):901 (1963).

118. E. Marler, C. A. Nelson, and C. Tanford, "The polypeptide chains of rabbit γ-globulin and its papain-cleaved fragments," Biochemistry 3(2):279 (1964).

119. G. Markus, A. L. Grossberg, and D. Pressman, "The disulfide bonds of rabbit γ-globulin and its fragments," Arch. Biochem. Biophys. 96(1):63 (1962).

120. E. Marler and D. Gitlin, "Comparison of the peptides released from human and rabbit γ_2-globulins by enzymatic digestion," Nature 198(4887):1304 (1963).

121. C. Merryman and B. Benacerraf, "Studies on the structure of mouse antibodies," Proc. Soc. Exp. Biol. Med. 114(2):372 (1963).

121a. H. Metzger, "Characterization of a human macroglobulin. V. A Waldenström macroglobulin with antibody activity," Proc. Nat. Acad. Sci. USA 57(5):1490 (1967).

122. H. Metzger and M. Mannick, "Recombination of antibody polypeptide chains in the presence of antigen," J. Exp. Med. 114(2):372 (1963).

123. H. Metzger and S. J. Singer, "Binding capacity of reductively fragmented antibodies to the 2,4-dinitrophenyl group," Science 142(3593):674 (1963).

124. H. Metzger, L. Wofsy, and S. J. Singer, "Affinity labeling of the active sites of antibodies to the 2,4-dinitrophenyl hapten," Biochemistry 2(5):979 (1963).

125. H. Metzger, L. Wofsy, and S. J. Singer, "The participation of A and B polypeptide in the active sites of antibody molecules," Proc. Nat. Acad. Sci. USA 51(4):612 (1964).

125a. S. Mihaescu, "The heterogeneity of the polypeptide L chain rabbit IgG. Electrophoresis and immunoelectrophoresis," Rev. Roumaine Biochim. 4(3):193 (1967).

126. F. Miller and H. Metzger, "Characterization of a human macroglobulin. I. The molecular weight of its subunit," J. Biol. Chem. 240(8):3325 (1965).

127. C. Milstein, "Chemical structure of light chains," Proc. Roy. Soc., Ser. B 166(1003):146 (1966).

128. C. Milstein, "Variations in the C-terminal half of immunoglobulin lambda chains," Biochem. J. 104(2):28c (1967).

128a. C. Milstein, "Variations in amino acid sequence near the disulfide bridges of Bence–Jones proteins," Nature 209(5021):370 (1966).

128b. E. Mozes, J. B. Robbins, and M. Sela, "Heterogeneity of the light chains of rabbit immunoglobulin G fractions and of a series of antibodies directed towards antigens of differing complexity," Immunochemistry 4(4):239 (1967).

129. R. L. Nachman and R. L. Engle, "Gamma-globulin: unmasking of hidden antigenic sites on light chains," Science 145(3628):167 (1964).

130. C. A. Nelson, "Isolation of a new intermediate in the papain cleavage of rabbit gamma-globulin," J. Biol. Chem. 239(11):3727 (1964).

131. C. A. Nelson, M. E. Noelken, C. E. Buckley, C. Tanford, and R. L. Hill, "Tryptic hydrolysis of the inactive fragments of rabbit gamma-globulin in antibodies," Federation Proc. 22(2):657 (1963).

132. C. A. Nelson, M. E. Noelken, C. E. Buckley, C. Tanford, and R. L. Hill, "Comparison of tryptic peptides from rabbit gamma-globulin and two specific rabbit antibodies," Biochemistry 4(7):1418 (1965).

133. A. Nisonoff, "Resynthesis of precipitating antibody from univalent fragments," Biochem. Biophys. Res. Commun. 3(5):466 (1960).

134. A. Nisonoff and D. J. Dixon, "Evidence for linkage of univalent fragments of half-molecules of rabbit gamma-globulin by the same disulfide bond," Biochemistry 3(9):1338 (1964).

135. A. Nisonoff and W. J. Mandy, "Quantitative estimation of the hybridization of rabbit antibodies," Nature 194(4826):355 (1962).

136. A. Nisonoff, G. Markus, and F. C. Wissler, "Separation of univalent fragments of rabbit antibody by reduction of a single, labile disulfide bond," Nature 189(4761):293 (1961).

137. A. Nisonoff and J. L. Palmer, "Hybridization of half molecules of rabbit gamma-globulin," Science 143(3604):376 (1964).

138. A. Nisonoff and D. Pressman, "Studies on the combining site of anti-p-azobenzoate antibody. Loss of precipitating and binding capacities through different mechanisms on acetylation," J. Immunol. 83(2):138 (1959).

139. A. Nisonoff and M. M. Rivers, "Recombination of a mixture of univalent antibody fragments of different specificity," Arch. Biochem. Biophys. 93(2):460 (1961).

140. A. Nisonoff, F. C. Wissler, and L. N. Lipman, "Properties of the major component of a peptic digest of rabbit antibody," Science 132(3441):1770 (1960).

141. A. Nisonoff, F. C. Wissler, L. N. Lipman, and D. L. Woernley, "Separation of univalent fragments from bivalent rabbit antibody molecule by reduction of disulfide bonds," Arch. Biochem. Biophys. 89(2):230 (1960).

142. A. Nisonoff, F. C. Wissler, and D. L. Woernley, "Mechanism of the formation of univalent fragments of rabbit antibody," Biochem. Biophys. Res. Commun. 1(6):318 (1959).

143. A. Nisonoff, F. C. Wissler, and D. L. Woernley, "Properties of univalent fragments of rabbit antibody isolated by specific adsorption," Arch. Biochem. Biophys. 88(2):241 (1960).

144. A. Nisonoff and D. L. Woernley, "Effect of hydrolysis by papain on the combining sites on an antibody," Nature 183(4671):1325 (1959).

145. M. E. Noelken and C. Tanford, "Unfolding and renaturation of a univalent antihapten antibody fragment," J. Biol. Chem. 239(6):1828 (1964).

146. C. Nolan and E. L. Smith, "Glycopeptides. III. Isolation and properties of glycopeptides from a bovine globulin of colostrum and from fraction II-3 of human globulin," J. Biol. Chem. 237(2):453 (1962).

146a. "Notation for human immunoglobulin subclasses," Bull. World Health Organ. 35(6):953 (1966).

147. V. Nussenzweig and B. Benacerraf, "Studies on the properties of fragments of guinea pig γ_1 and γ_2 antibodies obtained by papain digestion and mild reduction," J. Immunol. 93(6):1008 (1964).

148. V. Nussenzweig and B. Benacerraf, "Electrophoretic patterns at acid and alkaline pH of reduced guinea pig 7 S gamma-globulin and antihapten antibodies of different specificities," Intern. Arch. Allergy Appl. Immunol. 27(4):193 (1965).

149. V. Nussenzweig, E. C. Franklin, and B. Benacerraf, "Immunologic properties of products obtained by mild reduction of human and guinea pig γ-globulins," Immunol. 93(6):1015 (1964).

150. D. E. Olins and G. M. Edelman, "The antigenic structure of the polypeptide chains of human γ-globulin," J. Exp. Med. 116(5):635 (1962).

151. D. E. Olins and G. M. Edelman, "Reconstitution of 7 S molecules from L and H polypeptide chains of antibodies and γ-globulins," J. Exp. Med. 119(5):789 (1964).

152. K. Onoue, Y. Yagi, P. Stelos, and D. Pressman, "Antigen-binding activity of 6 S subunits of β_2-macroglobulin antibody," Science 146(3642):404 (1964).

153. E. F. Osserman and K. Takatsuki, "Clinical and immunochemical studies of four cases of heavy ($H\gamma_2$) chain disease," Am. J. Med. 37(3):351 (1964).

154. J. Oudin, "On the associated state of rabbit allotypes, the existence of rabbit antibody molecules against two allotypes, and the dissociation of human γ-globulin antigens into smaller molecules," Biochem. Biophys. Res. Commun. 5(5):358 (1961).

155. R. H. Pain, "The molecular weight of the peptide chains of γ-globulin," Biochem. J. 88(2):234 (1963).

156. J. L. Palmer, W. J. Mandy, and A. Nisonoff, "Heterogeneity of rabbit antibody and its subunits," Proc. Nat. Acad. Sci. USA 48(1):49 (1962).

157. J. L. Palmer and A. Nisonoff, "Reduction and reoxidation of a critical disulfide bond in the rabbit antibody molecule," J. Biol. Chem. 238(7):2393 (1963).

158. J. L. Palmer and A. Nisonoff, "Dissociation of rabbit γ-globulin into half molecules after reduction of one labile disulfide bond," Biochemistry 3(6):863 (1964).

159. J. L. Palmer, A. Nisonoff, and K. E. Van Holde, "Dissociation of rabbit γ-globulin into subunits. Reduction and acidification," Proc. Nat. Acad. Sci. USA 50(2):314 (1963).

160. I. A. Parfentjev, US Patents 2,065,196 (1936), 2,123,198 (1938), 2,175,000 (1939), cit. R. R. Porter, Plasma Proteins, 1:257 (1960).

161. J. F. Pechere, G. H. Dixon, R. H. Maybury, and H. Neurath, "Cleavage of disulfide bonds in trypsinogen and α-chymotrypsinogen," J. Biol. Chem. 233(6):1364 (1958).

162. M. L. Petermann, "The action of papain on beef serum pseudoglobulin and on diphtheria antitoxin," J. Biol. Chem. 144(3):607 (1942).

163. M. L. Petermann, "The splitting of human γ-globulin antibodies by papain and bromelin," J. Am. Chem. Soc. 68(1):106 (1946).

163a. P. J. Piggot and E. M. Press, "Cyanogen bromide cleavage and partial sequence of the heavy chain of a pathological immunoglobulin G," Biochem. J. 104(2):616 (1967).

164. C. G. Pope and M. Healey, "The preparation of diphtheria antitoxin in a state of high purity," Brit. J. Exp. Pathol. 20(3):213 (1939).

164a. R. R. Porter, "The formation of a specific inhibitor by hydrolysis of rabbit antiovalbumin," Biochem. J. 46(4):479 (1950).

165. R. R. Porter, "The structure of the heavy chain of immunoglobulin and its revevance to the nature of antibody-combining site," Biochem. J. 105(2):417 (1967).

166. R. R. Porter, "The hydrolysis of rabbit γ-globulin and its antibodies with crystalline papain," Biochem. J. 73(1):119 (1959).

167. R. R. Porter, "The structure of γ-globulins and antibodies," in: Basic Problems in Neoplastic Disease, Columbia University Press, New York (1962), p. 177.

167a. R. R. Porter, "The structure of antibodies," Sci. Am. 217(4):81 (1967).

168. R. R. Porter and E. M. Press, "Immunochemistry," Ann. Rev. Biochem. 31:625 (1962).

169. M. D. Poulik, "F'c fragment of immunoglobulins," Nature 210(5032):133 (1966).

170. M. D. Poulik, "Heterogeneity of L(B) chains of γ-globulins," Nature 202(4938):1174 (1964).

171. M. D. Poulik and G. M. Edelman, "Comparison of reduced alkylated derivatives of some myeloma globulins and Bence–Jones proteins," Nature 191(4795):1274 (1961).

172. M. D. Culik and J. Schuster, "Heterogeneity of H chains of myeloma proteins: susceptibility to papain and trypsin," Nature 204(4958):577 (1964).

172a. J. Prahl, "Enzymic degradation of the Fc fragment of rabbit immunoglobulin IgG," Biochem. J. 104(2):647 (1967).

173. E. M. Press, "The amino acid sequence of the N-terminal 84 residues of a human heavy chain of immunoglobulin G (Daw)," Biochem. J. 104(2):30c (1967).

174. D. Pressman, A. L. Grossberg, O. Roholt, P. Stelos, and Y. Yagi, "The chemical nature of antibody molecules and their combining sites," Ann. N. Y. Acad. Sci. 103(582) (1963).

175. D. Pressman, A. Nisonoff, and G. Radzimski, "Specific anion effects with anti-benzoate antibody," J. Immunol. 86(1):35 (1961).

176. D. Pressman and G. Radzimski, "Increased precipitability of antibody as a result of iodination," J. Immunol. 89(3):367 (1962).

177. D. Pressman and O. Roholt, "Isolation of peptides from an antibody site," Proc. Nat. Acad. Sci. USA 47(10):1606 (1961).

178. D. Pressman, P. Stelos, and A. Grossberg, "Retention of rabbit antibody activity during acetylation," J. Immunol. 86(4):452 (1961).

179. D. Pressman and L. A. Sternberger, "The nature of the combining sites of antibodies. The specific protection of the combining site by hapten during iodination," J. Immunol. 66(5):609 (1951).

180. F. W. Putnam et al., "The cleavage of rabbit γ-globulin by papain," J. Biol. Chem. 237(3):717 (1962).

181. F. W. Putnam, "Structural relationships among normal human γ-globulin myeloma globulins and Bence–Jones proteins," Biochem. Biophys. Acta 63(3):539 (1952).

182. F. W. Putnam and C. W. Easley, "Structural studies of the immunoglobulins. I. The tryptic peptides of Bence–Jones proteins," J. Biol. Chem. 240(4):1626 (1965).

183. F. W. Putnam, K. Titani, M. Wikler, and T. Shinoda, "Structure and evolution of kappa and lambda light chains," Cold Spring Harbor Symp. Quant. Biol., p. 6 (1967).

184. F. W. Putnam, C. W. Easley, and L. T. Lynn, "Site of cleavage of γ-globulins by papain," Biochim. Biophys. Acta 58(2):279 (1962).

184a. J. Reynek, O. Kotynek, and J. Kostka, "Contribution to the structural characterization of γ-globulin from pig colostrum," Folia Microbiol. 10(6):327 (1965).

184b. R. A. Reisfield and P. A. Small, "Electrophoretic heterogeneity of polypeptide chains of specific antibodies," Science 152(3726):1253 (1966).

185. O. Roholt, A. Grossman, K. Onoue, and D. Pressman, "Relative contributions of H and L chains to the binding site of antibody molecules," Abstr. VI Intern. Biochem. Congr. New York 2:176 (1964).

186. O. Roholt, K. Onoue, and D. Pressman, "Specific combination of H and L chains of rabbit γ-globulins," Proc. Nat. Acad. Sci. USA 51(2):173 (1964).

186a. O. Roholt and D. Pressman, "A discontinuous heterogeneity of heavy chains of rabbit antibody molecules," Science 153(3741):1257 (1966).

187. O. A. Roholt, G. Radzimski, and D. Pressman, "Polypeptide chains of antibody: effective binding sites require specificity in combination," Science 147(3658):613 (1965).

188. O. A. Roholt, G. Radzimski, and D. Pressman, "Antibody combining site: The B polypeptide chain," Science 141(3582):726 (1963).

188a. O. A. Roholt, G. Radzimski, and D. Pressman, "Preferential recombination of antibody chains to form effective binding sites," J. Exp. Med. 122(4):785 (1965).

188b. O. A. Roholt, G. Radzimski, and D. Pressman, "Specificity in the combination of Fd fragments with L-chains to form hapten-binding sites," J. Exp. Med. 123(5):921 (1966).

188c. O. A. Roholt, G. Radzimski, and D. Pressman, "Recovery of activity from inactive hybrids of H and L chains," J. Exp. Med. 125(1):191 (1967).

189. O. Roholt, A Shaw, and D. Pressman, "Preferential recombination of antibody fragments as shown by paired label studies," Nature 196(4856):773 (1962).

190. J. H. Rockey, N. R. Klinman, and F. Karush, "Equine antihapten antibody. I. 7 S β-2A- and 10 S γ-globulin components of purified anti-β-lactoside antibody," J. Exp. Med. 120(4):589 (1964).

191. H. J. Sage, C. F. Deutsch, G. D. Fasman, and L. Levine, "The serological specificity of the polyalanine immune system," Immunochemistry 1(2):133 (1964).

192. I. Schechter and M. Sela, "Combining sites of antibodies to L-alanine and D-alanine peptide determinants," Biochim. Biophys. Acta 104(1):298 (1965).

192a. J. Schechter and A. Berger, "On the size of the active site in proteases. I. Papain," Biochem. Biophys. Res. Commun. 27(2):157 (1967).

193. R. E. Schrohenloher, "The degradation of human γ-globulin by trypsin," Arch. Biochem. Biophys. 101(3):456 (1963).

194. R. E. Schrohenloher, H. G. Kunkel, and T. B. Tomasi, "Activity of dissociated and reassociated 19 S anti-γ-globulins," J. Exp. Med. 120(6):1215 (1964).

195. P. H. Schur and G. D. Christian, "The role of disulfide bonds in the complement-fixing and precipitating properties of 7 S rabbit and sheep antibodies," J. Exp. Med. 120(4):531 (1964).

196. J. H. Schwartz and G. M. Edelman, "Comparisons of Bence—Jones proteins and L polypeptide chains of myeloma globulins after hydrolysis with trypsin," J. Exp. Med. 118(1):41 (1963).

197. M. Sela, D. Givol, and E. Mozes, "Resolution of rabbit γ-globulin into two fractions by chromatography on diethylaminoethyl-sephadex," Biochem. Biophys. Acta 73:649 (1963).

198. M. Sela, "Immunological studies with synthetic polypeptides," Advan. Immunol. 5:30 (1967).

199. H. G. Seijen and M. Gruber, "Structure of γ-globulins and antibodies," J. Mol. Biol. 7(2):209 (1963).

199a. J. Sjöquist and M. H. Vaughn, "Heterogeneity of H and L chains of normal and myeloma γG-globulin," J. Mol. Biol. 20(3):527 (1966).

199b. S. J. Singer and R. F. Doolittle, "Antibody active sites and immunoglobulin molecules," Science 153(3731):13 (1966).

200. F. Skvaril, "Changes in outdated human γ-globulin preparations," Nature 185(4711):475 (1960).

201. F. Skvaril, "Inhibitions of spontaneous splitting of γ-globulin preparations with ε-aminocaproic acid," Nature 196(4853):481 (1962).

201a. F. Skvaril and J. Radl, "The fragmentation of human IgD during storage," Clin. Chim. Acta 15:544 (1967).

202. F. Skvaril, V. Brummelova, and F. Franek, "Isolation of kappa and lambda chains from normal human γG-globulin by ion-exchange chromatography," Biochim. Biophys. Acta 140(2):371 (1967).

203. A. Small, J. E. Kehn, and M. E. Lamm, "Polypeptide chains of rabbit γ-globulin," Science 142(3590):393 (1963).

203a. P. A. Small and M. E. Lamm, "Polypeptide chain structure of rabbit immunoglobulins. I. γG-immunoglobulin," Biochemistry 5(1):259 (1966).

203b. D. S. Smyth and S. Utsumi, "Structure of the hinge region in rabbit immunoglobulin G," Nature 216(5113):332 (1967).

204. S. R. Stein, J. L. Palmer, and A. Nisonoff, "Reformation of interchain bonds linking half-molecules of rabbit γ-globulin," J. Biol. Chem. 239(9):2872 (1964).

205. A. G. Steinberg, "Genetic variations in human immunoglobulins: the Gm and Inv types," in: Symposium on immunogenetics, Lippincott, Philadelphia (1967).

206. P. Stelos and D. Pressman, "Papain digestion of antigen–antibody precipitates," J. Biol. Chem. 237(12):3679 (1962).

207. P. Stelos, G. Radzimski, and D. Pressman, "Heterogeneity of rabbit antibody fragments," J. Immunol. 88(5):572 (1962).

208. P. Stelos, Y. Yagi, and D. Pressman, "Multiple discrete components of the univalent fragments of rabbit antibody," J. Immunol. 93(1):106 (1964).

209. D. Stolinsky and H. Fudenberg, "On the univalent fragments of human 7 S γ-globulin, Nature 200(4909):856 (1963).

210. J. M. Swan, "Thiols, disulfides, and thiosulphates: some new reactions and possibilities in peptide and protein chemistry," Nature 180(4587):643 (1957).

211. K. Takatsuki and E. F. Osserman, "Structural difference between two types of "heavy chain" disease proteins and myeloma globulins of corresponding types," Science 145(3631):499 (1964).

212. M. Tan and W. V. Epstein, "Purification of γ-globulin fragments by gel filtration," Science 139(3549):53 (1963).

213. W. D. Terry, "Subclasses of human IγG molecules differing in heterologous skin sensitizing properties," Proc. Soc. Exp. Biol. Med. 117(3):901 (1964).

214. W. D. Terry and J. L. Fahey, "Subclasses of human γ-globulin based on differences in the heavy polypeptide chains," Science 146(3642):400 (1964).

215. K. Titani and F. W. Putnam, "Immunoglobulin structure: amino and carboxyl terminal peptides of type I Bence–Jones proteins," Science 147(3663):1304 (1965).

215a. K. Titani, M. Wikler, and F. W. Putnam, "Evolution of immunoglobulins' structural homology of kappa and lambda Bence–Jones proteins," Science 155(3764):828 (1967).

215b. H. Ungar-Waron, J. C. Jaton, and M. Sela, "Action of papain on normal rabbit immunoglobulin M," Biochim. Biophys. Acta 140(3):542 (1967).

216. S. Utsumi and F. Karush, "Peptic fragmentation of rabbit γG-immunoglobulin," Biochemistry 4(9):1766 (1965).

217. S. Utsumi and F. Karush, "The subunits of purified rabbit antibody," Biochemistry 3(9):1766 (1965).

217a. S. Utsumi and F. Karush, "Chemical characterization of the peptic fragments of rabbit γG-immunoglobulin," Biochemistry 6(8):2313 (1967).

218. M. H. Winkler, "On the structure of specific antibody site," J. Theoret. Biol. 4(3):237 (1963).

219. P. L. Whitney and C. Tanford, "Recovery of specific activity after complete unfolding and reduction of an antibody fragment," Proc. Nat. Acad. Sci. USA 53(3):524 (1965).

220. S. F. Velick, C. Parker, and H. N. Eisen, "Excitation energy transfer and the quantitative study of the antibody hapten reaction," Proc. Nat. Acad. Sci. USA 46(11):1470 (1960).

220a. R. C. Weir and R. R. Porter, "The antigen-binding capacity of the peptide chains of horse antibodies," Biochem. J. 100(1):69 (1966).

221. R. C. Williams, "Heterogeneity of L-chain sites on Bence–Jones proteins reacting with anti-γ-globulin factors," Proc. Nat. Acad. Sci. USA 52(1):60 (1964).

222. L. Wofsy, H. Metzger, and S. J. Singer, "Affinity labeling – a general method for labeling the active sites of antibody and enzyme molecules," Biochemistry 1:1031 (1962).

223. L. Wofsy and S. J. Singer, "Effects of the amidination reaction on antibody activity and on the physical properties of some proteins," Biochemistry 2(1):104 (1963).

224. L. Wofsy, N. R. Klinman, and F. Karush, "Affinity labeling of equine anti-β-lactoside antibodies," Biochemistry 6(7):1988 (1967).

225. T. J. Yoo, O. A. Roholt, and D. Pressman, "Specific binding activity of isolated light chains of antibodies," Science 157(3789):707 (1967).

226. J. Zikan, B. Skvarova, and J. Rejnek, "Two structurally different types of rabbit light chains," Folia Microbiol. 12(2):162 (1967).

Chapter V

Antibody Biosynthesis

A considerable amount of research has been done on antibody synthesis. This is first and foremost a result of its practical importance, which is reflected in the fact that a considerable portion of the population is vaccinated each year with the most varied antigens. It is also the reason for the comprehensive literature on the productive phase of antibody synthesis, i.e., the period commencing at the moment antibodies appear in the blood. However, at the present, many investigators are concentrating on the elucidation of processes which are initiated directly after administration of antigen to the organism and which ultimately lead to formation of antibody molecules. As in the biosynthesis of other proteins, nucleic acids can be expected to play an important role in the mechanism of antibody formation. Hence, it is not surprising that much attention is being devoted at the present time to synthesis of nucleic acids in antibody-forming cells and tissues.

This chapter will be devoted principally to a discussion of data which are important for understanding processes taking place at the moment of induction of antibody synthesis, and which deal, for example, with the fate of antigens in the organism, RNA synthesis during antibody formation, etc. In vitro methods for studying antibody-synthesizing cells, which offer considerable advantages in biochemical investigations, are also examined in detail.

THE FATE OF ANTIGEN

The fate of an antigen in an animal after immunization is of fundamental importance for an understanding of antibody biosynthesis. If antigen or its fragments are not retained in the organism, their direct participation in antibody biosynthesis can be at once ruled out. Therefore, it is quite natural that many investigations have had as their object an inquiry into the pathways and transformations of antigen in the organism [9, 95, 144]. Soluble antigen injected intravenously is gradually eliminated from the blood over a period of several days. As Talmage et al. [267] showed on the basis of elimination of ^{131}I-labeled bovine γ-globulin, several stages can be distinguished in this process: 1) rapid decrease in the concentration of the injected antigen during the first day as a result of establishment of equilibrium with extracellular fluids; 2) gradual decrease in the antigen content due to degradation, which takes place at the same rate as degradation of homologous protein administered simultaneously with the antigen; 3) a phase of rapid disappearance of antigen from the blood as a result of initiation of antibody biosynthesis. The last phase is not actually observed in the case of homologous protein or when antigen is eliminated from tolerant animals with no antibody-forming capacity. Weigle presents similar data [279].

Antigen apparently penetrates rapidly into cells. This was established by Harris [143], who studied antibody synthesis by lymph node cells exposed to *Shigella* antigen in vitro and then transferred to irradiated animals. It was found that a 30-minute contact of antigen with the cells was sufficient to stimulate antibody synthesis and that addition of antiserum to *Shigella* at the end of incubation had no effect on this process. However, if the antiserum was added to the suspension before incubation, antibody synthesis did not take place. Five minutes after the start of incubation of antigen with cells the antiserum no longer inhibited antibody synthesis. Autoradiographic and electron-microscopic studies of ^{125}I-ferritin disclosed the presence of this antigen in small lymphocytes within five minutes [140b]. Thus, a few minutes was sufficient for antigen to penetrate into antibody-synthesizing cells.

How does antigen enter cells? Cellular antigens (bacteria, erythrocytes) usually penetrate into cells by means of phagocytosis. Soluble antigens evidently penetrate by pinocytosis, i.e., by absorption of drops of the surrounding fluid [23, 88, 229, 268] by cells. This process can be observed with a light microscope. Sometimes absorption of drops of antigen-containing fluid can only be observed with an electron microscope, in which case the process is called ropheocytosis [88]. In either case, first invaginations of cell membranes are observed. These become increasingly deeper and finally are transformed into intracellular vacuoles. Evidently most cells are capable of pinocytosis, including lymphocytes and plasma cells [88], although according to some reports the latter possess this property only in the early stages of their development [268].

After addition of labeled antigen (^{125}I, ^{131}I, ^{3}H) or antigen visible under an electron microscope, as in the case of ferritin, an Fe-containing protein, the label is detectable primarily in macrophage cells [97, 160, 244]. However, it is not completely clear whether antigen is always present in antibody-synthesizing cells. After administration of microgram quantities of labeled antigen (^{131}I-flagellin) to rats, Nossal et al. [208, 209] radioautographically detected the label only in macrophages and similar cells of lymphoid follicles, and not in plasma cells. Analogous results were obtained by McDevitt et al. [185d] who used synthetic polypeptide as the antigen with its antigenic determinants labeled with ^{125}I or ^{125}I-hemocyanin. No radioactive label could be detected in antibody-containing cells, although the macrophages contained considerable labeled antigen. These results are scarcely attributable to insufficient sensitivity of the method since, according to calculations, 15 molecules or less of undegraded antigen were present in a cell. However, it should be borne in mind that it is impossible to obtain information on antigenic fragments containing no label in this way [161a].

Wellensick and Coons [281] obtained different results in an electron microscope study of the distribution of ferritin in lymph node cells of presensitized rabbits. After administration of antigen in even greater amounts than used in experiments with radioactive iodine, these investigators consistently found ferritin in plasmoblasts and in immature (and sometimes in mature) plasma cells. It is noteworthy that ferritin was found not only in cytoplasm, but also in the nuclei of cells, including plasma cells. In the cytoplasm, ferritin was bound to membranes and ribonucleoprotein particles, while in nuclei it was associated with dense chromatin material. The electron diffraction patterns obtained by de Petris et al. [218] after injection of ferritin into rabbits showed a similar picture. Using large doses of γ-globulin with ^{3}H-labeled hapten as the antigen, Roberts and Haurowitz [228] discovered many grains above the cytoplasm of mature and immature plasma cells in radioautograms. It is at present difficult to explain differences in the results of these two groups of investigators, one of which found antigen in antibody-synthesizing cells, while the other did not, but they may possibly be due to the different amounts of antigen administered. In any event, the presence of antigen or its fragments in antibody-synthesizing cells cannot as yet be ruled out.

The question as to which cell components antigen is bound after its penetration into cells also lends itself to experimental solution. If components isolated by differential centrifugation are used in the study, it is difficult to avoid adsorption of antigen on them when the tissue is homogenized. Haurowitz et al. [145, 146] studied the fate of intravenously injected ovalbumin labeled with ^{131}I and other isotopes and found the tag in all cell components: first somewhat more in microsomes, then in mitochondria, but considerable quantities were also found in nuclear fractions and in the supernatant. Franzl [127] found antigen in microsomes. Coons et al. [185] detected a predominant accumulation of antigen in cell nuclei, but this was not confirmed in all subsequent investigations. However, Coons [185] in experiments with ferritin consistently found this protein in the nuclei of lymph cells.

The time of residence of antigen in the body has been determined by several investigators; it is now believed that antigens can remain in tissue for very long periods of time. Such studies have principally employed radioactively labeled antigens.

Ingraham [164-166] used bovine γ-globulin conjugated with sulfanilic acid as antigen. The hapten was labeled with ^{35}S. Since sulfanilic acid was metabolically inert, the label found in the tissues was attributed to the antigen or its fragments. It was found that of the amount of label detected 24 hr after injection into mice, about 1% and 4% were still in the liver and spleen, respectively, 200 days later. ^{35}S-Sulfanilic acid and its compounds with tyrosine, histidine, or glycine-tyrosine are very rapidly eliminated from the body.

Haurowitz and Crampton [145] found ^{131}I-labeled ovalbumin in the liver in an amount of $2 \cdot 10^{14}$ molecules, i.e., 2000 molecules per cell, on the 29th day after intravenous injection to rabbits. An extrapolation of these data showed a decrease in the number of molecules per cell from 2000 to 200 within 300 days, and to 20 per cell within 3000 days. Although such calculations are hardly distinguished by their accuracy, they do show that antigen remains in the body for a long time.

Data of Garvey and Campbell are very instructive; these authors studied the retention of a label in the liver of rabbits after single or multiple injections of bovine albumin and hemocyanin labeled with ^{35}S-sulfanilic acid [95, 134, 135]. From Fig. 33, which presents results of these experiments, it is evident that antigen is retained for considerable periods of time. Thus, 140 days after injection of ^{35}S-labeled bovine serum albumin, 0.02% of the injected quantity (10 μg or 10^{14} molecules) was still present, while 0.05% or 25 μg of the injected quantity of ^{35}S-hemocyanin was found 330 days after injection. The elimination rate during this time was such that considerable quantities of antigen were still to be found after several years (for example, after three years, 0.75 μg albumin, or $5.79 \cdot 10^{12}$ molecules, may still be present).

The experiments of Speirs, who detected ^{3}H-toxins 270 days after their injection to the organism, also give evidence of the prolonged retention of antigen by tissues [242].

In what form are antigens retained in tissues? A large body of evidence indicates that in the organism protein antigens are split into small fragments. Within a relatively short time after elimination of labeled protein antigens, the label ceases to precipitate with the corresponding antiserum. This was demonstrated with hemocyanin [134] and bovine sulfanilic acid-azo-γ-globulin [164]. Hawkins and Haurowitz [147] detected γ-globulin and albumin precipitable by bovine antiserum in saline extracts from rat spleen only during the first seven and four days, respectively, after injection of the antigens.

Many tissues have the capacity to degrade protein antigens; this is especially true of spleen tissue [180]. In a study of cleavage of human serum albumin by rabbit spleen extracts it was found that the protein molecule was split into three distinct fragments, for each of which a corresponding antibody was found in the serum of rabbits immunized with this same protein.

If it is indeed true that the molecules of protein antigens are degraded in the organism, then antibodies should be formed not only specific for antigen determinants on the surface of native molecules, but also for "internal" antigenic determinants, exposed

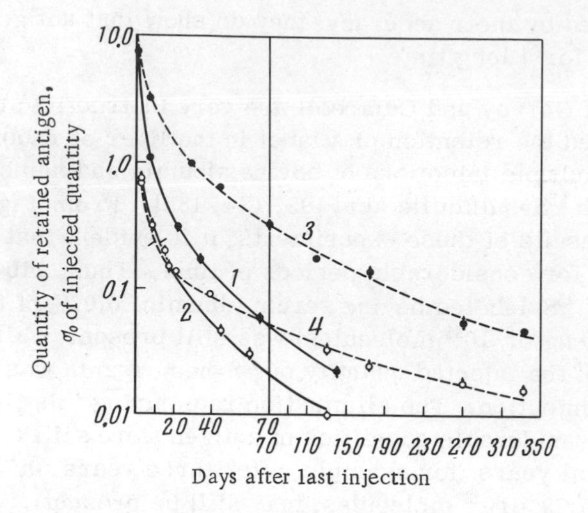

Fig. 33. Semilogarithmic curves showing retention of radioactive antigens in rabbit liver tissue [135]. Injections of ^{35}S-sulfanyl-azo-albumin: 1) one; 2) several. Injections of ^{35}S-sulfanyl-azo-hemocyanin: 3) one; 4) several.

after degradation into fragments. Some investigators have experimentally demonstrated that antibodies to such "latent" determinants actually do exist. Thus, Ishizaka et al. [163] found that some antibodies still remained in antiserum to bovine albumin after absorption by native antigen and that these antibodies were able to react with pepsin-cleaved fragments of the same albumin. Results of Sorkin and Boyden [241] provide a direct confirmation of antigen degradation in cells; these investigators isolated a labeled peptide from cells of a peritoneal exudate after their contact with ^{131}I-albumin. Thus, in discussing the time of retention of antigens in tissues it must be kept in mind that antigens are most likely retained as small fragments.

An extremely important question concerns the properties of these antigen fragments, which as seen in the foregoing, are long retained by tissues. Some investigators have been able to detect biologically active antigens for a long time in the organism.

McMaster and Kruse [188] employed a very sensitive method of passive anaphylaxia to study this question. Results were assessed on the basis of the response of mouse ear veins under a microscope. It was found that protein antigens are retained for a prolonged period in mice. For example, bovine γ-globulin remained in the liver for 101 days, while human serum albumin was retained for 36 days. These results were later corroborated in experiments on rabbits, in which bovine globulin was found two months after injection by the same method [189].

Garvey and Campbell also used a very sensitive anaphylactic method to determine biologically active substances in rabbit liver after a single injection of bovine sulfanilic acid azo albumin [135]. Their data indicated that antigen can be retained for a long time without loss of biological activity. However, it should be borne in mind that the methods used in these investigations were not direct methods, and hence the information obtained on the properties of the antigenic fragments studied cannot be regarded as definitive.

Evidence indicating that antigenic fragments can be retained in cells as complexes with RNA is illuminating. Saha et al. [230] isolated several fractions containing RNA and a radioactive label from extracts of liver of rabbits immunized with bovine ^{35}S-sulfanilic-azo-albumin. In one of these series of experiments, RNA was extracted from the supernatant left after centrifugation of the extract at 78,000 *g* for two hours. The label in these RNA preparations was freed by weak alkaline hydrolysis. In other experiments, radioactivity was found in different soluble ribonucleoprotein fractions of a supernatant obtained after protracted high-speed centrifugation and in nucleoprotein complexes isolated from a supernatant remaining after treating the extract with 0.6*M* perchloric acid. The nucleoprotein complexes contained up to seven amino acid residues. It has recently been shown that a definite portion of the antigen eliminated with the urine is bound with an RNA-like substance; this is especially true after the second antigen injection [135a]. RNA preparations extracted with phenol from macrophages of a mouse peritoneal cavity incubated with ^{131}I-hemocyanin or phage T2 contained small quantities of antigen which, however, were able to evoke a secondary immune response in previously immunized mice [70, 132]. It is still too early to draw any conclusions as to the role of such antigen-RNA complexes in the biosynthesis of antibodies, but pertinent studies will yield extremely important information, considering the contemporary views on the role of nucleic acids in protein biosynthesis.

Thus, a survey of the experimental literature creates the impression that antigens or their fragments can be retained in the organism for a long time. Of course, the methods used to demonstrate this view had certain shortcomings. Thus, when radioactively labeled antigens are used, there is the possibility that the label will split off and combine with other proteins. In experiments in which passive anaphylaxia was used, possible transfer, not only of the antigen itself, but also of other secondary factors produced by the antigen, must be taken into account. Nevertheless, the principal point, namely, that antigens remain for a long time in the organism, seems at the present time sufficiently substantiated.

Of course, this still does not prove that antigenic fragments participate directly in antibody synthesis. It is possible that these fragments are retained in cells without exerting any effect on this process. However, whenever the theoretical problems of antibody biosynthesis are disçussed, the possibility of prolonged retention of biologically active antigen fragments in the organism should always be taken into account.

IN WHICH ORGANS AND CELLS ARE ANTIBODIES SYNTHESIZED?

The site of antibody synthesis depends on the dose and the site of injection of antigen and on the properties of the antigen itself. McMaster was the first to show that antibodies are synthesized in the lymph nodes nearest to the site of antigen injection. After intradermal injection of a suspension of dead microbes into mice or rabbits, the highest titer of the corresponding agglutinins was found in extracts from neighboring lymph nodes [186, 187]. Later, this observation was repeatedly verified for different types of antigens [247]. Fontalin's experiments, in which neighboring lymph nodes were removed, provide a good example. Thus, after a tetanus anatoxin was injected into the hip muscle and the neighboring lymph nodes were removed, the average antitoxin titer in the blood of the animal was 0.4 ± 0.5 AE on the 21st day as compared with 1 ± 0.17 AE for animals from which the contralateral lymph nodes had been removed. In intact animals, the comparative figure was 0.8 ± 0.22 AE [48, 49]. Neighboring lymph nodes also exhibited the most pronounced cytological alterations. As Gurvich and Shumakov (Zdorovskii laboratory) showed, it is precisely in these neighboring lymph nodes that considerable accumulation of plasmoblasts takes place after immunization with a soluble antigen; moreover, these nodes acquire the ability to synthesize antibody when transferred to a nonimmune animal [18, 22, 66].

Humphrey and Sulitzeanu [161] made a systematic study of the site of synthesis of antibody to pneumococci in rabbits by determining the amount of ^{14}C-labeled antibody extractable from different organs of intravenously immunized animals. The results varied from experiment to experiment, but in most cases a considerable percentage of antibody was synthesized in the lungs, followed by the bone marrow, spleen, and lymph nodes, and finally the appendix. The leading participation of the lungs in the biosynthesis of antipneumococcal antibodies was probably attributable to the type of antigen used and to the mode of injection, whereby a considerable portion of the bacteria was retained in lung capillaries.

A study of the antibody-forming activity of different organs in an in vitro culture gives a clear picture of the relative participation of these organs in antibody formation. The very first investigations of this type showed that antibody synthesis by cells of a given organ in vitro is also dependent on the site of antigen injection [247]. For example, after a single intravenous injection of alien erythrocytes in vitro, hemolysins were synthesized by spleen cells in vitro, but not by mesenteric lymph nodes or the thymus. In contrast, after intraperitoneal injection of antigen, lymph nodes synthesized more antibodies [227].

Askonas and Humphrey made a systematic study of the synthesizing capacity of various organs in rabbits [68, 72] by determining incorporation of ^{14}C-amino acids into antibodies and γ-globulins by tissue slices from different lymph nodes, spleen, liver, and other organs after in vitro incubation. After injection of ovalbumin in Freund's adjuvant, it was found that the tissue

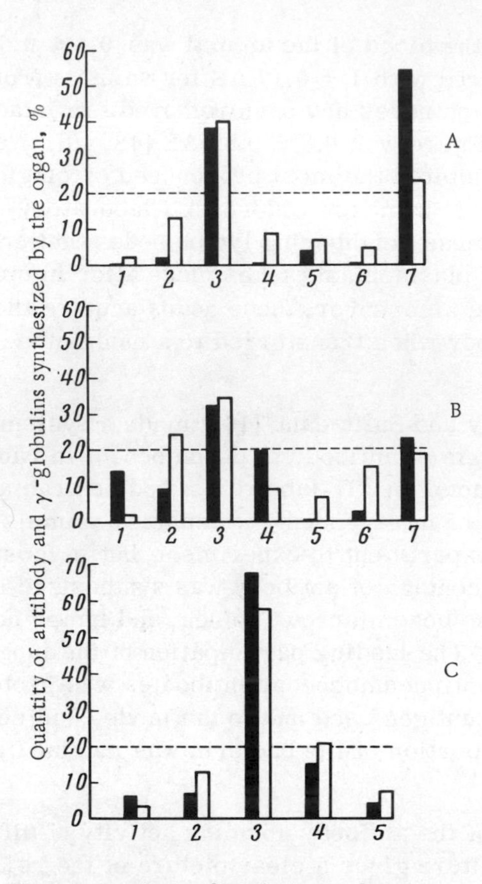

Fig. 34. Synthesis of antibodies and γ-globulins by tissue slices from rabbits after different methods of immunization with ovalbumin [69]. A) Intramuscular immunization in an adjuvant; B) immunization first intramuscularly, then intravenously; C) repeated intramuscular immunization. 1) Spleen; 2) lymph nodes; 3) bone marrow; 4) lungs; 5) liver; 6) kidney; 7) granuloma. Shaded columns – antibody; white columns – γ-globulins.

which grew at the site of a granuloma injection synthesized more antibody. When sometime later this same antigen was injected intravenously, the relative participation of the lungs and lymph nodes in antibody synthesis had increased considerably. After

TABLE 35. Formation of Hemolysis Plaques by Cells from Different Organs of Mice Immunized with Sheep Erythrocytes [131]

Mouse strain	Mean hemolysin titer	Number of hemolysis plaques per 10^6 cells									
		spleen	lymph nodes	peripheral leukocytes	thymus	bone marrow	lungs	liver	brain	kidneys	skeletal muscles
NIH	1 : 825	786	451	38	5	3	16	14	2	3	1
C3H	1 : 461	314	270	21	2	5	18	10	1	1	2
C57B	1 : 385	270	184	14	3	6	7	14	1	1	3
AKR	1 : 135	94	48	7	3	3	8	2	2	1	2

intravenous injection, the lungs, spleen, and bone marrow are the chief antibody-producers (Fig. 34). The bronchial lymph nodes and spleen were the most active (per unit weight). The liver is most likely not capable of antibody synthesis. Immunization with pneumococci sharply enhances the participation of the lungs in antibody formation. At the same time, γ-globulin synthesis by the lungs also increases (this γ-globulin could not be identified as antibody). Data from studies of Uchitel' and Konikova provide evidence of an increase in synthesis of nonspecific γ-globulins after immunization [44-47].

Thus, these investigations have shown that, as could be expected, antibody synthesis is most intense at the site of antigen accumulation.

The formation of hemolysins by cells from organs of mice intravenously immunized with erythrocytes was studied by counting the hemolytic plaques obtained by incubation of lymphoid cells with antigen by the Jerne method [131]. As is evident from Table 35, the greatest number of antibody-forming cells is found in the spleen and lymph nodes on the fourth day after immunization by this method. On the whole, analogous data were obtained in a study of the antibody-synthesizing capacity of tissues after transfer to nonimmunized, usually preirradiated animals [99]. In several cases, the transferred cells were placed in diffusion chambers. Thus, Holub [151] showed that spleen, lymph node, and lymph cells of normal rabbits which had been incubated with antigen in chambers placed in the intestinal cavity of newborn rabbits synthesized antibody. Thyroid gland cells did not have this capacity, and in bone marrow cells it was only slight.

Thus, the principal antibody-forming organs are the spleen [8], lymph nodes, and in some cases, the bone marrow. When antigen is injected in adjuvants, a considerable portion of antibody is synthesized in the local granuloma.

Antibody-Synthesizing Cells. The contemporary classification of lymphoid tissue cells participating in immune responses is based principally on morphological criteria. In its general features it was formulated by the International Commission of Scientists in 1959 in Prague [220]. In the slightly expanded version of Pokrovskaya et al. [38, 39] three main groups of immunologically competent cells of lymph tissue are distinguished:

I. Reticular cells (dormant reticular cells, basophilic reticular cells, transitional reticular cells, plasma cells).

II. Cells of plasma series (plasmoblasts, immature and mature plasma cells).

III. Cells of lymphoid series (lymphoblasts, prolymphocytes, and basophilic lymphocytes).

Various methods have been used to determine which of these cells participates directly in antibody synthesis.

Fagraęus [116] concluded from a study of the histological picture of the spleen of animals during immunization that plasma cells are the prime factor in antibody biosynthesis. These data were fully confirmed with the aid of an immunofluorescent method developed by Coons et al. [3, 57, 102, 185], who found antibodies to proteins in groups of plasma cells in the red pulp of spleen and the medullary regions of lymph nodes in hyperimmune rabbits. The antibodies appeared first in the cytoplasm of immature plasmoblastic cells, from which colonies of plasma cells later developed [185].

Lebedev carried out a detailed investigation of the morphology of antibody-synthesizing cells by the Coons method after primary and secondary immunizations [27, 28]. Table 36 presents results of a study of lymph nodes after secondary immunization.

It may be observed that the majority of cells producing a

maximum of antibodies are plasma cells, although a small number of lymphocytes also contain antibodies.

Methods employing radioactive isotopes [80] or elements with a high atomic weight, such as mercury and iodine, which can be easily detected with the electron microscope [26, 31, 174] are evidently suited for detection of antibody-synthesizing cells.

Data obtained from the study of antibody formation by single cells are very illuminating for determining the nature of antibody-synthesizing cells [210]. When single cells from lymph nodes are incubated, antibodies are secreted into the surrounding medium where they can be detected by immobilization of bacterial flagella by the adherence of bacteria to a cell, or by neutralization of a bacteriophage. Using bacterial antigens, Nossal and Mäkelä found that almost all antibody-secreting cells belonged to the plasma group. Another group of investigators, using a bacteriophage as antigen, found that cells of the lymphoid series were antibody-producers in a high percentage of cases [73-77]. Differences in results were possibly due to the different experimental conditions (antigens, antibody detection, cell description, etc.). Still other data obtained by the Coons method indicate that antibodies are also found in cells similar to mature lymphocytes [277]. The main quantity of antibody is apparently elaborated by plasma cells, although cells of the lymphocyte type are also able to synthesize antibody under certain conditions. The antibody-synthesizing capacity of macrophage cells still remains to be demonstrated [159, 162].

The special role of plasma cells in the elaboration of the greater percentage of antibody is indicated by their structure. The principal morphological characteristic of these cells is a well-developed endoplasmic network (reticulum) filling the cytoplasm. It consists of parallel double membranes with ribonucleoprotein particles distributed on the outer surface as is readily detectable with an electron microscope (Fig. 35) [85, 86, 88, 119]. This explains the pronounced pyroninophilicity of plasma cell cytoplasm.

The same picture can be observed in a study of the secretory cells of other organs, for example, the pancreas [86]. The plasma cell cytoplasm, which is rich in RNA and ribosomes, is apparently equipped to elaborate a large quantity of protein for "export." Indeed, a large quantity of antibody is found in the caniculi of the

TABLE 36. Percent Composition of Different Antibody-Synthesizing Cells in Slices of Nearby Lymph Nodes of Rabbits Immunized Twice with Diphtheria Anatoxin [28]

Time after second immunization	Hemocytoblasts	Plasma cells		Lymphocytes	
		immature	mature	small	medium and large
2	37.9 ± 3.6	40.35 ± 3.6	10.25 ± 0.33	0	12.16 ± 0.53
3	25.4 ± 4.5	45.5 ± 4.05	10.7 ± 1.8	5.1 ± 4.8	13.65 ± 2.56
4	11.92 ± 0.89	46.2 ± 2.92	29.08 ± 2.22	2.6 ± 0.67	11.16 ± 0.86
5	10.77 ± 1.39	41.1 ± 2.88	37.2 ± 4.4	4.9 ± 1.61	5.47 ± 2.94
7	7.1	37.3	48.5	4.5	3.0
8	3.2	7.7	73.1	11.5	0
12	0	2.1	81.9	16.0	0
14	0	2.05	91.9	6.05	0

endoplasmic reticulum of plasma cells. This is quite evident in the electron microscope studies of de Petris et al. [218], who found antibody specific for ferritin in the caniculi of the cell reticulum in rabbits immunized with this antigen (Fig. 36).

The structure of lymphocytes differs appreciably from the structure of plasma cells. Cytoplasm occupies only a small part of the whole cell. It is practically impossible to detect the endoplasmic reticulum, and the number of ribonucleoprotein particles is very low. Such cells probably synthesize only a very small quantity of protein for "export" [86, 88]. However, a study of plaque-forming cells by the Jerne method showed that plasmacytes and lymphocytes were represented in equal numbers among these cells. The lymphocytes had an endoplasmic reticulum which was, however, less pronounced [143a].

The number of plasma cells depends on the nature of the antigen, its dose, and the site of injection, and on whether the immunization is the first or second. In several cases, the number of plasma cells may be very high. Thus, after the first injection of *Salmonella* N-antigen Nossal and Mäkelä found about 3% plasma cells in the neighboring lymph node, while after a second injection 20%, and sometimes 30%, plasma cells was found [210]. In

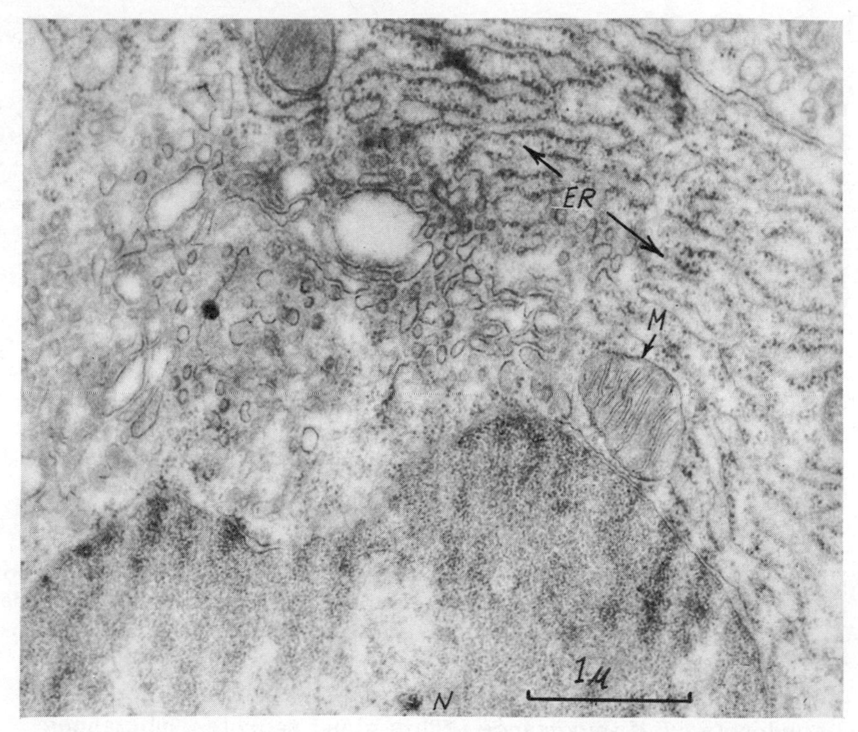

Fig. 35. Electron micrograph of a plasma cell from an inguinal lymph node of a rat (Preparation V. M. Manteifel, Institute of Molecular Biology, Academy of Science of the USSR). N, nucleus; M, mitochondria; ER, endoplasmic reticulum.

experiments with incubation of single cells in microdrops, these same investigators showed that up to 60% plasma cells elaborate antibody at the peak of the second response. However, lower figures were obtained by the Coons method [27, 28].

The thymus plays an important role in the origin of lymphocytes. These cells are probably formed in the thymus and then distributed throughout the organism [29, 269, 280a]. The migration of lymphocytes from one organ to another may be regarded as proven [50]. It is not surprising that blood cells are also able to synthesize antibody. According to rough data obtained by counting hemolysis plaques, 110,000 out of 600 million rabbit blood lymphocytes are able to synthesize antibody at the peak of the immune response [179].

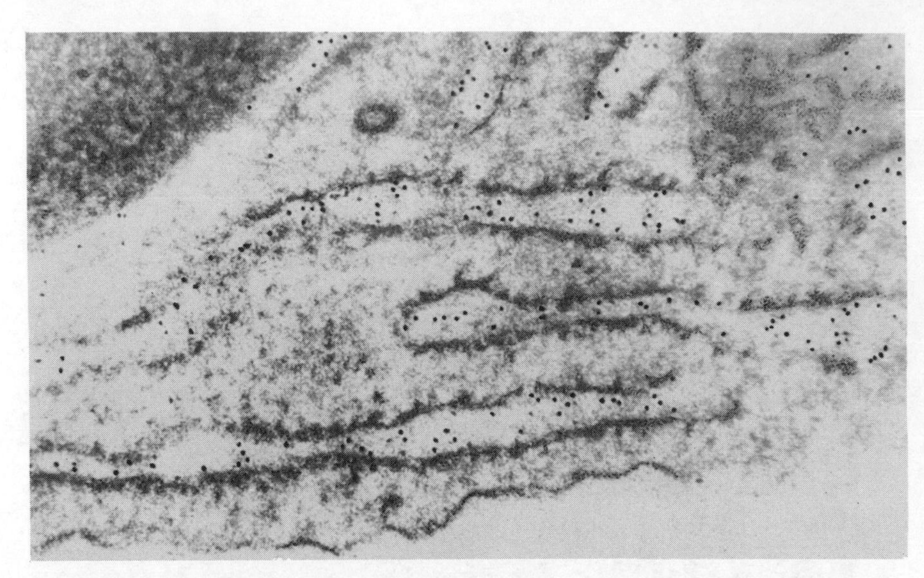

Fig. 36. Electron micrograph of a lymph node plasma cell of a rabbit immunized with ferritin [218]. The cells were treated with ferritin, the molecules of which (black dots) are only visible in cisternae of the reticulum and are not associated with ribosomes.

Another aspect of thymus activity – the hormone activity – is undoubtedly of importance. This gland secretes substances which simulate lymphopoiesis. For example, thymus tissue exhibits such an activity on being enclosed in diffusion chambers [29]. Extracts from thymuses of cows, rats, or mice stimulated incorporation of labeled precursors into DNA and total protein of lymph node cells in vivo [177]. An active factor – thymosine, a glycoprotein with a molecular weight of about 10,000 [138b] – was isolated from these extracts. It is currently assumed that the action of the thymus is necessary for at least some lymphoid cells to acquire the capacity to respond to antigen, i.e., to be converted into antigen-competent cells.

The antibody-synthesizing capacity varies during an animal's lifetime [191]. Thus, during the first month after birth it increases sharply, reaching a maximum in 40-week-old mice. During this period, the antibody-synthesizing capacity of spleen cells transferred to other mice of the same strain increased 600 times, while the body weight and spleen weight increased only 8 and 4 times, respectively. Then the activity decreased, and in very old mice (120 weeks) it was only 25% of the level found in 40-week-old mice.

The question of the immediate precursors of antibody-synthesizing cells will be examined in more detail in the chapter on the genetic aspects of antibody formation. At the present it is sufficient to point out that according to data from several investigations, these cells are formed from large, rapidly dividing lymphocytes. According to other data, however, they are formed from other types of lymphocytes, dormant cells, reticular cells [101], and even by direct transformation from small lymphocytes [152].

In recent years, increasing attention has been devoted to the role of phagocytes in antibody synthesis [132c, 199]. There is hardly any doubt that phagocytosis resulting from injection of corpuscular antigens is of primary importance. However, there is also evidence that when soluble antigens are injected molecular aggregates are the most important factor; it is possible that these aggregates are also quickly phagocytized [128].

A morphological investigation of the spleen disclosed, curiously enough, the distribution of plasma cells around a macrophage [268]. Moreover, cytoplasmic bridges between the macrophage and one or two lymphocytes or plasma cells immediately adjacent to it were observable under the electron microscope [233]. It is not impossible that RNA and ribosomes are transmitted across these bridges [64]. These data are revealing in connection with the data of Fishman et al. [122-125] which suggest that antibody synthesis has two phases: the first phase involves the capture of antigen by phagocytes, and then immediately afterwards a kind of stimulator (see below) is transferred into cells already directly engaged in antibody synthesis from these phagocytes.

The question of the cellular origin of different types of immunoglobulins is very interesting. Since these types are distinguished by their heavy (γ, μ, α) or light ($\varkappa$, λ) chains, this question reduces in substance to a determination of the cellular localization of synthesis of these types of chains. It would be of paramount interest to ascertain whether different immunoglobulins are synthesized in the same cells.

A large number of human lymphoid cells were investigated with the aid of a fluorescent label [195]. All three types of immunoglobulins – γG, γM, and γA – were found in cells having the same morphological characteristics. Sometimes γG- and γM-globulins were found simultaneously in the same cell by using two stains of different fluorescence. Plasma cells more often contain γG-globulin. Elsewhere [98], both γG- and γM-globulins were found in human spleen plasma cells, but most often each individual cell contained only one type of globulin. Only cells in mitosis sites contained both types of globulins. Bernier and Cebra [87a] found only one type of heavy chain in each cell. Bussard and Binet [92] studied the electron-microscope structure of hemolysin-producing rabbit cells individualized by the local hemolysis gel technique (Jerne). It was found that almost all cells had the structure of plasma cells. They synthesized 19 S hemolysins, since it was just this type of antibody which was found in the serum of the test animals.

Nossal et al. [212] obtained very interesting results in a study of the type of antibodies produced by individual cells of rat lymph nodes after immunization with bacterial antigens. Figure 37 presents schematically the results of these experiments. No morphological differences were found between cells synthesizing antibody sensitive to mercaptoethanol (γM), and cells insensitive to the latter (γG). In several cases, only some of the antibodies were sensitive to mercaptoethanol or to treatment with serum to rat γG-globulins. These cells were regarded as producers of both types of antibodies, and they were encountered more often when the process of synthesis had switched from γM-antibody to γG-antibody.

In certain pathological conditions, when only synthesis of γM-globulins is observed, plasma cells cannot be found in lymphoid tissue. In these cases, synthesis of γM-globulin takes place in large and medium-sized lymphocytes, as was confirmed with a fluorescent label test [290]. According to other data, in dysgammaglobulinemia γM-globulins are synthesized in large pyroninophilic cells, resembling plasmoblasts, with a high cytoplasm-to-nucleus ratio [105]. In patients with multiple myeloma or macroglobulinemia, Solomon et al. [240] were unable to establish a link between synthesis of γG-, γA- and γM-globulins and any definite morphological cell types: both plasma cells as well as other cells could contain any of these proteins.

On the other hand, Stavitsky et al. [234] found that γM-globulins and antibodies were formed principally only in nonphagocytizing

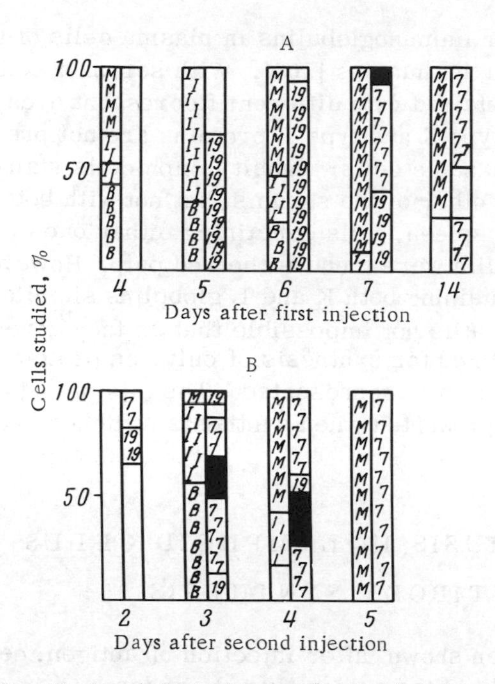

Fig. 37. Morphological types of cells of rat lymph nodes producing γG- or γM-antibodies to *Salmonella* [212]. A) Primary response; B) secondary response. Left side of column: morphology of the cells (M mature and I immature plasma cells, B blasts). Right side of column: type of antibody produced by cells (7 γG-antibodies; 19 γM-antibodies; black rectangle, both types of antibodies at the same time).

mononuclear cells in the red pulp of spleen, while γG-proteins were synthesized predominantly in plasma cells in nonfollicular white pulp. But these investigators still do not deny the possibility of simultaneous synthesis of both types of immunoglobulins in the same cells. It is evidently difficult to associate definitely the synthesis of γG- and γM-globulins and antibodies with any specific morphological cell types. In contrast, when γM-antibodies are synthesized (third day after immunization) a large number of plasmoblasts are found among spleen cells.

The site of γA-globulin synthesis has been little studied. There is evidence that this type of protein is found in larger quantities than other immunoglobulins in plasma cells of the small intestinal mucosa in humans [103]. With serum specific for $\varkappa$ and λ chains, stained with different fluorescent dyes, it was ascertained that type K and type L proteins are not present simultaneously in the same cells: rabbit lymph cells stained either with one or the other of the serums, but not with both [87]. In a study of human spleen, cells containing either one or the other type of immunoglobulin were found in the red pulp. However, in mitosis sites, cells containing both K and L globulins simultaneously can be found [217]. It is not impossible that as they develop cells become specialized for synthesis of only one protein and the site of the other protein is represented. The site of γM-globulins may thus be represented when synthesis switches over to γG-globulins.

RNA SYNTHESIS IN LYMPHOID CELLS DURING ANTIBODY SYNTHESIS

As has been shown, after injection of antigen, cells begin to multiply rapidly in the spleen, lymph, nodes, and in certain other organs, and this multiplication is accompanied by the appearance of a large number of plasmocytic cells. These cells, which have a high ribosome count, are responsible for antibody synthesis. It is reasonable to assume that after immunization, the quantity of nucleic acids in antibody-synthesizing organs increases. Indeed, many investigators have been able to show that when antibody synthesis is most intense, the nucleic acid content in the spleen and lymph nodes is also highest.

This was first demonstrated in 1949 by Ehrich et al. [114] and Harris [142]. In the former study, it was found that after injection of typhoid antigen into rabbits' paws, both the RNA and the DNA contents in the popliteal lymph nodes increased. The RNA/DNA ratio increased from the fourth to sixth day, during which time antibody synthesis was also at a maximum. Harris did not observe an increase in DNA content, but found that the RNA content had doubled by the second to fifth day after immunization with dysentery antigen, influenza virus, or sheep erythrocytes. The maximum RNA content either coincided with or somewhat preceded the maximum antibody concentration in the lymph node studied.

These data agreed well with cytological studies in which it was shown that during the same period pyroninophilic cells, which had a high RNA content in their cytoplasm, accumulated in the spleen and lymph nodes [114, 142, 283].

The increase in RNA content depends on several factors, in particular, on the method of immunization and on the antigenic properties. Consequently, absolute figures may vary. Thus, in some cases the observed RNA increase was less than that found in the aforementioned experiments [6, 25, 51, 172].

In isolating RNA from lymphoid tissue, it should be borne in mind that when lymphocytes are destroyed, a large quantity of RNA-ase is liberated. If, however, the tissue is homogenized in phenol solutions, high polymerized RNA preparations can be obtained. In our experiments, we were able to isolate 1.5 mg high-polymer RNA, precipitated with 1.5*M* NaCl solution, per gram moist spleen or lymph node tissue from normal rabbits (the amount of RNA was determined from the P content), while 1.9 mg high-polymer RNA was isolated from 1 g of the same tissue from immune rabbits on the fourth day after secondary immunization with human serum albumin.

Figure 38 gives results of a determination of the viscosity of high-polymer RNA from lymphoid tissue of immune rabbits and, for comparison, results of a determination of the viscosity of high-polymer RNA from rat liver (carried out in our laboratory by K. Vaptsarova, Bulgaria). From Fig. 38 it is seen that in both cases the viscosity increases in steps with rise in temperature. This indicates that the polynucleotide chains of RNA are continuous. This "temperature effect" is evidence of a high degree of polymerization of the isolated RNA preparations [41].

A study of the quantity of RNA in different subcellular fractions during immunization revealed variations in the microsome fraction. Thus, on the third to fourth day after primary immunization of rats with sheep erythrocyte membranes the RNA content in this fraction increased by approximately 1.3 times [172]. This effect was attributable to an increase in the number of cells having a high ribosome content; such cells are necessary for biosynthesis of antibody molecules.

In a study of incorporation of nucleic acid precursors into different fractions of spleen tissue, it was found that nucleic acid

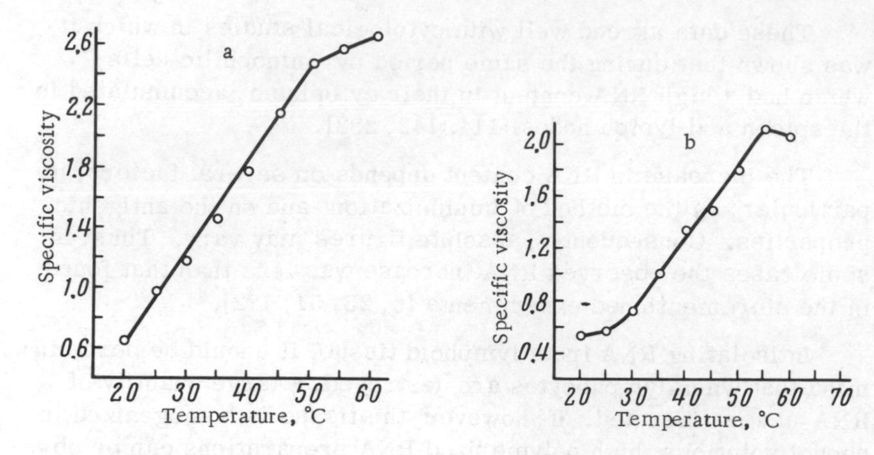

Fig. 38. Temperature dependence of the viscosity of high-polymer RNA from lymphoid tissue of immune rabbits (a) and rat liver (b).

metabolism intensified considerably after immunization. After addition of ^{32}P to a suspension of spleen cells from an immunized rabbit, a considerable increase in incorporation of the label into all cell fractions (i.e., acid-soluble, fat-soluble, DNA and RNA fractions, etc.) was observed [110, 285].

A detailed study of rapidly labeled RNA fractions in lymph nodes of immunized rats [185a] showed that the bulk of RNA was newly formed ribosomal RNA (r-RNA). In the first 10-20 minutes after adding the label (^{3}H-uracil) a considerable amount of labeled RNA was detected in fractions whose sedimentation constant was higher than that of the heavy component of ribosomal RNA, i.e., >30 S. After longer time intervals, the sedimentation properties of rapidly labeled RNA were similar to those of r-RNA. It is evident that r-RNA of lymph tissues, as of other animal tissues, is synthesized as large precursors. It is possible that it is transported to the cytoplasm as a component of certain particles, since some of the RNA in the sediment obtained by centrifugation of a postmicrosomal cytoplasmic fraction at 140,000 *g* for four hours resembles r-RNA in a sucrose gradient.

Newly formed r-RNA appears quite quickly in microsomes, but a fraction with properties similar to messenger RNA (m-RNA) appears even sooner in microsomes [185a]. By the use of the thermal method of Georgiev and Mant'ev [185b] it was possible to isolate an RNA fraction with a very high template activity in the Nirenberg system [185b]. This fraction, which is probably nuclear m-RNA, contained molecules having a wide range of sedimentation properties. It was found that light (10 S-23 S) as well as heavy (24 S-40 S) fractions had similar template activities. Since m-RNA of much smaller dimensions was found in microsomes, it can be assumed that during transport to microsomes m-RNA decomposes into smaller fragments. However, the relatively small size of microsomal m-RNA may also be the result of decomposition as it is isolated. m-RNA is probably transported in particulate form since it is found in the sediment after centrifugation of a postmicrosomal cytoplasmic fraction [185a].

Since the spleen and lymph nodes have a very heterogeneous cell composition and antibodies are synthesized in only a relatively small percentage of cells, data on antibody-forming cells themselves would be of considerable interest. Preparative separation of cells has not as yet been accomplished, but such a study would most probably be carried out with radioautography.

Schooley [235] was one of the first to perform such experiments. ^{3}H-Labeled uridine was injected into mice on the day following their immunization, and then the distribution of the label among cells of the plasma series was determined at different times thereafter. It was found that this precursor of nucleic acids is not incorporated into plasmocyte cells, i.e., antibody-synthesizing cells. Australian immunologists in Nossal's laboratory confirmed this result and showed that a considerable portion of the RNA in plasmocytes (immature and mature) is stable [201]. In these experiments, incorporation of ^{3}H-uridine into rat lymph node cells on the fourth day after immunization was studied in vitro. Since this precursor was incorporated into DNA and RNA, some of the samples were digested with DNA-ase or RNA-ase in order to discover which of the nucleic acids had incorporated the label. It was found that only a very small amount of the label was incorporated into plasma cells and small lymphocytes, while incorporation into blasts was many times greater (Fig. 39).

In the latter case, the bulk of the labeled cytidine was presumably incorporated into RNA, since RNA-ase digestion considerably reduced the grain count on the radioautogram. During the first four hours, almost 90% of the incorporated uridine was found in RNA of blast nuclei.

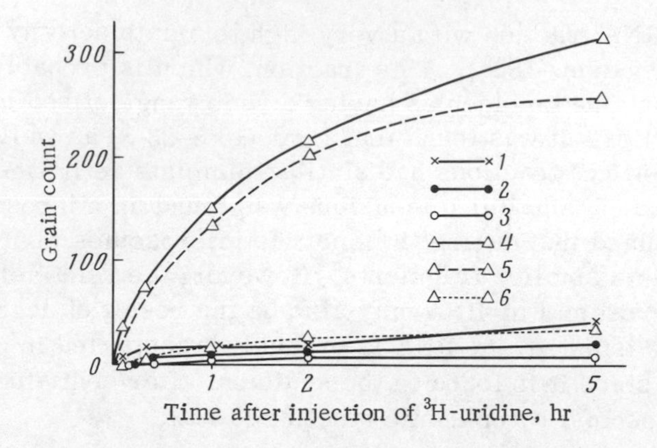

Fig. 39. Incorporation of ^{3}H-uridine into rate lymph node cells on the fourth day after second immunization [201]. Grain count over the cells: 1) immature plasma cells; 2) mature plasma cells; 3) small lymphocytes; 4) blasts; 5) blasts after treatment with DNA-ase; 6) blasts after treatment with RNA-ase.

In Mitchell's experiments [200], ^{3}H-uridine incorporation into plasma cells was even lower. Taking incorporation into blasts as 100%, incorporation into small lymphocytes was 2-4%, whereas into mature plasma cells it was only 0.8%. A concurrent study of incorporation of ^{3}H-leucine showed that blasts synthesized stable proteins predominantly for cell needs, whereas mature plasma cells synthesized proteins for "export," i.e., antibodies. This is evident from the nature of the changes in the curves in Fig. 40: curves showing incorporation of amino acids into blast proteins are almost continuously ascending, while the analogous curves for mature lymphocytes reach a certain level and then continue horizontally. A count disclosed that blast cells incorporated 90 times more uridine than plasma cells and 12 times more than lymphocytes per unit incorporated leucine.

In other experiments [200], further conversion of ^{3}H-cytidine incorporated into blast cells was studied. It was found that the amount of label decreased in the nucleus and increased in the cytoplasm (Fig. 41). This would seem to indicate that the RNA synthesized in the nucleus of the precursor cells to antibody-synthesizing cells is gradually transported into the system.

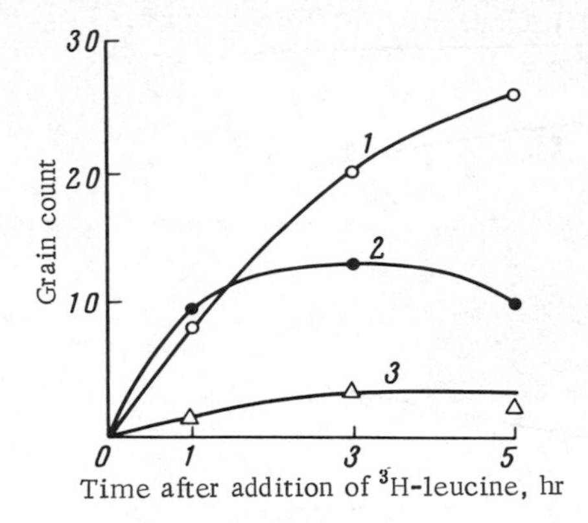

Fig. 40. Incorporation of ^{3}H-leucine into lymph node cells in vitro [200]. Grain count over: 1) blasts; 2) mature plasma cells; 3) small lymphocytes.

Thus, the results of these experiments give the impression that RNA is stable in cells responsible for the bulk of antibody production. Miller [198] came to the same conclusion, showing that plasma cells retain a label in RNA incorporated during the proliferative phase of a secondary immunization for periods of 30 days or more.

One of the conclusions was that RNA was stable and had activity necessary for antibody biosynthesis. Such a phenomenon would be biologically very expedient: indeed, the bulk of antibody is synthesized during the first few days following injection of antigen. Cases have been known in which stable messenger RNA has been found in the cells of mammals [4]. However, the experiments under discussion can also be interpreted in another way; namely, the bulk of RNA is indeed stable, but this RNA is only ribosomal RNA, while the small quantity of RNA which plasma cells do produce is messenger RNA. Consequently, further experiments with other methods are necessary for a final assessment of the stability of RNA.

Experiments with actinomycin D could provide some additional information on this topic. The effect of this antibiotic on cells, in which it suppresses RNA synthesis, is usually explained as follows: combining with DNA, it thereby prevents RNA-polymerase from acting as a catalyst for the formation of new RNA molecules. Hence, if protein synthesis takes place in the presence of actinomycin D, a stable informational RNA probably exists for this protein, and vice versa. Mitchell [200] found that addition of actinomycin D in concentrations sufficient to inhibit cytidine incorporation completely diminished incorporation of leucine in blasts as well as in plasma cells by only 33%. This would appear to indicate the presence of a stable RNA template. However, the experiments lasted only three hours.

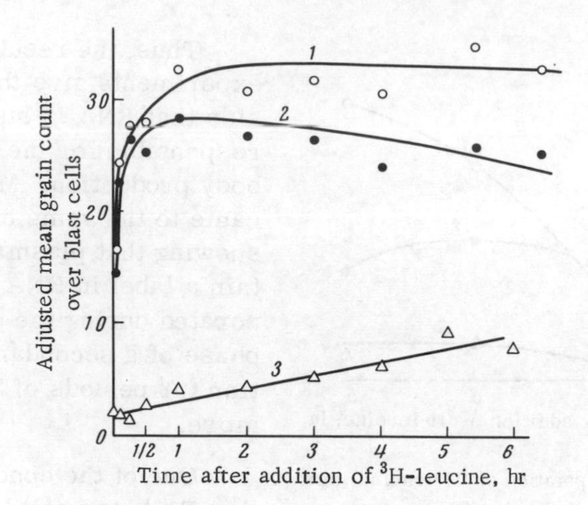

Fig. 41. Change in the grain counts over blast cells and their components after 3.5 minutes of contact with ^{3}H-uridine [200]. Grain count over: 1) whole cells; 2) nuclei; 3) cytoplasma.

A longer exposure to actinomycin D suppressed antibody biosynthesis. It was found that this antibiotic inhibited the primary and secondary responses to bacteriophage T2 in vitro in those cases where the antibiotic was added to the culture together with antigen [271]. But antibody synthesis was only partially depressed when the second immunization with a phage in vitro was carried out four days before addition of actinomycin. Apparently, synthesis of new RNA must precede antibody synthesis. On the other hand, an initiated antibody synthesis no longer requires RNA formation; this fact is evidence in support of the existence of sufficiently stable messenger RNA [184a].

Roughly analogous results were obtained in a study of the effect of actinomycin D on synthesis of antibody to bovine albumin in a suspension of spleen cells from an immunized rabbit [237]. Small doses of antibiotic (1 μg/ml) inhibited incorporation of ^{3}H-uridine into RNA by about 90%, but at the same time had no inhibitory effect on antibody synthesis. Large doses suppressed antibody synthesis, but production of other proteins by cells was suppressed to the same degree. This gives rise to the suspicion that in this case the toxic effect of the antibiotic is exerted on the cell as a whole. Perhaps it is precisely the use of relatively large doses (8 μg/ml) which explains the inhibition of biosynthesis of antibody to the poliomyelitis virus observed when actinomycin D is added to a culture of spleen cells from an immune rabbit [261]. According to Geller and Speirs [136], actinomycin has a weak inhibitory effect on biosynthesis of antibody to a tetanus anatoxin in vivo. Even high toxic doses did not cause complete inhibition. In the experiments of Hanna and Wust [141, 286, 287], injection of actinomycin D delayed an immune response in mice by two days, although thereafter the antibody titers were the same as in the control animals. Results of Stavitsky's experiments with the same antibiotic [247a] also confirm the presence of stable messenger RNA.

Thus, changes in RNA metabolism in lymphoid organs after injection of antigen may be represented schematically as follows: initially there is a sharp intensification of synthesis of all types of RNA necessary for protein production, but when antibody synthesis reaches a peak, RNA synthesis in the antibody-producing cells almost stops. The bulk of experimental data gives good grounds for assuming that messenger RNA specific for antibody is quite stable.

Attempts to Induce Antibody Synthesis by RNA Fractions Isolated from Lymphoid Tissue of Immune Rabbits

If the building up of peptide chains of antibody is totally dependent on a special messenger RNA synthesized after immunization and is independent of the presence of antigen, the question then arises as to whether it might not be possible to induce antibody synthesis in cells unexposed to antigen with the aid of RNA from the spleen or lymph nodes of immune rabbits. The solution of this question would not only be of fundamental importance for an understanding of antibody biosynthesis, but would have a wide general biological significance, since such experiments would be a very convenient direct test for the biological activity of messenger RNA.

Sterzl and Hrubesova [256] were the first to explore this problem when they studied the capacity of nucleoprotein fractions from the spleen of rabbits immunized with *Salmonella* to induce antibody synthesis. In these experiments, nucleoproteins were transferred to young rabbits incapable of synthesizing antibody after injection of antigen. After injection of nucleoproteins, agglutinating antibodies were found in the rabbits' blood. However, later experiments to verify this result by attempts to transmit to young rabbits the capacity to synthesize antibodies to other antigens (*Brucella*) [153] or to use as recipients chicken embryos into which had been injected nucleoproteins or purified nucleic acids from the spleen or immune chickens [156] were unsuccessful. Antibody biosynthesis could not be detected with the aid of a radioactive label in rabbits after transfer of nucleoproteins containing 15-30% nucleic acids [155]. In these experiments, only an intensification of incorporation of ^{14}C-amino acids into a γ-globulin fraction after injection of nucleoproteins could be observed. Negative results were also obtained by several other investigators [169, 205], but the possibility of transfer of trace amounts of antigen with nucleoproteins was demonstrated [205].

In recent years, attempts have been made to detect antibody synthesis by transfer of highly purified RNA fractions. Mannick and Egdahl [193] isolated RNA from the lymph node of a rabbit which had been given a skin homotransplant. Then this RNA was incubated with cells from a normal rabbit in a hypertonic sucrose solution, i.e., under conditions which facilitated the penetration of infectious virus RNA into the cells. These incubated cells acquired the capacity to transmit transplantation immunity.

An examination of a series of investigations by Fishman et al. [122-125] is relevant at this point. These investigators proposed the following two-step mechanism of antibody synthesis. First, the antigen penetrates into RNA-synthesizing macrophages. This RNA initiates antibody biosynthesis in lymphoid cells. To confirm this view, macrophages from the intestinal cavity of rats were incubated with antigen (bacteriophage T2, hemocyanin). Then the cells were homogenized, the homogenate was filtered, and the filtrate was incubated in vitro with lymph node cells. After incubation the culture medium was able to neutralize phage or agglutinate erythrocytes with adsorbed hemocyanin. Treatment of the filtrate with RNA-ase or streptomycin eliminated this effect. In subsequent experiments it was found that incubation of lymphocytes with RNA isolated from a homogenate of macrophages exposed to antigen gave an analogous result. In these experiments, lymphocytes were incubated with RNA in vivo in diffusion chambers. At the same time, it was ascertained that macrophage RNA labeled with ^{3}H-uridine did indeed penetrate into lymphocytes [124]. In the last of these investigations, Fishman showed that macrophage RNA in diffusion chambers with 0.1-μ -diameter pores was also capable of inducing antibody synthesis after injection into the abdominal cavity of rats [125].

The data of Askonas and Rhodes [70], however, call for great caution in the interpretation of such experiments. These investigators studied the fate of antigen (hemocyanin labeled with radioactive iodine) after contact with macrophages in the organism of a mouse or after addition directly to macrophages before extraction of RNA from them. In both cases, the RNA isolated from the macrophages contained antigen and hence was able to stimulate antibody formation. This stimulation was apparently not to be attributed to the presence of messenger RNA in the preparations, despite the fact that treatment with RNA-ase diminished their immunogenicity.

Friedman et al. [132] obtained analogous results in reproducing Fishman's experiments. After incubation with phage T2, RNA preparations from macrophages did actually acquire the capacity to induce synthesis of neutralizing antibodies. However, they contained all types of phage antigens. Treatment with RNA-ase did not completely suppress the activity of RNA preparations.

In the USSR, the capacity of RNA from immune lymph nodes and spleen to induce antibody synthesis has been studied by Fuks and Konstantinova [52, 53]. Rabbits were immunized with bovine albumin, and, on the third day after the second injection, RNA was isolated from lymph nodes and spleen. This RNA was either injected into the foot pad of a rabbit or incubated with lymph node cells from a normal rabbit after which this suspension was injected intraperitoneally into a normal rabbit. Within 12 hr, antibody titers of 1:10-1:80 were detected in the rabbits' blood by hemagglutination. Treatment with RNA-ase completely suppressed the capacity to stimulate antibody synthesis in seven of nine rabbits. By histochemical methods it was found that after injection of RNA into the rabbit's foot pad, the cells of the corresponding popliteal lymph node contained a large quantity of RNA. Antigen could not be detected in RNA preparations using an immunological method. In considering these experiments, it should be recalled that antigen or its fragments might have been transferred together with RNA, although antibodies appeared too early to admit the possibility of simple immunization by antigen impurities.

Several recently published reports deal with the induction of antibody synthesis by RNA by the Jerne method. In these experiments, RNA from the spleen of animals immunized with alien erythrocytes was incubated with spleen cells from normal animals to which erythrocyte antigens were later added. In the first report of this series, by Cohen and Parks [100], it was shown that the number of hemolytic plaques due to formation of hemolysins in cells incubated with "immune" RNA was 94 ± 21.1, while without RNA the number of plaques was 20 ± 5.4. Friedman [130] studied this problem in more detail in an investigation of the induction capacity of RNA as a function of the time elapsed after immunization of donor mice with RNA (Table 37). After treatment with RNA-ase, RNA preparations were no longer capable of increasing the number of hemolytic plaques. RNA from a mouse which had been given an injection of actinomycin D was also incapable of stimulating plaque formation. According to Stanchev [42a], the RNA fractions extracted with phenol at 55°C had the greatest stimulatory effect.

However, the Jerne method has one fundamental shortcoming: some hemolysin-forming cells are always found in the spleen of normal animals. Consequently, it is not impossible that injection of RNA activates some of the cells which have been, as it were, conditioned for antibody synthesis or those cells which had already synthesized antibody in a quantity too small to be detected by this method.

Attempts to induce synthesis of myeloma immunoglobulin by normal lymphoid cells after treating them with RNA isolated from plasma tumor cells have been unsuccessful [98a]. Although some RNA penetrated into the cells, it was not possible by means of radio-immunoelectrophoresis to demonstrate the synthesis of proteins possessing the antigenic determinants of pathological immunoglobulins.

In concluding, it can be stated that there exists evidence that antibody synthesis can be induced by RNA, but it cannot be re-

TABLE 37. Formation of Hemolysis Zones by Spleen Cells of Mice Immunized with Sheep Erythrocytes and Spleen Cells of Nonimmunized Mice after Incubation with RNA Isolated from Spleen Cells of Donor Mice at Different Times after Immunization [130]

Donor mice*	Mean hemolysin titer in donor	Number of hemolytic plaques per 10^6 spleen cells from immune mice		Number of hemolytic plaques per 10^6 normal spleen cells after incubation with RNA	
		mean	range	mean	range
Nonimmune	<1:10	2	1-7	1	0-3
Immune:					
+24 hr	1:13	38	21-96	11	8-17
+48 hr	1:46	138	101-260	19	10-31
+72 hr	1:230	342	180-460	33	19-39
+96 hr	1:423	612	380-790	40	23-58
+144 hr	1:450	401	29-640	38	19-52
+240 hr	1:122	17	11-80	2	0-11

*The hours indicate the time elapsed from injection of the antigen until killing of the animal from which the spleen cells were taken for determination of the number of hemolysin-forming cells or for isolation of RNA.

garded as conclusive. A more substantial proof would be the induction of synthesis of radioactive antibodies by addition of "immune" RNA, containing absolutely no traces of antigen, to normal cells. But even in this case it would be difficult to ascertain the actual mechanism of this process.

It should also be pointed out that no rigidly demonstrated experiment in which biosynthesis of any protein has been induced by RNA has as yet been carried out [4]. An exception is the capacity of virus RNA to induce cells to synthesize virus proteins; however, it is still unclear whether this effect is due to the high concentrations of messenger RNA in the preparations or whether the explanation of this effect is to be found in the specific characteristics of virus RNA, for example, the ease of cell penetration, or the capacity to combine with ribosomes. The absence of rigid proofs of induction of antibody synthesis by RNA does not conflict with data obtained from the study of biosynthesis of other proteins in animal cells.

DYNAMICS OF ANTIBODY SYNTHESIS

Antibody synthesis initiated after injection of antigen into an animal takes place in several stages. In the first of these — the latent, or induction period — antibody synthesis is not discernible. As soon as the first antibodies begin to appear, the productive phase of antibody synthesis begins; this phase can, in turn, be divided into several stages: an exponential stage of antibody formation, a stage of maximum antibody synthesis, and a stage of diminished antibody synthesis. The distinguishing features of each of these stages will be examined below.

Latent Period

After injection of antigen, a certain time elapses during which antibody is not detectable in the serum. The duration of this period depends on several factors, in particular, on the antigenic properties, on the method of immunization, on whether the antigen has been injected for the first time or is a repeat, and, of course, on the method of determination of the antibody content. Data obtained by the most sensitive methods are, of course, especially valuable. One such method involves determination of the beginning of the immune phase of elimination of a labeled antigen. In a study of elimination of bovine albumin and globulin labeled with radioactive iodine, Dixon et al. [106] discovered that antibody synthesis began on the seventh to ninth day after primary injection of albumin, and on the fourth to fifth day after primary injection of globulin. Antibodies appeared on the second to fourth day, but not earlier, after the secondary injection of these proteins.

Sterzl et al. studied the duration of the latent period in detail by very sensitive methods, involving determination of immune lysis of bacteria and determination of hemolysis of erythrocytes under a microscope. These methods were sensitive to quantities of antibody on the order of 10^{-6} μg antibody N. The serum of newborn pigs, maintained under sterile conditions on a synthetic diet, was used as the source of complement necessary for initiation of the antibodies studied [257]. In immunization of sheep with *Salmonella*, antibodies were not detectable in either lymph or blood during the first two days after injection. In other experiments, lymphoid cells stimulated by antigen in vitro were transferred to newborn rabbits. It was found that antibodies usually appeared no earlier than the third day after transfer of $50 \cdot 10^6$ cells per ml, the concentration usually used in such experiments, or after transfer of a 10 times greater cell concentration [255].

However, reports published in recent years have indicated much shorter durations for the latent period. In these investigations, the immune elimination of viruses from the blood of animals was studied. Thus, Uhr et al. [273, 274] in experiments using the bacteriophage ØX174 as antigen found that the initiation of immune elimination depended on the amount of bacteriophage injected. At a dose of $6 \cdot 10^{10}$ plaque-forming particles it began within 50-60 hr, and at doses of $6 \cdot 10^8$ and $6 \cdot 10^6$ it began within approximately 45 and 35 hr, respectively. In one of these experiments even smaller amounts of bacteriophage ($3 \cdot 10^6$) were injected and immune elimination began within 24 hr.

Svehag and Mandel [262] used other antigens — poliomyelitis virus and Coksackie virus. In some instances, virus-neutralizing antibodies were found in the serum of the rabbits on the first day after injection of the virus, although the serum possessed no neutralizing activity before the virus injection. In an experiment in which a mixture of virus and serum was held at 3°C for an extended period, it was found that inactivation was more appreciable when the virus was incubated with serum taken eight and even four hours after virus injection than with serum taken before immunization. These investigators conjecture that antibody elaboration begins soon after immunization.

In interpreting investigations of this type, it should be borne in mind that a biological test such as virus neutralization is far from being quantitative and specific. There is most likely no single cause of virus inactivation. Hence, it cannot as yet be stated with certainty that the latent period is short and that antibody synthesis can be initiated by some cells immediately after contact with antigen. It is interesting that no antibodies were detected during the first two days after primary immunization by a method of flagellae immobilization, which is just as sensitive as the method of bacteriophage neutralization [211].

It is very important to acquire some conception of the processes taking place during the latent period. It could be assumed simply that the delay of appearance of antibody in the blood is due to the fact that antibodies or their protein or polypeptide precursors synthesized during this time do not pass into the blood. Some investigators have suggested this as a possibility [140]. However, this assumption could not be confirmed in further experiments.

Gurvich [10] showed that injection of ^{14}C-glycine shortly before injection of antigen after immunization led to formation of antibody containing only a small amount of isotope. However, after injection of radioactive amino acid during the peak period of antibody synthesis, antibody with a very high activity was detected. Dixon et al. [107] obtained analogous results. Taliaferro and Talmage [266] injected a ^{35}S-labeled hydrolyzate of yeast protein three days after a second injection of bovine serum albumin into rabbits when the blood contained no precipitins. Shortly thereafter, the animals were killed, the spleen was removed and minced, and the spleen cells were injected into another rabbit. Antibody formation commenced in the recipient and reached a peak on the second day, but the content of radioactive label in these antibodies was very low. However, when labeled amino acids were injected into the recipient rabbit after the cell transfer, the antibody radioactivity was very high. Evidently no antibody precursors constituted of amino acids are formed during the first three days after antigen injection.

Thus, it may be assumed that formation of antibody or antibody precursors does not take place during the latent period.

The nature of processes taking place during the first moments after entrance of antigen into the organism remains unclear. A large body of evidence indicates that the latent period is exceptionally sensitive to various external factors. Thus, for example, it is susceptible to X-rays. In contrast, during the period of intensive antibody biosynthesis, X-rays no longer had an inhibitory effect [36, 37, 43, 265]. For example, in the experiments of Dixon et al. [108] it was found that irradiation had the greatest inhibitory effect on antibody synthesis when carried out 12-48 hr before immunization. Three days after antigen injection, even doses of 800 r had no effect (Table 38).

Generally analogous results were obtained by Taliaferro et al. [263-265] in a study of the effect of radiation on formation of hemolysins by rabbits. Hemolysin synthesis was sharply suppressed when the antigen was injected 12 hr to 21 days after irradiation, but only retarded when immunization was performed from four days before to six hours after irradiation.

TABLE 38. Effect of X-Ray Irradiation on Synthesis of Antibodies to Bovine γ-Globulin [108]

Radiation dose	Time of radiation	Quantity of antibody nitrogen precipitated by 1 ml serum to bovine γ-globulin at 80% precipitation of antigen, μg
Control	–	16.4 ± 10
400	6 hr after immunization	14.2 ± 7
400	Simultaneous with immunization	12.7 ± 5
400	5 hr before immunization	7.7 ± 4
400	12 hr before immunization	2 ± 2
400	48 hr before immunization	0 ±
800	3 hr after immunization	16.1 ± 21
800	1 hr after immunization	6.4 ± 6

The Swedish investigators Fagraeus and Berglund discovered that the inhibitory effect of cortisone on antibody synthesis was quite pronounced when cortisone was injected a short time before immunization [117]. The most pronounced inhibition was observed as a result of injection of prednisolone over a period of four hours beginning from the tenth hour before antigen injection [83, 84]. The effect of hormones was considerably diminished when injected at later times.

Thus, irradiation and hormone injections inhibited antibody synthesis especially if the animals were treated before antigen injection. After such treatment the mitotic capacity of cells, which is usually manifested by antibody-synthesizing tissues during the latent period, should decrease. It is known that irradiation and hormone injections inhibit mitosis [20]. In particular, cortisone has an especially pronounced inhibitory effect on mitosis in lymphoid tissue [1]. The impression is that irradiation and hormone injections damage cells in such a way that they become less susceptible to antigen or incapable of elaboration of an immune response.

It is interesting to compare the effect of 6-mercaptopurine on antibody synthesis with the effect of irradiation and hormone injections. The inhibitory effect of this analog of purine bases

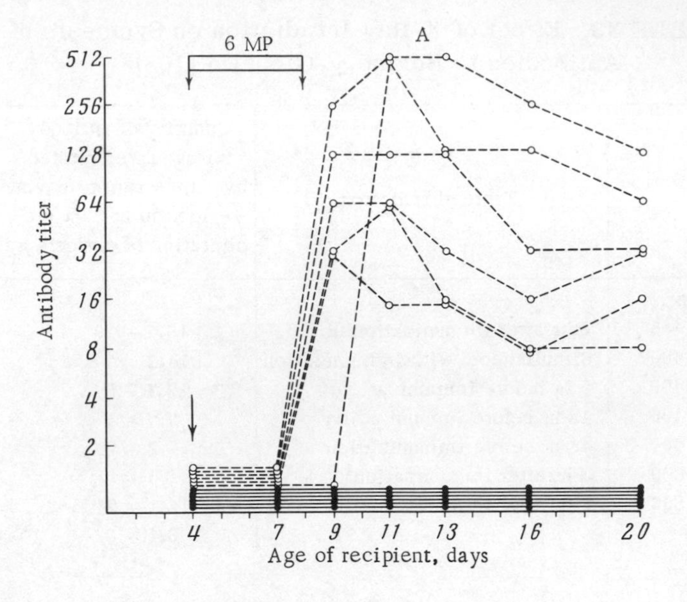

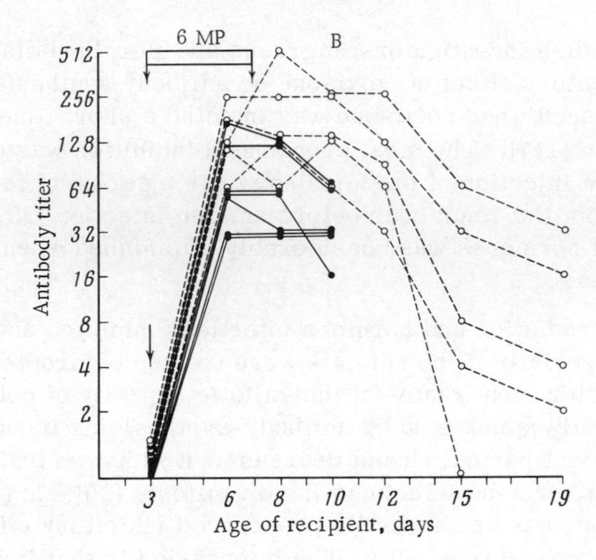

Fig. 42. Effect of 6-mercaptopurine (6MP) on antibody synthesis by adult rabbit lymph cells transferred to newborn rabbits [253a]. A) The cells were mixed with antigen before precipitation; B) injection of cells from an immune adult rabbit. Solid line, experiments with 6MP injection; broken line, control experiments.

was studied by Sterzl [253a] on the following model. A mixture of *Brucella* and spleen cells from adult rabbits was injected into newborn rabbits incapable of independent antibody formation. Antibodies were detected in the recipients after approximately seven days.

However, when 6-mercaptopurine was injected during the first days after transfer of the cells, antibody synthesis was completely suppressed. On the other hand, this effect did not occur if the cells from a preimmunized donor were injected (Fig. 42).

This special sensitivity of antibody synthesis to external factors during the latent period was the basis for Sterzl's view that the induction period of antibody synthesis, beginning immediately after antigen injection, differed qualitatively from the subsequent productive period of antibody synthesis [252]. It is this period which is most critical for an understanding of the process of antibody synthesis as a whole. Soon after antigen injection, the number of lymphoid cells increases sharply; however, cell division is hardly the most essential aspect of the latent period. This is clear, in particular, from the previously mentioned experiments of Sterzl and Trnka [259], who injected different quantities of spleen cells from an immunized adult rabbit into newborn rabbits. It was found that the sharp increase in the number of cells did not appreciably shorten the latent period.

It is obvious that the crux of the matter does not lie simply in the small number of antibody-synthesizing cells during this period. But at present, the processes taking place at the subcellular level after antigen injection remain unclarified.

Antibody Production

Antibodies appear in the blood of an animal within a certain time, known as the latent period, after injection of antigen. The antibody content quickly reaches a peak and then decreases again rapidly. A small quantity of antibody is synthesized over a very prolonged period. Each of these stages will be examined in more detail below.

Kinetics of Antibody Formation. The antibody content in the blood rises very rapidly within a short time after their appearance. Gurvich [11, 12] studied this phase of antibody

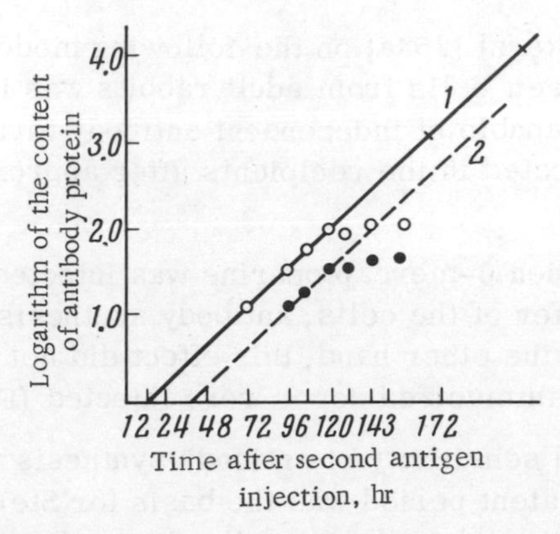

Fig. 43. Curves showing the change in the absolute contents of two antibodies in rabbit blood. Repeated injections of human serum albumin and horse serum albumin were made simultaneously into the rabbit [12]. 1) Anti-human serum albumin; 2) anti-horse serum albumin.

formation in rabbits immunized with one or two protein antigens. Results indicated that the antibody content in the blood increased exponentially for a certain time (Fig. 43). According to Ingraham's data [167] the antibody titer doubled within 2.2–3.7 hr during the first period of the exponential stage, and within 3.6–6.2 hr during the second period. By extrapolation of the curves, it can be calculated that if antibody synthesis continued to accelerate at such a rate for two days or more, the antibody content would be equal to the content of all proteins in the serum after two days and would be more than the total quantity of proteins in the animal after four days [12]. Gurvich also found that synthesis of one antibody did not interfere with contemporaneous synthesis of another antibody. For example, it is evident from Fig. 44 that at the moment when the content of the one antibody in the blood was at its peak, the other antibody was still being synthesized intensively. Apparently, each antibody is synthesized independently of the synthesis of other γ-globulins. The content of radioactive amino acid in the antibodies depended on the rate of synthesis alone and not on any other factors.

As was stated above, no protein or peptide precursors of antibodies are formed during the latent period. Several investiga-

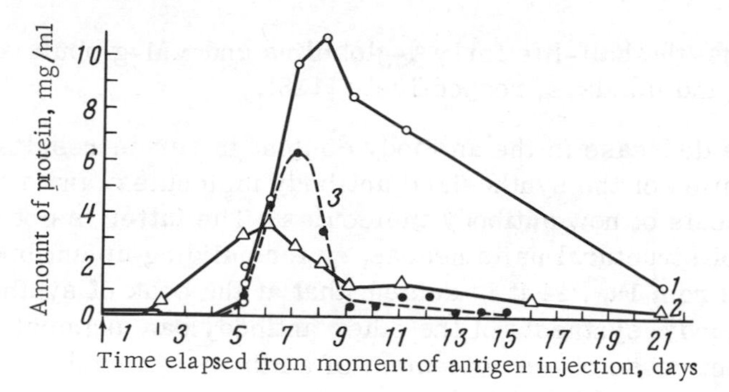

Fig. 44. Changes in the content of antibody protein in the serum of rabbits immunized with horse serum albumin (second day) and cat serum γ-globulin (zero day) [17]. Content of antibody protein: 1) antialbumin; 2) anti-γ-globulin; 3) amount of antialbumin antibody newly formed each day.

tors studying incorporation of radioactive amino acids into antibody have shown that antibodies are synthesized from amino acids [10-12, 17, 140]. According to current conceptions, this is also the origin of all other proteins [42].

The amount of antibody formed depends on many factors, e.g., on the type of antigen, the animal species, and the method of immunization. Under optimum conditions of immunization (hyperimmunization) of rabbits, the quantity of antibody protein formed can attain a very high level. Thus, Gurvich [17] observed the synthesis of 0.5 g antialbumin antibody protein per kg body weight during exponential phase. Thus, the quantity of antibody in serum can reach 10-15 mg/ml and more in 6-10 days after secondary immunization.

After antibody synthesis reaches its peak, it begins to diminish at an increasing rate. Dixon et al. [107] demonstrated that antibody synthesis diminished abruptly after primary immunization, while after secondary immunization, as well as in hyperimmune rabbits, a phase of more moderate antibody formation set in about five days after a sharp decrease. The half-life of antibody during the first phase of formation was 1.3 days for primary and secondary immunizations and two days for hyperimmune rabbits. The half-life in the retarded phase was 12 days for hyperimmune rabbits. Different classes of immunoglobulins differed in their catabolic properties.

Thus, the half-life for γA-globulins and γM-globulins of mice was 1.2 and 0.5 days, respectively [118].

The decrease in the antibody content in serum results from decomposition of the synthesized antibody molecules and a decrease in synthesis of new antibody molecules. The latter is not due to a deficit of structural units necessary for building up antibody molecules. From Fig. 44 it is evident that at the peak of synthesis of one antibody, synthesis of the other antibody can augment sharply. The exact cause of the abatement of antibody synthesis is unclear. Dixon et al. [107] offer interesting evidence that when elimination of antigen (bovine serum γ-globulin) from the blood ceases, the antibody content in the blood also begins to fall sharply.

Haurowitz has recently made similar observations [224]. However, it is not clear to what extent these processes are mutually related, although they may be regarded to a certain extent as indirect evidence of the direct participation of antigen in antibody biosynthesis.

From an analysis of the fluctuations of the antibody content in the blood, Gurvich concluded that the abatement of antibody synthesis was due to inhibition by agents specific for the given antibody [11, 12, 17]. This investigator has in recent years been engaged in a concentrated search for these inhibitors. One series of experiments was run to determine whether antibodies themselves inhibit their own synthesis. For this purpose, the effect of addition of immune antiprotein serums, or antibodies isolated from them, on the synthesis of radioactive antibody in a tissue culture was studied [13]. In some instances, the serums proved capable of inhibiting antibody synthesis. However, this effect was independent of the antibody concentration and was not observed when aged sera were used. In addition, antibodies isolated with immunoadsorbents had no inhibitory effect on antibody synthesis or antibody secretion from cells even in high concentrations (on the order of 1 mg/ml).

Nevertheless, it is noteworthy that in animal experiments it was possible to suppress partially or completely an already commenced antibody synthesis by addition of sera containing γG- or γM-antibody [121, 216, 272]. The detailed mechanism of this effect is not completely clear. Apparently, antigen is somehow blocked, but this blocking plays no essential role or does not take place at all in synthesis in vitro [165a, 282].

In other experiments, attempts were made to discover substances which inhibit antibody biosynthesis in different tissues or their subcellular fractions [14-16]. Of the subcellular granules, mitochondria had the greatest inhibitory effect. The inhibitor was insoluble and extractable with acetone. It was also contained in the insoluble fraction of microsomes. It was found that this factor was an unsaturated fatty acid. However, the inhibitory effect was nonspecific for antibodies, inasmuch as it was also observed in a study of synthesis of water-soluble proteins from liver and kidneys. It is not impossible that fatty acids may play some role in the regulation of protein synthesis, but their role in specific inhibition of antibody biosynthesis is at present unclear. Thus, the cause of the rapid abatement of antibody synthesis during immunization remains unexplained.

It should be stressed that this problem is closely connected with more general problems of antibody synthesis, and its clarification could shed some light on imperfectly understood processes leading to antibody synthesis after antigen injection.

Thus, after reaching a maximum, the antibody content in the blood decreases sharply. However, the continued presence of antibody for long periods can be proved with sensitive methods. Richter and Haurowitz detected nonprecipitating antiprotein and antihapten antibodies for several months after single or repeated injections of various antigens [223]. These antibodies were not simply stored in the blood, but were actually newly synthesized. Thus, 470 days after a second injection of bovine serum albumin into chicks, incorporation of ^{35}S-methionine into antibodies was observed [138].

Sequence of Synthesis of 19 S and 7 S Antibodies. In recent years, the study of the synthesis of 19 S (γM) and 7 S (γG) antibodies during immunization has received much attention. The chief accomplishment of these investigations was the demonstration in the majority of cases of a sequence in which formation of antibodies of the macroglobulin type was followed by synthesis of 7 S antibodies.

This was first observed by Kabat [173] who in experiments on prolonged immunization of horses with pneumococcal polysaccharides initially detected only macroglobulin antibodies, but thereafter observed formation of light antibodies in increasing quantities. Later it was found that this principle was characteristic of protein antigens as well as of polysaccharide antigens.

Stavitsky [78, 79] studied the sequence of 19 S and 7 S antibodies to different antigens in rabbits. Especially significant results were obtained from a single injection of antigen (proteins, bacteriophage T2, erythrocytes, H-antigens of *Salmonella*). By using different methods of identification of γM- and γG-antibodies he was able to show that in all cases a single immunization led to formation of heavy antibodies (macroglobulins). The 7 S antibodies began to appear in the serum only after a certain time had elapsed and as their content increased, the content of macroglobulin antibodies diminished increasingly rapidly until their complete disappearance.

In a study of the secondary response, results were not as clear-cut, partly because the antigen (protein) was injected while 7 S antibodies were still present in the blood. As a result, synthesis of 19 S antibodies recommenced and the amount of 7 S antibodies rose further. It was concluded that the primary and secondary responses were qualitatively different, and that in the secondary immunization 19 S antibodies were formed by cells which had not been in contact with the antigen. However, later experiments, as will be shown below, were not able to verify these assumptions.

Uhr et al. [270, 273] made a comprehensive study of the primary and secondary responses of guinea pigs to injection of the bacteriophage ∅X174. To distinguish 19 S antibodies from 7 S antibodies, the former were inactivated with mercaptoethanol (0.1*M* solution). The first injection of the bacteriophage led to formation of 19 S antibodies and after one week, to formation of 7 S antibodies. Formation of 19 S antibodies was independent of the bacteriophage dose, whereas synthesis of 7 S antibodies began only at a definite bacteriophage concentration. Synthesis of 7 S antibodies persisted for many months. The second injection did not induce to formation of 19 S antibodies, but the 7 S antibody content increased. From these and other results, the investigators concluded that an "immunological memory" for 19 S antibodies is not created.

For their continued synthesis, injection of new doses of antigen was necessary after the lapse of certain time intervals since the preceding immunization.

Analogous data were obtained using poliomyelitis virus as antigen [262]. Small doses of polio virus induced only a brief transitory synthesis of 19 S antibodies in rabbits, while synthesis of 7 S antibodies following formation of 19 S antibodies were detected after large doses. Repeated large doses again induced synthesis of 19 S and 7 S antibodies. After synthesis of 19 S antibodies following injection of small virus doses, the "immunological memory" persisted for only a short time (several days).

We investigated synthesis of 19 S and 7 S antibodies after a resolving dose of antigen in hyperimmune rabbits [33, 34]. Radioactive antibodies synthesized by spleen cells or lymph node cells in vitro in a special chamber (see below) were studied. In contrast to the experiments discussed above, a dextran gel Sephadex G-200, with which we were able to detect differences in the size of the antibody molecules, as well as their susceptibility to inactivation by mercaptoethanol, was used to separate the two types of antibodies. The experiments were set up as follows. On the third or fifth day after secondary immunization of rabbits with human serum albumin, the spleen and lymph nodes were minced, and suspensions of cells of these organs were cultured in vitro in the presence of ^{14}C-amino acids. Newly synthesized antibodies were extracted with the aid of albumin linked to cellulose and mixed with rabbit serum. This mixture was fractionated on Sephadex G-200. The radioactivity at the first protein peak (i.e., with 19 S globulins) was evidently attributable to macroglobulin antibodies, while the radioactivity at the second protein peak (i.e., 7 S proteins) was associated with 7 S antibodies.

A considerable portion of the antibodies synthesized on the third day after immunization were 19 S antibodies (Fig. 45). Almost all antibodies formed on the fifth day after immunization were 7 S antibodies. This rapid appearance and the correspondingly rapid abatement in synthesis of 19 S antibodies is evidence that synthesis took place in cells which had already been exposed to the antigen and had retained their "immunological memory" for 19 S antibodies. In our experiments, synthesis of 19 S antibodies evidently followed the second type of mechanism, since it is known that in rabbits antibody synthesis after a second injection

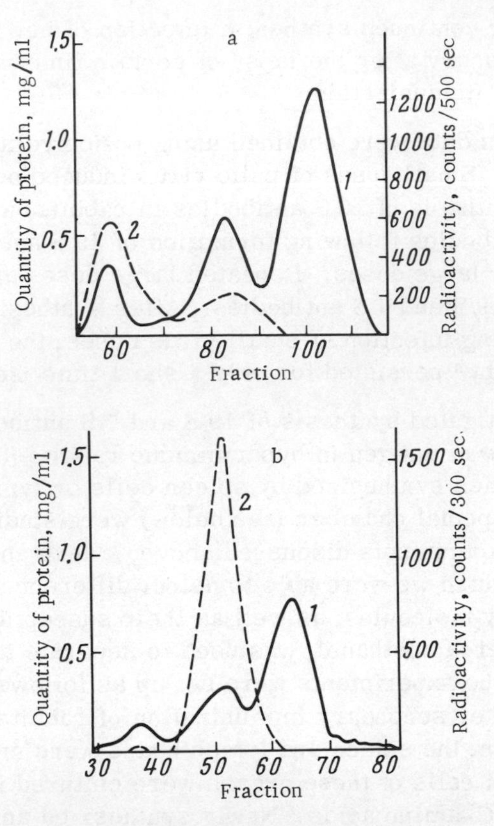

Fig. 45. Determination of the molecular size of radioactive antibodies synthesized in a culture of rabbit lymphoid cells with the aid of gel filtration on a column containing Sephadex G-200 [32]. A mixture of antibodies with normal rabbit serum was passed through. Antibodies synthesized on third day (a) and fifth day (b) after second immunization. I) Protein content; 2) radioactivity.

of serum albumin begins on the second to fourth day, whereas after the first injection of this same protein, antibodies are not detectable before the seventh to ninth day [106].

Thus, there was no appreciable difference noted between the first and second responses with respect to synthesis of 19 S and 7 S antibodies, other than the greater rapidity of the responses after the second immunization.

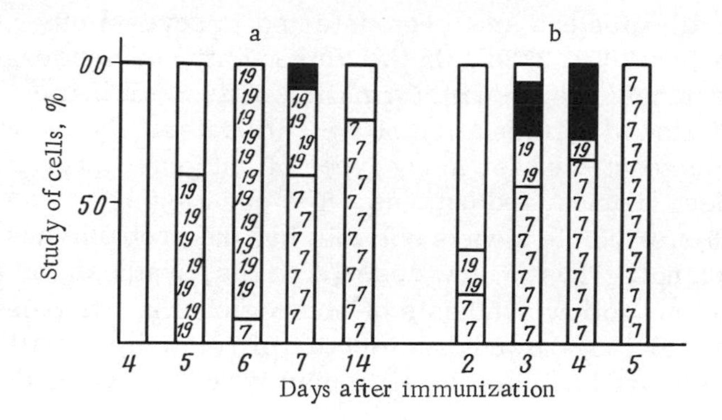

Fig. 46. Types of antibodies synthesized by rat lymph node cells on different days after first (a) and second (b) immunization [212]. For explanation of symbols, see Fig. 37.

Very similar results were obtained in the experiments of Nossal et al. [212] who studied antibody production by single cells of rat lymphocytes after immunization with *Salmonella* flagellae. The results are given in Fig. 46.

It can be observed that 19 S antibodies appeared as early as the second day after the immunization, while after the first antigen injection they could not be detected before the fifth day. In the secondary response, synthesis of macroglobulin antibodies had already ceased by the fifth day.

This rapid sequence of events confirms the presence of an "immunological memory" for 19 S antibodies.

It is as yet difficult to discern the reason for the divergence of the results of experiments with virus antigens [262, 273] on the one hand from the results of our experiments and the experiments of Nossal et al. on the other. The answer is possibly to be sought in differences in the properties and the doses of the antigens employed. A certain caution is to be exercised in interpreting experiments in which macroglobulin antibodies have been determined exclusively on the basis of their susceptibility to mercaptoethanol, since more and more evidence indicates that some 7 S antibodies, including antiphage antibodies, are susceptible to this reagent [260]. It should be pointed out that an "immunological memory" for 19 S antibodies has also been detected in several other experiments [89, 205a, 218b]. In the aforementioned experiments, antibody synthesis began with formation of 19 S antibodies which was then followed by formation of 7 S antibodies. This sequence has been demonstrated for many types of antigens: proteins, polysaccharides, viruses, and haptens. The question arises as to whether this principle always obtains, but the probable answer is that it does not. Thus, at low doses of virus, synthesis of 7 S antibodies may not follow synthesis of macroglobulin antibodies [262, 273]. Ada et al. [58] showed that injection of relatively small doses of nonpolymerized soluble flagellin may result in synthesis of only 7 S antibodies.

Formation of macroglobulin antibodies soon decreases after synthesis of 7 S antibodies has begun. However, if 7 S antibody synthesis is prevented by irradiation or injection of mercaptopurine, macroglobulin antibodies may continue to be produced even longer than usual. When 7 S antibodies are injected into animals treated in these ways, synthesis of 19 S antibodies ceases within 24 hr [231]. This effect is produced only by injection of 7 S antibodies specific for the same antigen.

Thus, it can be inferred from these experiments that the arrest of 19 S antibody synthesis is attributable to the appearance of antibodies of lower molecular weight in the blood.

The problem of switching over of synthesis from 19 S antibodies to 7 S antibodies is of fundamental importance, and an elucidation of the causes of this phenomenon can provide valuable information for an understanding of antibody biosynthesis in general.

It is at present difficult to give a completely satisfactory clarification of the biological significance of this switchover of synthesis from macroglobulin antibodies to γG-antibodies. Stavitsky et al. [234] presented some interesting views, according to which the system of macroglobulin antibody synthesis is phylogenically and ontogenically prior. And, indeed, the antibodies synthesized in the newborn are predominantly the γM-type [239]. This same antibody type precedes the lighter type not only in mammals but also in birds [225] and lower vertebrates [275]. On the other hand, the system of 7 S antibody synthesis, which plays the more important role in the immunological response, develops much later. In *Mustelus canis*, a very ancient species of fish, antibodies are found only in immunoglobulins resembling γM-globulins. No other immunoglobulins are found in this fish [194].

IN VITRO ANTIBODY SYNTHESIS

The study of antibody formation by tissues and cells isolated from the organism has a long history [40, 247]. More and more problems, on the most varied aspects of antibody biosynthesis, are now being solved with models of cultures of antibody-synthesizing tissues or cells, whereby the entire arsenal of modern biochemical methods is used. Hence, it would be appropriate to discuss in detail the culturing methods used and some of the results of in vitro studies of antibody synthesis.

Methods for Culturing Antibody-Synthesizing Tissues and Cells

Whole organs, through which suitable solutions are perfused, can be used to study many problems of antibody biosynthesis. Since lung tissue is known to synthesize a considerable quantity of antibody after intravenous immunization of rabbits with pneumococci, Askonas and Humphrey [69] employed isolated lungs through which, with the aid of a special apparatus, was perfused a Hanks saline solution containing glucose, a mixture of amino acids, radioactive lysine or glycine, and 10% rabbit serum. After a certain time, radioactive antibodies were determined in the perfused fluid and in the tissue extracts.

In most investigations of in vitro antibody synthesis, methods developed for culturing of different tissues and cells in vitro are usually employed [2, 7, 245].

In some instances, small pieces of tissue of spleen, lymph nodes, or other organs are cultured. To obtain fragments on the order of fractions of a millimeter in size, either a special knife or shears are used. The fragments may be incubated in a medium containing 10% serum for several hours [68]. Coons [214] cultured fragments of rabbit lymph nodes (1-2 mm) for a prolonged period in rotating test tubes, the inner surface of which was coated with heparinized rabbit plasma. The medium could be changed after a given period so that thereby the dynamics of antibody synthesis could be studied [196]. Tissue pieces have been cultured by immersion in agar [139]. A shortcoming of this method, however, is that synthesis indexes vary from test tube to test tube due to differences in the quantity of tissue in each.

For biochemical studies it is frequently necessary to determine accurately the number of cells in each of the incubated samples. In such cases, suspensions of spleen or lymph node cells are convenient. These suspensions can easily be prepared by passing the tissue through a stainless steel [148, 276] or capron [32] mesh after mincing it. Preparation of a suspension from compact lymph nodes is, however, more complicated than preparation of a spleen cell suspension. By Attardi's method [74] a suspension of lymph node cells can be prepared with relatively little cell damage. This investigator first minced the lymph nodes into 5 mm^3 pieces with sharp scissors in the cold under a layer of culture medium containing 50% serum. Then these were placed in a siliconized flask in a medium containing 50% serum and the mixture was blended with a magnet freely suspended in the mixture and rotating at approximately 20 rpm. After six hours of such treatment, 20-40% of the cells had entered the suspension.

The suspension may be cultured in conical flasks [253], Petri dishes placed in a steam-saturated desiccator [276], or in a siliconized Warburg [148]. In all these experiments, the atmosphere consisted of 95% O_2 and 5% CO_2.

The optimum culturing conditions were achieved by applying the cells in a thin layer to a semipermeable membrane separating the culture medium (under continuous agitation) below from the required gas mixture above. This type of chamber was proposed by Steiner and Anker [250]. Its convenience lies in the fact that the fluid can be added to the cells, i.e., to the top part of the chamber, as well as to the medium in the lower part. These chambers are presently finding effective application in the Gurvich's laboratory [14]. Ainis obtained good results from incubation of cells in a more complicated version of this type of chamber (Fig. 47). The cells are applied in a thin layer to the middle section of the chamber onto a millipore membrane with much larger pores than cellophane, but still too small for the cells to pass through.

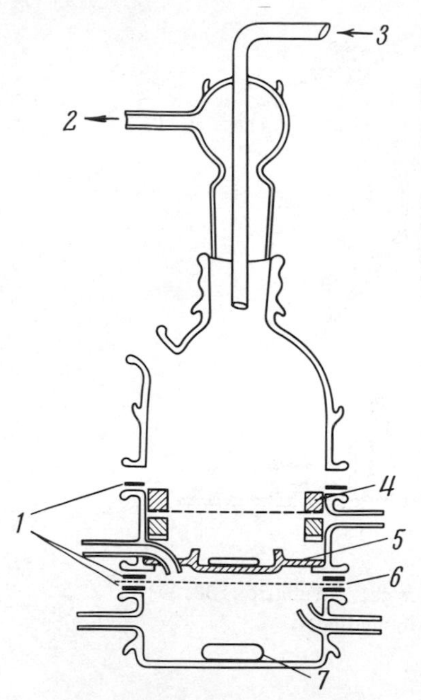

Fig. 47. Perfusion chamber for in vitro cell incubation [59]. 1) Rubber gaskets; 2) exit of gas mixture; 3) entrance of gas mixture; 4) ring with millipore filter; 5) support with magnet; 6) dialysis membrane; 7) magnet.

The fluid under the cells is agitated and renewed as required. The lower section of the chamber is separated from the middle section by a dialysis membrane and also contains the culture medium which is agitated by another magnet.

If it is necessary to incubate several samples at the same time under strictly identical conditions, chambers in which several vessels with membranes can be placed under one hood at the same time are convenient. This type of chamber has been constructed by La Via [181, 183]. Tissue pieces were placed on a millipore membrane placed over a Petri dish containing the culture medium agitated by a magnet. All the dishes were placed under the same hood.

We used another version of this type of chamber for incubation of suspensions of spleen and lymph node cells from hyperimmune rabbits (Fig. 48). The cells were applied to a cellophane membrane (3) stretched tightly over a glass ring with a ground flange (2). The ring was placed on a dish (1) also with a ground flange, and the two pieces were firmly connected by rubber bands. The culture medium had first been poured into the dish and the agitating magnet had been installed. The vessels, each standing in a small crystallizing or Petri dish (5), were placed on a glass plate covered with moist filter paper (6). Over the dishes was placed a hood to which was connected two tubes — one for the entrance and the other for the exit of the gas mixture (95% O_2 and 5% CO_2) which was supplied at the required rate from gas meters. The entire chamber was placed in a temperature-controlled incubator mounted on a special base, under

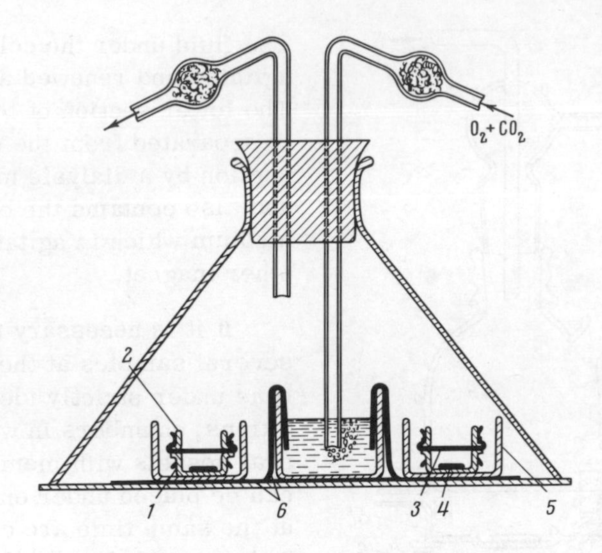

Fig. 48. Chamber and vessels for cell incubation [33, 34]. See text for symbols.

which were located motors with turning magnets (one for each vessel). To avoid overheating of the incubator, the motors were switched on only at certain time intervals by means of a special control device (KEP-12).

A procedure in which antibody synthesis by a single, separately incubated cell was studied proved useful for solving many problems of fundamental importance [210]. Nossal [206] was the first to develop a technique for incubating single antibody-synthesizing cells. The technique, briefly, involves the following: lymph node cells are suspended, washed with medium, and diluted to a concentration of 10^5-10^6 cells/ml. Further operations are carried out with micromanipulators in a Fonbrun oil chamber containing many large drops of medium for washing the individual cells, micro-drops for incubation (10^{-6} ml), and drops with a large number of lymph cells. The cells are first washed with a micromanipulator, after which the chamber is held in an incubator at 37° for three hours [210]. Then the presence of antibody in a microdrop is determined by one of the suitable methods (bacterial antigens are used).

Antibody formation by single cells has been studied by other investigators who used bacteriophages as antigens. A detailed description of a procedure for investigating single cells, consisting basically of the technique summarized above, has been published [76].

In the aforementioned experiments, the culture medium usually used was one of the synthetic media employed in incubating various tissues, e.g., Eagle medium Nos. 199 or 1066, with addition of 10-30% serum from the same species of animal from which the lymph cells have been taken. An Eagle medium containing only some amino acids is suited for experiments with radioactive amino acids.

Addition to the media of an ultrafiltrate of serum, embryo extracts, coenzymes, nucleic acid bases and their derivatives, or carbohydrates did not stimulate antibody synthesis [284]. Glucose, ascorbic acid, nucleosides, and glutamine also had no effect on in vitro antibody synthesis [276]. Certain vitamins and hormones were likewise without effect [196].

Ambrose, in Coons' laboratory [62], obtained interesting results in a study of a medium required for induction of antibody synthesis. In these experiments, the effect of the medium composition on the capacity of lymph node fragments to give a secondary response after in vitro contact with antigen was investigated. It was found that addition of physiological concentrations of hyprocortisone (0.01-1.0 μmole) to Eagle medium can replace serum. An optimum medium was obtained by adding serine (0.5 μg/ml) which in several instances had an even greater stimulatory effect on antibody synthesis. In a serum-free medium, antibody synthesis can take place in the presence of D-glucosamine or any of several mucopolysaccharides, but the rate of synthesis is slower [60]. The experimental development of a completely synthetic medium in which cells are able to produce antibodies over an extended period would be of considerable interest.

Methods for determination of the quantity of in vitro synthesized antibodies are partially dependent on the type of antigen used to immunize the donor animal. Various agglutination reactions (bacteria, erythrocytes), passive hemagglutination [247], neutralization of bacteriophages [73], immobilization of *Salmonella* flagellae or immune attachment of bacteria to antibody-synthesizing cells [210], and diffusion in agar [113] have been used. For biochemists, quantitative methods, such as the Heidelberger method [59], and methods employing radioactive isotopes are of especial value. The latter are very sensitive and are particularly accurate when immunoadsorbents are used to isolate newly synthesized radioactive antibodies from the culture medium [14, 32].

A method for determination of hemolysin forming lymphoid cells from the number of hemolysis plaques after incubation of these cells with erythrocytes used as antigens is becoming increasingly popular. Jerne and Nordin [170, 171], who were the first to propose this method, used agar as the supporting medium. In other cases, a suspension of carboxymethylcellulose [168] or agarose [258] has been used. A benzidine stain can be used for better detection of the hemolysis plaques appearing after addition of complement [171, 258], although within a short time after incubation of the cells with erythrocytes with adsorbed complement (20-30 minutes), the plaques are visible with the naked eye [168]. During incubation, it is the synthesis of hemolysins which actually takes place, rather than liberation of previously formed antibodies into the medium, since various factors harmful for the cell (e.g., heating for 30 minutes at 56°) almost completely prevent appearance of plaques. Although this method has proved to be very useful for shedding light on many problems, it nevertheless has one important disadvantage; namely, it is able to detect only must be borne in mind that erythrocytes possess determinants which are similar to the determinants of certain bacteria, and it is possible precisely for this reason that a small quantity of hemolysin-forming cells are found in nonimmune animals. The development of analogous methods for other types of antigens would be very valuable.

Primary and Secondary Responses in vitro

Attempts to induce antibody synthesis in cells to which antigen had been added in vitro date back to Karrel. Some investigators have published positive results, but usually these could not later be confirmed [247]. Stevens and McKenna [190] made such an attempt relatively recently. In their experiments, addition of protein antigen induced antibody synthesis in vitro in spleen cells of a peritoneal exudate, judging from agglutination of tanned erythrocytes with adsorbed antigen. A later attempt to confirm the results of these experiments was unsuccessful [252]. According to the data of Stavitsky [247], hemagglutination could possibly have been induced by certain nonspecific substances. Michaelides and Coons [196] were unable to obtain a primary response in a tissue culture. The method used in these experiments involved incubation of minced tissue in rotating test tubes and was capable of sustaining antibody synthesis for a prolonged period. Fishman [123], as mentioned above, worked out a system which, as it were, gave a primary response outside the organism. A primary immune response was induced in an organ culture. Before removing the spleen the animals received an injection of phytohemagglutinin to stimulate mitosis [138a].

It can now be regarded as certain that a second response can be induced by adding antigen to lymphoid cells from an immune animal. Stavitsky [246] has described several successful attempts to induce synthesis of radioactive diphtheria antitoxin in vitro after addition of toxin to spleen fragments from previously immunized animals.

In recent years, a procedure of secondary in vitro stimulation has been worked out in detail by Coons [196]. Lymph nodes were removed from rabbits several months after immunization with diphtheria toxin or bovine serum albumin, and their fragments were incubated in plasma clots in rotating test tubes together with antigen. Within four days after adding the toxin and seven days after adding albumin, antibody could be detected in the culture medium by passive hemagglutination (Table 39). The antibody content increased gradually, and synthesis continued for four or more weeks.

If the rabbits were pre-immunized with two antigens, both were able to cause a second response in the culture. In these experiments, very small quantities of antigens were able to stimulate antibody synthesis. Doses of 0.01-1.0 μg gave high titers. In some experiments, even doses of 0.001 μg protein stimulated production of a small amount of antibody. It should be mentioned, however, that in 66% of the experiments, antibody synthesis was observed even without addition of antigen in vitro. This dictates a certain caution in interpreting experiments in which high antibody

TABLE 39. Stimulation of Antitoxin Synthesis by Antigen in vitro by Lymph Node Fragments from Rabbits Immunized Six Months before the Experiment* [196]

Antigen	Antibody titer at different incubation times					
	3rd day	6th day	9th day	13th day	17th day	21st day
0.05 *Lf*	0	0	160	320	0	0
0.05 *Lf*	0	0	80	160	0	0
0.05 *Lf*	0	10	160	160	0	0
Without antigen	0	0	0	0	0	0

*The fragments were exposed to the antigen for two hours, after which the medium was removed and fresh medium was added without antigen.

titers were determined. Nevertheless, addition of sufficiently large antigen doses induced a much higher rate of synthesis than in the control test tubes.

The possibility of obtaining a secondary response in a suspension of spleen cells has recently been demonstrated [222]. In these experiments, the capacity of spleen cells of rabbits preimmunized with sheep erythrocytes to form hemolytic plaques was studied by Jerne's method. Whereas in the control 4-59 plaques per 10^7 cells were obtained, in the test samples up to 5000 plaques per 10^7 cells appeared by the 5th-6th day of incubation. The response was specific.

The possibility of a secondary response in vitro is of considerable interest, since it permits the tracing of biochemical processes taking place in cells immediately after contact with antigen, and ultimately leading to induction of antibody synthesis.

Course of Antibody Synthesis in vitro

After the first (and in most cases, the second) injection of antigen into an animal, two to three days should elapse before the spleen or lymph nodes begin to synthesize antibodies in in vitro incubation [247, 252, 259]. However, recently published experimental data indicate that antibody synthesis by lymph node fragments begins much sooner after a second antigen addition to incubated cells. In these experiments, tissues were incubated under conditions permitting a complete in vitro secondary response

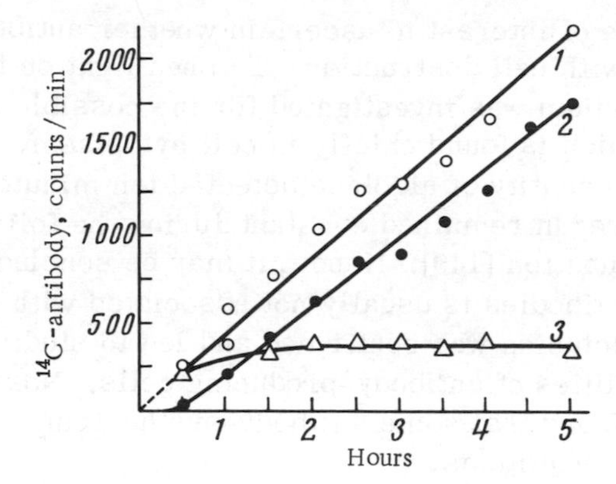

Fig. 49. Incorporation of radioactive amino acid into antibodies of lymphoid cells incubated in vitro [148]. 1) Total quantity of radioactive antibodies; 2) antibodies secreted into the medium; 3) antibodies isolated from cells.

(see above). Consequently, it is not surprising that antibodies were detected on the 7th to 10th day, i.e., the same time as after addition of antigen in vitro, in the culture medium when fragments of lymph nodes in a plasma clot were placed in rotating test tubes one day and even 2.5 hr after a second injection of antigen to rabbits [196]. Analogous results were obtained when the fragments were incubated in a medium with hydrocortisone [150].

Antibody secretion by cells incubated in vitro was traced with the aid of a radioactive label [68, 93, 148, 149]. Formation of intracellular antibodies began immediately after addition of the radioactive amino acids, and their quantity gradually rose. However, after a certain time (approximately 30 min), this augmentation was arrested, and with continued incubation the quantity of intracellular radioactive antibodies remained more or less constant (Fig. 49). Antibodies do not appear in the culture immediately, but only after a certain time (approximately 30 min) has elapsed. The length of this lag phase was the same in experiments with perfused lungs [69], incubation of tissue fragments [68], or incubation of a suspension of lymph cells [148]. Retardation of secretion of antibodies into the medium is not associated with adaptation of the cells to the new conditions since it is also observed when a label is added at different times after incubation has begun [148].

It would be of interest to ascertain whether antibody secretion is associated with cell destruction. To shed light on this question, the culture medium was investigated for the possible accumulation of aldolase, which is found chiefly in cell cytoplasm. It was found that the small quantity of aldolase detected ten minutes after incubation was begun remained constant during the following 50 minutes of incubation [149]. Hence, it may be concluded that secretion of antibodies is usually not associated with cell destruction. This conclusion was confirmed and led to studies of the mitotic capabilities of antibody-producing cells. Nossal and Mäkelä [210] found that some antibody-synthesizing cells are indeed capable of mitosis.

The degree of the dependence of antibody secretion on their continued synthesis was also studied [149]. For this purpose, synthesis was stopped in cells under incubation with radioactive amino acid by addition of puromycin. It was found that despite the presence of this inhibitor, some of the previously synthesized antibodies was secreted into the medium. However, for complete secretion of all antibodies formed, continued antibody synthesis, or some other undisturbed processes in the cell, is apparently necessary.

If labeled amino acid in the medium is replaced by unlabeled, about 80% of the radioactive antibodies is secreted into the medium in 30 minutes. Thus, the intracellular store of antibody is renewed at a rate of 3-4% per minute [149].

Quantity of Antibodies Synthesized in vitro

In some investigations the quantity of antibodies synthesized during incubation of lymphoid tissue in vitro has been calculated. Since the incubation conditions and the methods of antibody determination varied considerably, it is not surprising that the figures obtained by the different investigators also varied.

In instances where radioactive isotopes were used, the weight of the antibodies synthesized can be approximately calculated from the incorporation of the label. Thus, Askonas and Humphrey [68] calculated that in 3 hr 1 g of spleen tissue synthesizes 32 μg antibody to ovalbumin, while Vaughn et al. [276] found that 1 g of spleen synthesizes 400 μg of such antibodies in 24 hr.

In other experiments, a quantitative precipitation method was used. The results of these experiments were generally similar: in 24 hr 1 g of lymphoid nodes synthesized 400 μg diphtheria antitoxin [247], and the number of cells in one milliliter synthesized 4.5 mg antibody protein in 13 days [59]. The high figures obtained by Steiner and Anker [250] (on the order of 1-3 mg antibody protein per day, and more, per gram of spleen) were not found by other investigators.

We determined the quantity of antibodies formed by the spleen of immune rabbits with the aid of immunoadsorbents. When incubation of the tissue was begun on the 3rd day after a challenging dose of human serum albumin, 1 g of spleen formed 60 to 300 μg antibody protein in 40 hr. It should be emphasized that the quantity of synthesized antibody protein varies individually [32]. The quantity of antibodies formed is even greater if the incubation is begun on the 4th day after secondary immunization.

In experiments with isolated lungs of rabbits immunized with pneumococci, Askonas and Humphrey [69] found that perfusion of these organs resulted in formation of 8 mg antibody protein in one hour. It is interesting, that in in vivo experiments hyperimmune rabbits synthesized 0.5 g antibodies per kilogram body weight in one day. If it is assumed that the body of a rabbit contains $5 \cdot 10^{11}$ lymphoid cells, then each cell will have formed 8000 molecules per minute. Since not all cells synthesize antibodies, the actual number of molecules synthesized by an active cell is 10 50 times greater [12].

Another interesting question is the percent of antibodies synthesized of the total quantity of proteins formed during this time by lymphoid cells. An estimate of this figure can be obtained by comparing the amount of labeled amino acid incorporated into antibody with the quantity of label in other proteins. It was found that the antibody fraction of total proteins synthesized was considerable. Thus, in one experiment in which antibody synthesis was quite intense, Helmreich et al. [148] found that of the 154,000 counts/min in newly synthesized protein in the cells and medium, 32,000 counts/min or 20% was in antibody to the dinitrophenyl group.

Similar values were obtained in our experiments in which we studied the biosynthesis of all water-soluble proteins in lymph node cells and of antibodies to bovine globulin. With an immunoadsorbent (fixed globulin), antibody with a total activity 130,000 counts/min was isolated in one of the experiments. After adsorption, the total activity of the protein remaining in solution was 490,000 counts/min.

Recently, Drizlikh [19] made a comprehensive enquiry into this problem in which he determined not only antibody, but also nonspecific γ-globulins in the culture medium. It was found that on the 4th day after a secondary immunization, the total quantity of proteins newly formed by rabbit spleen cells comprised 65-75% γ-globulins, of which only 40% was antibodies. It is curious that addition of antigen had no effect on antibody synthesis.

Microglobulins Synthesized in a Tissue Culture

We studied the size of immunoglobulin molecules synthesized in a tissue culture with the aid of gel filtration [35]. Lymph node and spleen cells from rabbits immunized with human serum albumin were incubated in vitro in the presence of radioactive amino acids. After 18 hr, the cells were destroyed by homogenization in a hypotonic solution; after centrifugation of the culture medium and the homogenate at 105,000 *g* for 30 min, all antibodies were extracted from the supernatant with the aid of a fixed antigen. Then newly synthesized γ-globulin was adsorbed from this sample with the aid of an insoluble complex of fixed rabbit γ-globulin and its specific donkey antibody, i.e., an adsorbent-antibody (cf. Chapter I). Then the conjugated γ-globulin was eluted at pH 1.8 together with the donkey antibody, and the eluate was applied to a column containing Sephadex G-200 1*N* acetic acid. To determine the size of the newly synthesized γ-globulin, bovine serum and lysozyme, and, in other cases, rabbit γ-globulin, ovalbumin, or chymotrypsinogen were added to the eluate. Absorption at 280 mμ was measured in the effluent fractions. To each of the fractions was added 0.5 mg bovine globulin as the supporting protein, and all the fractions were precipitated with trichloracetic acid. Then, the protein, dissolved in formic acid, was applied to paper discs and the radioactivity was counted in a liquid scintillation counter.

The results showed the radioactivity distributed in three peaks (Fig. 50). The first of these was evidently γM-globulin and uncleaved complexes of γ-globulins with donkey antibody, and in its magnitude corresponded to polymer complexes of bovine albumin. Most of the radioactivity emerging from the column was accumulated at the second peak, which in magnitude fully corresponded with γG-globulin. The third peak (approximately 5% of the emerged radioactivity) was the most interesting. The molecule which emerged at this peak was the most interesting. The molecule was somewhat smaller than the ovalbumin molecule and similar in size to the chymotrypsin molecule. Its molecular weight was approximately 25,000. Antibodies synthesized in a tissue culture and isolated with immunoadsorbents did not give this third peak in gel filtration. This indicates that this peak was not an artifact. The protein of the third peak was also not the result of proteolysis of γG-globulin by lymphoid cell enzymes since it could not be isolated after incubation of pure radioactive antibodies with cells from a normal rabbit.

The third type of proteins, designated as "microglobulin," was identified by a method developed by us which involved fixed immunoglobulin subunits [35a, 35c].

After being passed through a Sephadex G-25 column in borate buffer, pH 8, the microglobulin solution, was divided into two equal parts. Nonspecific adsorbent (fixed HSA) was added to one part of the solution for the detection of nonspecific adsorption. After centrifugation a washed complex of fixed light chains and the corresponding donkey antibodies was added to the supernatant. The adsorption of radioactive microglobulin on a complex of fixed heavy chains and corresponding antibodies was studied. The bound radioactive protein was eluted with 0.1*N* HCl and measured with a scintillation counter. In control experiments with ^{14}C RGG the yield of the elution procedure was found to be about 84%. As shown in Table 39A, practically all microglobulin had adsorbed on the complex of the fixed light chains and antibodies.

To the second half of the microglobulin solution a complex of the fixed heavy chains and corresponding antibodies was added after the nonspecific adsorbent. In contrast to the previous experiment about 13% of the total radioactivity was adsorbed on the complex of the fixed heavy chains and antibodies.

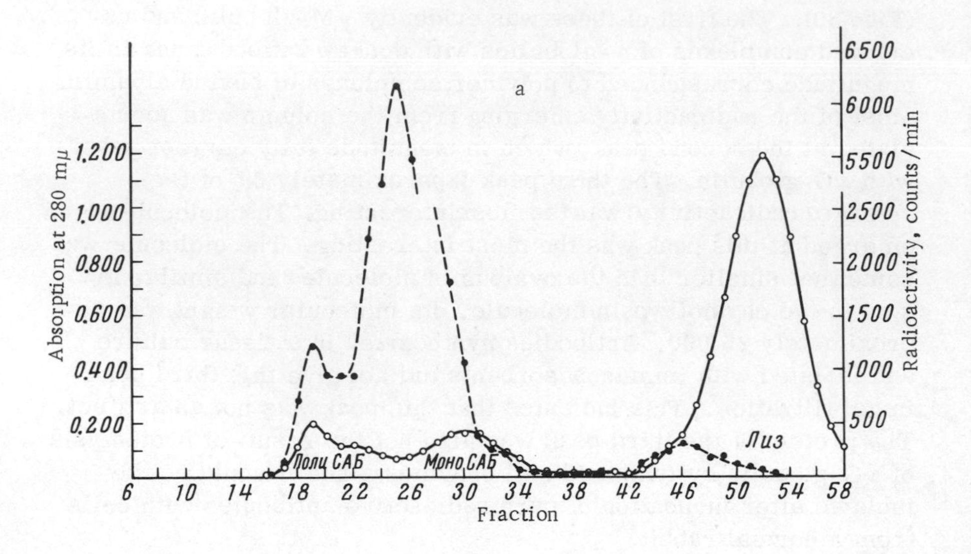
a
Absorption at 280 mμ
Radioactivity, counts / min
Полц САБ
Моно САБ
Лиз
Fraction

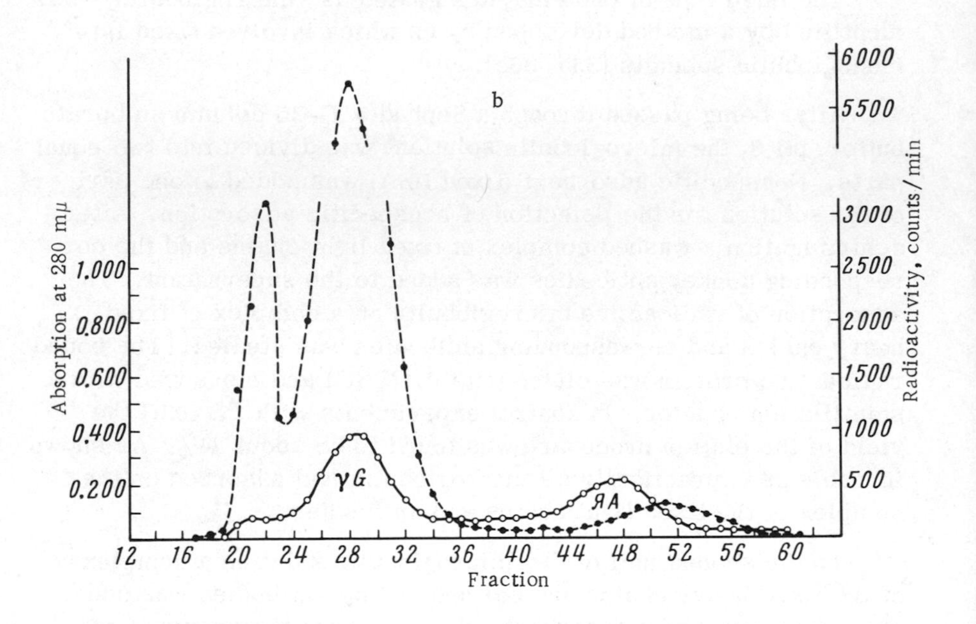
b
Absorption at 280 mμ
Radioactivity, counts / min
γG
ЯА
Fraction

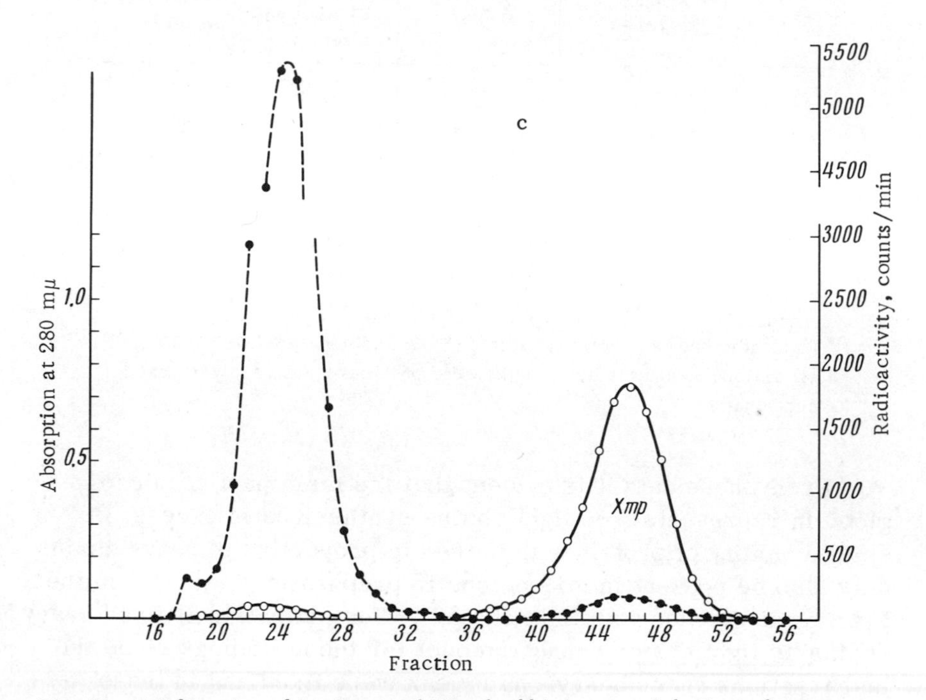

Fig. 50. Gel filtration of in vitro synthesized rabbit immunoglobulins and test proteins through a column containing Sephadex G-200. a) Bovine serum albumin (polymer complexes, polyBSA; monomer complexes, monoBSA) and lysozyme (lys); b) γG-globulin and ovalbumin (OA); c) chymotrypsinogen (Ch) [35]. The dashed curve and black circles represent the radioactivity, the solid curve and open circles gives absorption at 280 mμ.

TABLE 39A. Adsorption of the Microglobulin Synthesized in vitro by Different Antibody-Immunoadsorbents

Aliquot No.	Total radio-activity of aliquot, counts/min	Fixed proteins (4 mg)	Treatment of fixed proteins with 0.4 ml of anti-RGG anti-serum	Adsorbed radio-activity,* counts/min
1	752	HSA	Not treated	0
		Light chains of RGG	Treated	727
		Heavy chains of RGG	Treated†	7
2	752	HSA	Not treated	0
		Heavy chains of RGG	Treated†	95

* Total adsorbed radioactivity [eluted (84%) and noneluted with 0.1*N* HCl].
† All antibodies against light chains were previously adsorbed with fixed light chains.

From these data it is evident that the main part of microglobulin represents free light chains synthesized in excess. A small quantity of protein with antigenic properties of heavy chains may also be present in microglobulin preparations (at pH 8 in the form of complexes with light chains). Contamination of fixed heavy chains by light chains cannot account for these findings since the last fractions of the heavy-chain peak were fixed and in these light chains are present either only in trace amounts or are totally absent. Furthermore, the fixed heavy chains used for the preparation of the antibody-immunoadsorbent were treated with anti-RGG antiserum from which anti-light chain antibodies had been absorbed.

Askonas and Williamson [72a, 72b] demonstrated clearly the presence of a certain quantity of free light chains in cells. In their first experiments, they used plasma tumor cells as a model. These cells formed only complete molecules of mouse myeloma 7 S globulin and contained a small quantity of free light chains. The light-chain pool found was in a dynamic state since some of the light chains were continuously combining with heavy chains to form a complete immunoglobulin molecule. These results were confirmed in experiments with lymph nodes of immune mice [72b]. It was found that within five minutes after addition of a label to the culture the amount of label in the light-chain fraction and γG-globulin fraction was the same. However, after 40 minutes the amount of label in the fraction of whole immunoglobulin molecules was 6-7 times greater. Shapiro et al. [236a] also obtained evidence of the existence of free light chains using an electrophoretic method.

It is quite probable that light chains have a certain regulatory effect on the synthesis of immunoglobulin molecules [35a, 72a]. It may be assumed that a light chain can combine with a heavy chain while the latter is still on the ribosome. In this case, the ribosome is deprived not only of the heavy chain, which tends to form aggregates, but also of readily soluble complexes of light and heavy chains. As is known, light and heavy chains are readily able to form complexes with each other without the intervention of other factors. This way of forming immunoglobulin molecules is similar to the mechanism proposed by Colombo and Baglioni [100a] for formation of the hemoglobin molecule, according to which free β-chains remove α-chains from ribosomes.

Synthesis of the Carbohydrate Components of Immunoglobulins

The study of the synthesis of carbohydrate components in antibodies and γ-globulins is still essentially in its initial stages. Incorporation of ^{14}C-labeled sugars into γ-globulin carbohydrates of rabbits immunized by bovine serum albumin [238] has been determined in tissue culture [238]. After incubation of spleen of lymph node cells with labeled glucose, mannose, or galactose, the bulk of the isotope (70-80%) was found in γ-globulin protein. The carbohydrate component incorporated 20% mannose, 13% glucose, and 32% galactose of the total amount of incorporated isotope. It is still not known precisely where and how the carbohydrate component of antibodies and γ-globulins is formed, nor in what way it becomes bound to peptide chains. It is known only that microsomes contain molecules of immunoglobulins without carbohydrates, while the supernatant remaining after sedimentation of microsomes contains already completed immunoglobulin molecules with a carbohydrate component [262b]. It would be interesting to check the hypothesis that the carbohydrate component facilitates efflux of immunoglobulin molecules from cells.

PARTICIPATION OF RIBOSOMES IN THE BIOSYNTHESIS OF ANTIBODIES. CELL-FREE SYSTEMS. POLYRIBOSOMES

According to current conceptions of protein biosynthesis, ribonucleoprotein particles (ribosomes) participate directly in the formation of the polypeptide chains [5, 42]. There is reason to suppose that the polypeptide chains of antibodies are also formed with the aid of ribosomes found in the cytoplasm of plasma cells. As is known, these cells have the structure of secretory cells and are distinguished by a high content of ribosomes, which are concentrated in the endoplasmic reticulum. It is not surprising that a certain quantity of antibodies can be extracted from microsomes, which are experimentally obtained fragments of this ribosome-containing reticulum. Such experiments were carried out on microsomes of lymphoid tissues of animals immunized with different antigens. From the microsomes contained in 1 g of spleen or bone marrow of rabbits immunized with pneumococci, 180 μg antibodies was extracted with Tris phosphate buffer (pH 7.5), after which another 90 μg antibodies was extracted with Na_2CO_3 buffer (pH 9.5) [65-67, 71].

In another series of experiments, lymph node microsomes from guinea pigs immunized with bovine globulin conjugated with the 2,4-dinitrophenyl group (DNP) as hapten [175, 176] were studied. It was found that these ribosomes were capable of binding DNP albumin. This reaction was specific inasmuch as DNP albumin was eluted with DNP lysine, i.e., by the same means as from an antibody-DNP-albumin precipitate. Further, binding of DNP-albumin by microsomes was inhibited by the hapten. In control experiments with microsomes from nonimmunized animals this effect was not observed. The antibody activity in liver microsomes of immune pigs was 4% of the activity of lymph node microsomes. In other experiments, antibodies to DNP, detected by the passive skin anaphylaxia reaction, were extracted from microsomes at pH 2.5. Antibodies bound with microsomes and nucleoproteins have also been found by other investigators [120, 129, 288]. It has recently been shown that immunoglobulins first accumulate in microsomes [262a].

In recent years a large number of investigations have been devoted to the synthesis of proteins in cell-free systems. In several instances, formation of individual proteins in cell-free systems of both animal and bacterial origin [5, 157] has been demonstrated. In particular, the synthesis of albumin and several other proteins of microsomal preparations from rat liver has been shown [96, 133]. Some investigators have undertaken the study of synthesis of antibodies or γ-globulins in cell-free preparations of lymph nodes or spleen [115, 154, 215]. Ogata et al. [215] showed that the microsomal fraction of the spleen of hyperimmune rabbits was able to incorporate ^{14}C-leucine into the protein of γ-globulins and antibodies. However, these results were not later confirmed. It is quite possible that when newly synthesized proteins are determined by the Heidelberger method, which these Japanese investigators used, a large quantity of nonspecific labeled proteins passes into the precipitate. This could lead to the erroneous conclusion that new γ-globulins and antibodies are formed in the system used by these investigators.

Eisen et al. [115] have published evidence that ^{3}H-lysine is incorporated into proteins similar to γ-globulin in cell-free extracts of lymph nodes from rabbits or guinea pigs. However, this report, published in 1961, had only a general character, and subsequent studies by the same investigators as well as others were unable to verify its conclusions. Thus, for example, Askonas [67] obtained negative results in a study of incorporation of radioactive label into antibodies and γ-globulin by spleen and lymph node microsomes from immunized rabbits and into globulin proteins in a cell-free system of plasma cell tumors in mice.

The incorporation of radioactive amino acids into immunoglobulins is direct evidence of the synthesis of immunoglobulins in a cell-free system. Several investigations (those mentioned above and others) [140a, 262c, 284a, 289] have shown that microsomes of lymphoid tissue are capable of incorporating radioactive amino acids into proteins, although certain difficulties are encountered in the identification of radioactive proteins. Radioimmunoelectrophoresis [251a] does not give accurate quantitative results although an attempt has been made to count the radioactivity of precipitin lines [140a].

Immunoadsorbents are useful for the identification of radioactive soluble proteins. However, the first attempt to use this

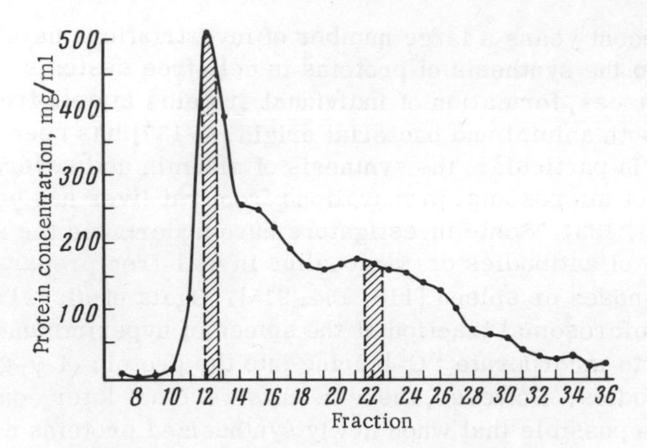

Fig. 51. Fractionation of a supernatant remaining after centrifugation of a cell-free microsomal system of rabbit lymphoid tissue on a column filled with Sephadex G-200. Fractions with radioactive protein are shown in the shaded columns (for explanation see text).

method was unsuccessful [34, 154]. In experiments with rabbit spleen microsomes, approximately 40-50% labeled proteins passed into the supernatant after treatment of the incubation medium with deoxycholate and precipitation of the ribosomes. In fractionation of labeled soluble proteins on Sephadex G-200 it was found that the molecular weight of some proteins was less than 200,000 since some of the label was retained by the gel. For example, in the experiment whose results are given in Fig. 51, radioactivity was found in the proteins of both fractions 12 and 22 [34].

However, in an attempt to identify labeled proteins with the aid of immunoadsorbents, no proteins with either antibody or antigenic properties of immunoglobulins could be found since the label was not adsorbed on either fixed antigens or on complexes of γG-globulins and antibodies.

Experiments with rat microsomes were more successful. We were able to show that with the use of immunoadsorbents it is possible not only to detect, but also to characterize quantitatively, the synthesis of immunoglobulins by these particles in vitro [35c].

The experiments were carried out with the spleen and liver of rats following two injections on successive days with 2 ml of a

10% suspension of sheep erythrocytes; the animals were killed on the third day after the second injection. In order to obtain microsomes, minced tissues were placed in a glass homogenizer and disrupted by four strokes of a loosely fitting spherically tipped Teflon pestle. The homogenization medium had the following composition: 0.035*M* Tris buffer (pH 7.8), 0.15*M* sucrose, 0.07*M* KCl, 0.01*M* $MgCl_2$, and 0.006*M* mercaptoethanol. To isolate the microsomes, the homogenates were centrifuged at 15,000 *g* for 20 minutes and then for 60 minutes at 105,000 *g*. The cell sap used to isolate the pH 5 fraction was obtained by centrifugation of the homogenates at 105,000 *g* for 90 minutes.

The microsomes and the pH 5 fractions were mixed with addition of the ATP-generating system and ^{14}C-labeled amino acids. The mixture was incubated for 25 minutes, then puromycin was added for the release of nascent polypeptide chains from polyribosomes and the incubation was continued for a further five minutes. Incorporation of ^{14}C-labeled amino acids into proteins was measured in a liquid scintillation counter by means of the disk method, and the protein concentration was determined by Lowry's method.

Under such conditions the spleen microsomes were found to be able to incorporate amino acids into proteins (Table 39B), the incorporation proceeding somewhat more actively in the presence of the pH 5 fraction from liver than with the same fraction from spleen. With liver microsomes incorporation was, however, always several times as great. (In the experiment described: 8640 counts/min per mg microsomal protein.)

After sedimentation of ribosomal particles, the supernatant contained 40-45% of the radioactive proteins. To find out whether or not these included proteins with the antigenic properties of γ-globulin, the supernatant was successively treated with antibody immunoadsorbents, i.e., complexes of cellulose-fixed antigens and their corresponding antibodies. These complexes are known specifically to bind free antigens in solution. First, to the supernatant was added rabbit γG-globulin fixed on cellulose with specific antibodies bound to it. After removal of this control adsorbent by centrifugation, the supernatant was treated with cellulose-fixed rat γG-globulin with corresponding bound rabbit antibodies. For the fixation on cellulose pure γG-globulins isolated

TABLE 39B. Incorporation of Radioactive Amino Acids into Proteins by Rat Spleen Microsomes*

Expt. No.	Microsomes	pH 5 fraction	Radioactivity incorporated into proteins, counts/min per mg microsome protein	Total counts/min: soluble protein	Total counts/min: sedimented proteins	Adsorption of the radioactive label from supernatants: on nonspecific adsorbent, counts/min	Adsorption of the radioactive label from supernatants: on specific adsorbent, counts/min	Soluble labeled proteins antigenically reacting as rat γG-globulin, counts/min
1	Spleen	Spleen	1240	8000	12,780	540	484	No
2	Spleen	Liver	1900	6800	7800	349	1000	651

*The reaction mixtures contained in a volume of 1 ml: microsomes (5.5 mg in Expt. 1 and 4 mg in Expt. 2), pH 5 fraction (25 mg in Expt. 1 and 11 mg in Expt. 2), 2 μmole ATP, 0.2 μmole GTP, 10 μmole phosphoenolpyruvate, 100 μg phosphoenolpyruvate kinase, 1 μC of a mixture of ^{14}C-labeled amino acids (glutamic acid, alanine, arginine, serine, and proline) with a specific activity of approximately 7 mC/mmole and 0.3 μmole of 15 other amino acids (nonlabeled). The mixtures were incubated for 25 minutes at 36.5°; puromycin was then added (final concentration $10^{-3}M$) and the incubation was continued for a further five minutes. After addition of deoxycholate (final concentration 0.5%) the mixtures were centrifuged for 90 minutes at 105,000 g.

by means of DEAE-Sephadex were used. The adsorbents were washed with saline and the adsorbed proteins were eluted with 0.1*N* HCl (the yield in the elution procedure was found to be 84%). The radioactivity of the eluates was estimated on a scintillation counter.

In the first experiment involving microsomes and a pH 5 fraction both isolated from spleen, the adsorption of the tracer was found to be similar for the two adsorbents. In the second experiment, involving spleen microsomes and pH 5 fraction from liver, adsorption on the complex of fixed rat γG-globulin and antibody exceeded that on the control adsorbent by a factor of three. After subtracting the control, the specific increase in radioactivity was 651 counts/min (Table 39 B). Hence about 10% of the soluble radioactive proteins exhibited the same antigenic properties as rat γG-globulin. In other similar experiments with the same quantity of spleen microsomes the values

TABLE 39C. Study of the Antigenic Properties of Protein Synthesized in a Cell-Free System from Rat Spleen

Adsorption sequence	Antibody-immunoadsorbent	Adsorption, counts/min	Specific adsorption, counts/min
1	Control adsorbent (complex of fixed rabbit γG-globulin and the corresponding antibodies)	239	–
2	Adsorbent binding light chains of rat γG-globulin (complex of fixed light chains of rat γG-globulin and the corresponding antibodies)	530	291
3	Adsorption binding heavy chains of rat γG-globulin (complex of fixed heavy chains of rat γG-globulin with the corresponding antibodies)	534	295
4	Adsorbent binding rat γG-globulin (complex of fixed rat γG-globulin and corresponding antibodies)	251	12

of 1070 and 648 counts/min were obtained. There was no specific adsorption in one experiment in which puromycin had not been added. The control experiments were performed with the supernatant of the incubation mixture which consisted of microsomes and pH 5 fraction both isolated from liver; the adsorption of isotopic label was similar for both control and specific immunoadsorbents.

In other experiments we showed that in a cell-free system from rat spleen, labeled amino acids are incorporated into proteins with the antigenic properties of both heavy and light chains. Specially prepared complexes of fixed peptide chains with corresponding antigens proved well-suited for identification of the radioactive proteins.

As an example, results are given of an experiment in which a supernatant obtained after centrifugation of an incubation medium consisting of 24 mg spleen microsomes and 60 mg pH 5 fraction from liver was studied. The supernatant was successively treated with different antibody-immunoadsorbents beginning with rabbit γG-globulin + antibody as a control (Table 39C).

Thus, we were able to show that in a cell-free system from rat spleen proteins can be formed which are similar in their antigenic properties to the heavy and light chains of γG-globulins. Apparently, spleen microsomes contain sufficient information to form γG-globulin peptide chains under the appropriate conditions.

Mach et al. [185c] chose a chemical method for identification of the substance synthesized in a cell-free system from mouse plasma tumor. The tryptic hydrolysis products of soluble proteins labeled in vitro and isolated light chains formed by the same tumors in vivo were comparatively analyzed (tumors of the type used formed only light chains). The chromatogram peaks were identical in the two cases. The substances at the peaks also had identical electrophoretic mobilities. These results confirm the possibility of biosynthesis of immunoglobulins in a cell-free system.

As is known, ribosome-polyribosome complexes participate directly in the building up of peptide chains [221]. A study of the synthesis of hemoglobin and liver proteins revealed that the size of polyribosomes is determined by the length of the messenger RNA it contains. Approximately 90 nucleotides are included between two ribosomes in an aggregate [249, 278]. The size of such a subunit is 300 Å. If similar relationships obtain in the syntehsis of any protein, the size of polyribosomes participating in the formation of peptide chains of antibodies may be calculated.

Since the code is triplet [24], the 500 amino acid residues of the γ-chain should correspond to the messenger RNA from 1500 nucleotides with a molecular weight of approximately 450,000. The size of the polyribosome engaged in γ-chain synthesis is consequently 4500 Å; the polyribosome is composed of 15 ribosomes. If the γ-chain actually consists of two separately synthesized chains half so large, then the polyribosomes should also be half so large. Light chains have 200 amino acid residues, and consequently the corresponding messenger RNA should consist of 600 nucleotides (molecular weight approximately 180,000), and the polyribosome should contain 6 ribosomes (1800 Å).

What are the experimental confirmations of these calculations? At the present time, polyribosomes of lymphoid tissues are being comprehensively studied by various investigators. However, considerable difficulties are being encountered, since the spleen and lymph nodes contain much ribonuclease, liberated when the tissue is homogenized, which degrades messenger RNA, thereby giving distorted figures of the true size of polyribosomes. To inhibit RNA-ase during isolation of microsomes, either a cell extract [232, 140a] or an extract from soluble liver proteins [205b] can be added. Numerous investigations [79a, 183a, 184b, 192, 203, 236a, 262c, 267b, 282a] have firmly established that antibody-synthesizing lymphoid cells or plasma tumor cells forming myeloma globulins contain polyribosome complexes. In several investigations it was possible to detect two types of polyribosome complexes — larger ones, in which heavy chains are probably formed, and smaller ones, corresponding to light chains. According to Becker and Rich [79a] the first type of complexes contains 16-20 ribosomes, while the second type contains 7-8 ribosomes. La Via et al. [183a] detected less ribosomes in polysomes: 10-12 and 3-5, respectively.

Polysome complexes actually participate in the formation of the peptide chains of immunoglobulins, since proteins which are immunologically similar to light and heavy chains may be found in them. Williamson and Askonas [282a] clearly demonstrated that heavy chains are associated with large polysomes, while light peptide chains are associated with the smaller polysomes. Since the size of polysome complexes closely agrees with the above calculations, it may be concluded that both heavy and light chains are synthesized whole at the translation level, and not from smaller subunits. This important conclusion is also confirmed by investigations carried out with other methods. Fleischman [126] in a study of incorporation of labeled leucine into the heavy chains of rabbit γG-globulin showed that the label concentration gradually decreased from the N-end of the chain to the C-end for short incubation periods. This is evidence of the build-up of chains from one initial point. Analogous data were obtained by Knopf et al. [178] who studied incorporation of labeled amino acid into Fc and Fab fragments of the heavy chain of myeloma immunoglobulin from mice in short periods of incubation.

INHIBITORS OF ANTIBODY SYNTHESIS

The considerable efforts devoted in recent years to the study of inhibitors of antibody synthesis are due to several factors. First, a specific inhibitor of antibody synthesis is needed in medical practice, since organ and tissue transplantations are considerably hampered by the immune response of the recipient. Furthermore, it is important to know what effect various drugs may have on the immunity of the organism. This refers especially to antibiotics used in different infectious diseases, in the pathogenesis of which immune mechanisms play an essential role. Finally, inhibitors having a certain mode of action could contribute considerably to an understanding of antibody formation.

In most studies alkylating analogs of bases of nucleic acids or amino acids or antibiotics which in some way inhibit biosynthesis of nucleic acids or proteins have been used as inhibitors. These substances are injected into animals at different times before and after immunization or are added to in vitro incubated cells of antibody-synthesizing organs [21, 30, 132b, 184, 243, 247].

Several groups of investigators have made systematic studies of a large number of compounds in order to find inhibitors of antibody synthesis. Results of these studies are given in Table 40, which lists substances exerting the strongest inhibitory effect on a primary immune response. The principal group in this table comprises analogs of bases of nucleic acids.

It must be observed that not all the inhibitors listed had the same effect. Thus, Myleran was the most effective when injected before immunization. In this respect, Myleran has an effect very similar to that of radiation [43, 81], which also inhibits antibody synthesis when administered before immunization. Many other inhibitors of the antimetabolite type were effective either when administered simultaneously with, or a short time after, antigen injection. The mechanisms of action of antimetabolites and alkylating agents are most likely different [30]. If an inhibitor is effective in experiments on some animals, it does not mean that it will act similarly on animals of other species. Thus, one of the most powerful inhibitors, 6-mercaptopurine, which is even used in clinical medicine, is completely ineffective for suppression of antibody synthesis in guinea pigs in a dose of 9 mg/kg intramuscularly, or 40 mg/kg intraperitoneally [137].

TABLE 40. Most Effective Antibody-Synthesis Inhibitors Discovered in Animal Experiments

Substance	Dose mg/kg wt	Animal and antigen	Duration of use (0, day of antigen injection)	References
6-Mercaptopurine	6	Rabbits, human serum albumin	From 0 to +5th day	[236]
6-Azauridine	5000	Transfer of a mixture of lymph cells from a grown rabbit with *B. suis* to new-born rabbits		[254]
6-Mercaptopurine	3-5			
6-Thioguanine	0.5-1			
Aminopterine	20			
Mylevan	1.5-3			
6-Mercaptopurine	75	Mice, sheep erythrocytes	From 0 to +4th day	[204]
6-Thioguanine	2			
6-(1-Methyl-4-nitro-5-imidazolyl)-thiopurine	25-100			
2-Amino-6-(1-methyl-4-nitro-5-imidazolyl)-thiopurine	12.5			
Mylevan	125-150	Mice, typhoid-para-typhoid vaccine	From −9th to −2nd day, and from 0 to +6th day	[81, 82]
Iprite nitrite	2.4			
Minopterine	17.5			
Triethylenemelamine	2			
6-Mercaptopurine	150			
6-Thioguanine	37.5			
Chloramphenicol	500-600	Rabbits, bovine γ-globulin	From −24 hr to +12th day	[280]
DL-Penicilamine	75	Rabbits, human serum albumin	From −1st to +14th day	[6]

A few inhibitors have been studied in more detail. In one investigation [280] it has been shown that chloramphenicol in a dose of 0.5 g/kg completely suppresses the primary immune response to albumin injection in rabbits. These same animals were able to develop a normal anamnestic response to a second antigen injection within 10-12 days. As these investigators pointed out, chloramphenicol inhibits linkage of messenger RNA to ribosomes in cell-free systems. Hence, it may be concluded that this substance inhibits the productive phase of the primary response, but has no effect on the capacity to develop an anamnestic response. However, in a later experiment [104] it was found that large doses (1.5 g/kg) of chloramphenicol also suppressed the capacity to develop a typical secondary response in mice, although only when injections were begun simultaneous with, or a few hours after, the first antigen injection. The mechanism of the effect of chloramphenicol is obviously very complicated.

It is interesting that several other substances are able partially or completely to suppress the secondary immune response capacity which usually appears after primary immunization. Such substances are triethylthiophosphamide, 6-mercaptopurine, 8-azaguanine, and versenate [94]. At the same time, these substances, like chloramphenicol, had no effect on the secondary response if injected simultaneously with the secondary immunization. From this observation, Coons et al. drew the very important conclusion that there is a fundamental difference between induction leading directly to antibody synthesis and the process ("priming") which takes place directly after the first contact of the antigen with the cell and renders the cell capable of recognizing the antigen on second contact. This process takes place immediately after antigen injection, since the injection of inhibitors even as soon as three hours after immunization does not suppress the capacity to develop a typical secondary response.

In most in vivo investigations, the secondary response was considerably more resistant to inhibitors (injected during the secondary immunization) than the primary response. The results of the first studies of Schwartz et al. [236], as well as several other investigations, were similar to those obtained by Coons in the study mentioned above.

One of the most active immunosuppressants is cyclophosphamide [132b]. This substance suppresses the primary immune response whether injected before or after immunization, and is even capable of suppressing antibody synthesis already begun. In repeated injections cyclophosphamide considerably prolongs the survival of transplants in different species of animals.

Nitroso-alkyl ureas, which are strong alkylating agents, also exhibit a unique immunosuppressant effect. Rokhlin [40a] showed that N-nitrosoethylurea injected 40, 48, and 54 hr after immunization suppresses formation of antibodies to sheep erythrocytes. However, this substance is not effective when injected before immunization, simultaneously with immunization, or 6, 24, or 30 hr after immunization. A single injection of nitrosoethylurea 48 hr after immunization with a 40% suspension of erythrocytes considerably depressed the secondary response (in secondary immunization after 21 days). N-nitrosopropylurea had a similar effect, while N-nitrosomethylurea inhibited antibody formation whether injected before immunization or 2.5 days after immunization. Nitrosoethylurea has a low toxicity and causes only negligible lympholysis.

Numerous investigations of inhibition of antibody synthesis have been carried out with in vitro cultures of lymphoid cells.

Ambrose and Coons [63] studied the effect of chloramphenicol on the secondary immune response in tissue culture. It was found that this antibiotic is able completely to suppress antibody synthesis in a dose of 50 μg/ml. Even the relatively small doses of 5 μg/ml gave 80% inhibition.

Again, the earliest periods of the immune response were the most susceptible to chloramphenicol: during the first six days synthesis was 90% inhibited, whereas when this inhibitor was injected on later days inhibition was only 40% (the bulk of antibodies is produced after the sixth day).

In most other experiments on inhibition of antibody synthesis in tissue culture, the effect of various substances on the productive phase of antibody formation was studied. Some of the results of such experiments are given in Table 41, from which it is evident that the most varied substances are capable of inhibition of in vitro antibody synthesis.

TABLE 41. Inhibition of Antibody Biosynthesis in a Tissue Culture by Different Inhibitors

Antigen	Substance	Concentration	Inhibition, %	Reference
Diphtheria toxin	DL-ethionine	1.0	82	248
	p-fluorphenylalanine	1.0	85	248
	γ-ethylamidoglutamic acid	1.0	80	248
	8-azaguanine	2.5	60	247
	4,5,6-trichlor-1-(β-D-ribofuranosyl)-benzimidazole	0.3	90	247
	puromycin	0.3	95	247
	5-bromdeoxypuridine	0.2	80	247
Ovalbumin	DL-ethionine	8.0	93	276
	p-fluorphenylalanine	5.5	93	276
	8-azaguanidine	1.7	40	109
		4.0	80	109
	5-bromdeoxyuridine	0.02	60	111
Alcohol dehydrogenase	6-mercaptopurine	0.5	40	282
	8-azaguanine	0.5	9	282
	5-iododeoxypuridine	1	57	282
	5-fluoruracil	1	27	282
	5-fluordeoxyuridine	1	17	282
	6-azathymine	10	15	282
	5-iodouracil	1	0	282
	6-bromuracil	1	0	282
	6-azauracil	1	0	282
	5-iododeoxycytidine	1	0	282
	puromycin	25*	100	282
	actinomycin D	1*	100	282
	mitomycin	10*	58	282
Poliovirus	actinomycin	8*	100	261
	chloramphenicol	50*	90	261

*Concentration in μg/ml.

To understand the mechanism of action of inhibitors it is important to know to what extent the substance inhibits biosynthesis of other proteins in a cell. Dutton and Pearce [112] carried out experiments on this problem (Table 42).

According to the data in Table 42, some inhibitors inhibited synthesis of all cell proteins. Others, for example colchicin and 6-azauridine, had a stronger inhibitory effect on antibody synthesis than on synthesis of other lymphoid cell proteins. However,

TABLE 42. Effect of Different Inhibitors on Incorporation of Radioactive Amino Acids into Antibodies and Other Cell Proteins [112]

Substance	Concentration, μg/ml	Duration of effect, hr	Inhibition of amino acid incorporation, %	
			antibodies	other cell protein
5-Bromuracil	50	8-20	—	2
Ethidium bromide	11	20-28	15	20
" "	11	0-48	73	34
Mitomycin C	20	20-28	20	31
Colchicin	1.2	20-28	16	1
	2.0	0-48	66	15
	0.4	0-48	39	2
	0.08	0-48	−22	0
	0.016	0-48	−7	−1
6-Mercaptopurine	465	20-28	59	23
5,6-Dichlorbenzimidazol reboside	32	20-28	52	59
6-Azauridine	32	0-48	92	84
"	2000	7-15	(−2)	7
"	250	7-15	(4)	−4
"	1000	0-48	2(21)	1
"	200	0-48	43(11)	−4

Note: In parentheses are given results of addition of 6-azauridine to cells taken on the second day after immunization. The other experiments began on the third day after immunization.

even in these instances the mechanism of action is unclear. It should be observed that if a substance specifically inhibiting only antibody synthesis exists, considerable research would be required to ascertain the site of action of such an inhibitor. There is always the possibility that an inhibitor exerts its effect on very secondary processes, which are nevertheless essential for antibody formation. Indeed, post hoc does not mean propter hoc.

Several inhibitors are able to suppress cell proliferation usually taking place after antigen injection and, in consequence, inhibit antibody synthesis. Examples of this type of inhibitors are analogs of DNA bases [111, 213]. In particular, 5-bromdeoxyuridine inhibits antibody synthesis in vitro if the cells are isolated within two to three days after immunization [111]. Other inhibitors of DNA synthesis, such as iprite nitrite, mitomycin C, are also able to suppress antibody synthesis. The effect of these substances can hardly be called specific.

LITERATURE CITED

1. I. A. Alov, "Neurohumoral regulation of cell division in normal and cancer tissues," Tr. Khabarovsk. Med. Inst. 15:31 (1957).
2. O. G. Andzhaparidze, V. I. Gavrilov, B. F. Semenov, and L. G. Stepanova, Tissue Culture and Virological Studies [in Russian], Medgiz, Moscow (1962).
3. N. M. Balaeva, "Luminescent-serological method for detecting globulins," in: Current Problems of Immunology [in Russian], Meditsina, Moscow (1965), p. 336.
4. I. A. Bass and V. A. Gvozdev, "Messenger RNA and protein synthesis," in: Biosynthesis of Proteins and Nucleic Acids [in Russian], Nauka, Moscow (1965), p. 50.
5. A. A. Bogdanov and R. S. Shakulov "Ribosomes. Structure, biosynthesis, and role in the synthesis of proteins," in: Biosynthesis of Proteins and Nucleic Acids [in Russian], Nauka, Moscow (1965), p. 86.
6. R. B. Chvarov, L. Venkov, M. Iomtov, and T. Nidolov, "Changes in nucleic acid and ribonuclease activities in guinea pig spleen at different times after immunization," Izv. Tsentr. Lab. Biokhim. Bulgar. Akad. Nauk. 2:71 (1964).
7. V. I. Gavrilov, The Transplantable Cell in Virology [in Russian], Meditsina, Moscow (1964).
8. V. I. Goncharova, "On the problem of the role of lymphoid organs in antibody synthesis," Byul. Eksperim. Biol. i Med. 48(12):85 (1959).
9. A. E. Gurvich, "Fate of an intravenously injected protein," Usp. Sovrem. Biol. 37(1):94 (1954).
10. A. E. Gurvich, "Ratio between synthesized specific and nonspecific serum γ-globulins," Vopr. Med. Khim. 1(3):169 (1955).
11. A. E. Gurvich, "Analysis of the mechanism of antibody formation," in: Proceedings of the All-Union Scientific Technical Conference on the Application of Radio-Isotopes in the National Economy and in Science [in Russian], Moscow (1958), p. 55.
12. A. E. Gurvich, "On the biosynthesis of antibodies," in: Virology and Immunology [in Russian], Nauka, Moscow (1964), p. 243.
13. A. E. Gurvich and G. I. Drizlikh, "Effect of antibody present in the medium on the biosynthesis of similar antibodies by spleen cells," Dokl. Akad. Nauk SSSR, 155(2):482 (1964).
14. A. E. Gurvich and E. V. Sidorova, "Study of inhibition of antibody biosynthesis," Biokhimiya 29(3):556 (1964).
15. A. E. Gurvich, E. V. Sidorova, Hsü Fen, and A. E. Tumanova, "The presence in mitochondria of a factor which inhibits biosynthesis of antibodies and other cell proteins," Biokhimiya 30(2):429 (1965).

16. A. E. Gurvich, E. V. Sidorova, A. E. Tumanova, and Hsü Fen, "Effect of different subcellular granules and lipids isolated from them on the synthesis of antibodies and other proteins," Biokhimiya 30(5):1044 (1965).

17. A. E. Gurvich and N. P. Smirnova, "Changes in the antibody content and rate of incorporation of ^{14}C-glycine into antibody during immunization of an animal with two antigens," Biokhimiya 22(4):626 (1957).

18. G. A. Gurvich and G. V. Shumakova, "Immunological activity of lymphoid organs and general principles of immunogenesis," Vestn. Akad. Med. Nauk SSSR No. 1, p. 57 (1960).

19. G. I. Drizlikh, "Mutual relationship between biosynthesis of nonspecific γ-globulins and antibodies and the effect of antigen on these processes," Biokhimiya 30(4):743 (1965).

20. O. I. Epifanova, Hormones and Cell Proliferation [in Russian], Nauka, Moscow (1965).

21. E. E. Zhuravleva and I. P. Gorchakova, "Effect of antibiotics on antibody formation," Zh. Mikrobiol. Epidemiol. i Immunobiol. 30(6):14 (1959).

22. P. F. Zdorovskii, Problems of Infection and Immunity [in Russian], Medgiz, Moscow (1961).

23. A. V. Zelenin, "Pinocytosis," Usp. Sovrem. Biol. 53(3):364 (1962).

24. Yu. N. Zograf and M. D. Frank-Kamenetskii, "Nucleic acids and protein specificity. The problem of the code," in: Biosynthesis of Proteins and Nucleic Acids [in Russian], Nauka, Moscow (1965), p. 7.

25. L. I. Krasnoproshina and M. V. Dalin, "Content of nucleic acids in lymphoid tissue in immunization of rabbits with heated *Salmonella typhosa* vaccine," Byul. Eksperim. Biol. i Med. 58(9):92 (1964).

26. A. Ya. Kul'berg and N. B. Azadova, "Specific contrasting in electron microscopy with the aid of mercury-tagged antibodies," Vopr. Virusol. 8(1):100 (1963).

27. K. A. Lebedev, "Primary immunological response (study of the dynamics of the occurrence and the morphology of antibody-containing cells by Coons' indirect method)," Arkh. Patol. 27(2):6 (1965).

28. K. A. Lebedev, "Secondary immune response (study of the dynamics of the occurrence and the morphology of antibody-containing cells by Coons' indirect method)," Arkh. Patol. 27(5):30 (1965).

28a. E. A. Luriya, A. T. Nikolaeva, A. E. Gurvich, and A. Ya. Fridenstein, "Antibody synthesis induced in organ cultures of lymph nodes," Bull. Exp. Biol. Med. 64(7):73 (1967).

29. E. A. Luriya and A. Ya. Fridenstein, "Role of the thymus in immunity," Usp. Sovrem. Biol. 57(2):269 (1964).

30. V. A. Lyashenko, "Effect of cytostatic substances on the immunological reactivity of the organism," Vestn. Akad. Med. Nauk SSSR 20(7):72 (1965).

31. L. B. Mekler, S. M. Klimenko, G. E. Dobrezov, W. K. Naumova, J. P. Goffman, and V. M. Zhdanov, "Cytochemical and immunochemical analysis at the electron microscopy levels: obtaining contrasting antibodies by use of iodine," Nature 203(4096):717 (1964).

32. R. S. Nezlin, "Synthesis of 19 S and 7 S antibodies in a culture of rabbit lymph node and spleen cells," Biokhimiya 29(3):548 (1964).
33. R. S. Nezlin, "The change in the molecular weight of antibodies synthesized in tissue culture after second immunization," in: Molecular and Cellular Basis of Antibody Formation, Academic Press, New York (1965), p. 595.
34. R. S. Nezlin, "Incorporation of amino acids into proteins by microsomes of rabbit lymph nodes and spleen," Biokhimiya 31(3):516 (1966).
35. R. S. Nezlin and L. M. Kulpina, "Microglobulin synthesized in cell culture of the lymph nodes and spleen of the rabbit," Nature 212(5064):845 (1966).
35a. R. A. Nezlin and L. M. Kulpina, "Microglobulin synthesized in vitro and its identification by means of complexes of cellulose-fixed peptide chains of γG-globulin and corresponding antibodies," Biokhimiya 31(6):1257 (1966).
35b. R. S. Nezlin and L. M. Kulpina, "Cellulose-fixed immunoglobulin subunits," Immunochemistry 4(4):269 (1967).
35c. R. S. Nezlin and L. M. Kulpina, "Incorporation of radioactive amino acids into γ-globulin in a cell-free system from rat spleen," Biochim. Biophys. Acta 138(3):654 (1967).
36. R. V. Petrov, "Significance of radiology for the development of knowledge on infection and immunity," Usp. Sovrem. Biol. 50(2):174 (1960).
37. R.V. Petrov, "Contemporary trends in radiation immunology," Usp. Sovrem. Biol. 58(2):262 (1964).
38. M. P. Pokrovskaya, N. A. Kraskina, V. I. Levenson, N. M. Gutorova, and N. I. Braude, "Morphology and nomenclature of immunologically competent cells of lymphoid tissue," Zh. Mikrobiol. Epidemiol. i Immunobiol. 42(3):8 (1965).
39. M. P. Pokrovskaya, I. G. Makarenko, N. A. Kraskina, N. I. Braude, and N. M. Gutorova, "Morphological foundations of immunogenesis," in: Contemporary Problems of Immunology [in Russian], Institute of Experimental Medicine, Leningrad (1959), p. 20.
40. L. G. Prokopenko, "Synthesis of antibodies in transplants and tissue culture," Usp. Sovrem. Biol. 56(1):44 (1963).
40a. O. W. Rokhlin, "Effect N-nitrosoalkylurea antibody formation," Dokl. Akad. Nauk SSSR (1968 in press).
41. A. S. Spirin, Certain Problems on the Macromolecular Structure of Nucleic Acids, Academy of Sciences of the USSR, Moscow (1963).
42. A. S. Spirin, "The problem of protein biosynthesis," Vestn. Akad. Nauk. SSSR, No. 4, p. 51 (1965).
42a. B. D. Stanchev, "Antibody production in vitro in nonimmune mouse spleen cells induced by treatment with fractions of homologous RNA from immunized donors," Mol. Biol. 1(1):3 (1967) (USSR).
43. V. L. Troitskii, D. R. Kaulen, M. A. Tumanyan, A. Ya. Fridenstein, and O. V. Chakhava, Radiation Immunology, Meditsina, Moscow (1965).
44. I. Ya. Uchitel' and A. S. Konikova, "Certain data on antibody formation," Byul. Eksperim. Biol. i Med., No. 12, p. 35 (1955).
45. I. Ya. Uchitel' and A. S. Konikova, "Comparison of synthesis of antibodies and nonspecific proteins in the organism," Byul. Eksperim. Biol. i Med., No. 7, p. 85 (1957).

46. I. Ya. Uchitel' and A. S. Konikova, "Contribution to the problem of the mechanism of antibody formation," in: Mechanisms of Antibody Formation, Academic Press, New York (1960), p. 195.
47. I. Ya. Uchitel' and E. L. Khasman, "Rates of synthesis of specific and non-specific proteins of the organism at different stages of immunogenesis," in: Problems of Infectious Pathology and Immunology [in Russian], Medgiz, Moscow (1963), p. 55.
48. L. N. Fontalin, "On the relationship between the site of action of antigen in the organism and the site of antibody production (under conditions of immunization with tetanus antitoxin)," Com. IV, Zh. Mikrobiol. Epidemiol. i Immunobiol., No. 4, p. 66 (1960).
49. L. N. Fontalin, The Immunological Function of Lymphoid Organs and Cells, Meditsina, Leningrad (1967).
50. A. Ya. Fridshtein, "Lymphoid tissue as an organ of immunity," in: Current Problems of Immunology [in Russian], Meditsina, Moscow (1964), p. 97.
51. M. A. Frolova, M. V. Dalin, and N. P. Perepechkina, "Changes in the content of nucleic acids during immunogenesis," Zh. Mikrobiol. Epidemiol. i Immunobiol. 41(6):70 (1964).
52. B. B. Fuks, I. V. Konstantinova, and A. P. Tsygankov, "Immunological and histochemical study of biosynthesis of specific antibodies induced by RNA from cells of an immunized animal," Vestn. Akad. Nauk SSSR, 153(2):485 (1963).
53. B. B. Fuks, I. V. Konstantinova, L. E. Stefanovich, I. G. Lik'yanova, L. I. Tsygankobs, S. G. Kolaeva, I. M. Kross, and L. V. Van'ko, "Specific biosynthesis of antibodies induced by ribonucleic acid from lymph nodes and spleen of immune rabbits," Dokl. Akad. Nauk SSSR 153(2):485 (1963).
54. E. V. Chernokhvostova, "On a procedure of differentiation of macroglobulin (19 S) and γ-globulin (7 S) antibodies," Lab. Delo, No. 6, p. 323 (1965).
55. R. S. Shakulov, M. A. Aitkhozhin, and A. S. Spirin, "On the latent degradation of ribosomes," Biokhimiya 27(4):744 (1962).
56. G. V. Shumakova, "Role of lymphoid organs in the formation of antitoxin immunity," Byul. Eksperim. Biol. i Med. 51(3):80 (1961).
57. N. V. Engel'gardt, "Use of antibodies to γ-globulin in an indirect method of fluorescent antibodies," Byul. Eksperim. Biol. i Med. 57(1):67 (1964).
58. G. L. Ada, G. J. V. Nossal, and C. M. Austin, "Studies on the nature of immunogenicity employing soluble and particulate bacterial proteins," in: Molecular and Cellular Basis of Antibody Formation, Academic Press, New York (1965), p. 31.
59. H. Ainis, "Antibody production in vitro by means of a new perfusion chamber," Nature 194(4824):197 (1962).
60. L. J. Alfred, P. Corvazier, and P. Grabar, "Formation of antibody in tissue culture, effect of certain mucopolysaccharides and hexosamine," Nature 200(4907):698 (1963).
61. K. Altman and M. S. Tobin, "Suppression of the primary immune response induced by dl-penicillamine," Proc. Soc. Exp. Biol. Med. 118(2):554 (1965).
62. C. T. Ambrose, "The requirement for hydrocortisone in antibody-forming tissue cultivated in serum-free medium," J. Exp. Med. 119(6):1027 (1964).

63. C. T. Ambrose and A. H. Coons, "Studies on antibody production. VII. The inhibitory effect of chloramphenicol on the synthesis of antibody in tissue culture," J. Exp. Med. 117(6):1075 (1963).

64. M. Aronson, "Bridge formation and cytoplasmic flow between phagocytic cells," J. Exp. Med. 118(6):1083 (1963).

65. B. A. Askonas, "Protein synthesis in mammalian cells with particular reference to antibody formation," Rec. Trav. Chim. 77:611 (1958).

66. B. A. Askonas, "Studies on the mechanism of antibody secretion," in: Mechanism of Antibody Formation, Academic Press, New York (1960), p. 234.

67. B. A. Askonas, "Formation of globulins by plasma cell tumors transplantable in mice," in: Protein Biosynthesis, Academic Press, London-New York (1964), p. 363.

68. B. A. Askonas and J. H. Humphrey, "Formation of specific antibodies and γ-globulin in vitro. A study of the synthetic ability of various tissues from rabbits immunized by different methods," Biochem. J. 68(2):252 (1958).

69. B. A. Askonas and J. H. Humphrey, "Formation of antibody by isolated perfused lungs of immunized rabbits," Biochem. J. 70(2):212 (1958).

70. B. A. Askonas and J. M. Rhodes, "Immunogenicity of antigen-containing ribonucleic acid preparations from macrophages," Nature 205(4970):470 (1965).

71. B. A. Askonas, J. L. Simkin, and T. S. Work, "Protein synthesis in cell-free systems," Proc. 4th Internat. Congr. Biochem. Vienna 8:181 (1958).

72. B. A. Askonas and R. G. White, "Sites of antibody production in guinea pig. The relation between in vitro synthesis of antiovalbumin and γ-globulin and distribution of antibody containing plasma cells," Brit. J. Exp. Pathol. 37(1):64 (1956).

72a. B. A. Askonas and A. R. Williamson, "Biosynthesis of immunoglobulins. Free light chains as an intermediate in the assembly of G-molecules," Nature 211(5047):369 (1966).

72b. B. A. Askonas and A. R. Williamson, "Balanced synthesis of light and heavy chains of immunoglobulin G," Nature 216(5112):264 (1967).

73. G. Attardi, M. Cohn, K. Horibata, and E. S. Lennox, "Antibody formation by rabbit lymph node cells. I. Single cell responses to several antigens," J. Immunol. 92(3):335 (1964).

74. G. Attardi, M. Cohn, K. Horibata, and E. S. Lennox, "Antibody formation in rabbit lymph node cells. II. Further observations on the behavior of single antibody-producing cells with respect to their synthetic capacity and morphology," J. Immunol. 92(3):346 (1964).

75. G. Attardi, M. Cohn, K. Horibata, and E. S. Lennox, "Antibody formation by rabbit lymph node cells. III. The controls for microdrop and micropipet experiments," J. Immunol. 92(3):356 (1964).

76. G. Attardi, M. Cohn, K. Horibata, and E. S. Lennox, "Antibody formation by rabbit lymph node cells. IV. The detailed methods for measuring antibody synthesis in individual cells, the kinetics of antibody formation by rabbits and the properties of cell suspensions," J. Immunol. 92(3):373 (1964).

77. G. Attardi, M. Cohn, K. Horibata, and E. S. Lennox, "Antibody formation of rabbit lymph node cells. V Cellular heterogeneity in the production of antibody to T5," J. Immunol. 93(1):94 (1964).

78. D. C. Bauer, M. J. Mathies, and A. B. Stavitsky, "Sequences of synthesis of γ_1 macroglobulin and γ_2 globulin antibodies during primary and secondary responses to proteins, *Salmonella* antigens and phages," J. Exp. Med. 117(6):889 (1963).

79. D. C. Bauer and A. B. Stavitsky, "On the different molecular forms of antibody synthesized by rabbits during the early response to a single injection of protein and cellular antigens," Proc. Nat. Acad. Sci. USA 47(10):1667 (1961).

79a. M. J. Becker and A. Rich, "Polyribosomes of tissues producing antibodies," Nature 212(5058):142 (1966).

80. M. C. Berenbaum, "The antibody content of single cells," J. Clin. Pathol. 11(6):543 (1958).

81. M. C. Berenbaum, "The effect of cytoxic agents on the production of antibody to T. A. B. vaccine in the mouse," Biochem. Pharmacol. 11(1):29 (1962).

82. M. C. Berenbaum and I. N. Brown, "Dose-response relationships for agents inhibiting the immune response," Immunology 7(1):65 (1964).

83. K. Berglund, "Inhibition of antibody formation by prednisolone: location of a short sensitive period," Acta Pathol. Microbiol. Scand," 55(2):187 (1962).

84. K. Berglund, "Studies on the inductive phase of antibody formation: effects of corticosteroids and lymphoid cells," in: Molecular and Cellular Basis of Antibody Formation, Academic Press, New York (1965), p. 405.

85. W. Bernhard, Fine Structure of Normal and Malignant Human Lymph Nodes, Pergamon Press, New York (1965).

86. W. Bernhard and N. Granboulan, "Ultrastructure of immunologically competent cells," in: Ciba Foundation Symposium on Cellular Aspects of Immunity, Churchill, London (1960), p. 92.

87. G. M. Bernier and J. J. Cebra, "Polypeptide chains of human γ-globulin: cellular localization by fluorescent antibody," Science 144(3626):1590 (1964).

87a. G. M. Bernier and J. J. Cebra, "Frequency distribution of α, γ, κ, and λ polypeptide chains in human lymphoid tissues," J. Immunol. 95(2):246 (1965).

88. M. C. Bessis, "Ultrastructure of lymphoid and plasma cells in relation to globulin and antibody formation," Lab. Invest. 10(6):1040 (1961).

89. J. Borel, M. Tauconnet, and P. A. Miescher, "7 S versus 19 S anamnestic response in rabbits," Proc. Soc. Exp. Biol. Med. 117(2):603 (1964).

90. W. Braun and M. Nakano, "Influence of oligodeoxyribonucleotides on early events in antibody formation," Proc. Soc. Exp. Biol. Med. 119(3):701 (1965).

91. A. E. Bussard, "Biosynthesis of antibodies. Facts and theories," Ann. Rev. Microbiol. 13:279 (1959).

92. A. E. Bussard and J. L. Binet, "Electron micrography of antibody producing cells individualized by the technique of local hemolysis gel," in: Molecular and Cellular Basis of Antibody Formation, Academic Press, New York (1965), p. 477.

93. A. Bussard and V. A. Huynh, "Incorporation of labeled amino acids in antibodies synthesized in vitro by cells of immunized rabbits," Biochem. Biophys. Res. Commun. 3(5):453 (1960).

93a. A. E. Bussard and M. Lurie, "Primary antibody response in vitro in peritoneal cells," J. Exp. Med. 125(5):873 (1967).

94. W. T. Butler and A. H. Coons, "Studies on antibody production. XII. Inhibition of priming by drugs," J. Exp. Med. 120(6):1051 (1964).

95. D. H. Campbell and J. S. Garvey, "Nature of retained antigen and its role in immune mechanisms," Advan. Immunol. 3:261 (1963).

96. P. N. Campbell and B. A. Kernot, "The incorporation of ^{14}C-leucine into serum albumin by the isolated microsome fraction from rat liver," Biochem. J. 82(2):262 (1962).

97. H. F. Cheng, M. Dicks, R. H. Shellhamer, and F. Haurowitz, "Localization of antigens by autoradiography," J. Histochem. Cytochem. 6(6):400 (1958).

98. G. Chiappino and B. Pernis, "Demonstration with immunofluorescence of 19 S macroglobulins and 7 S γ-globulins in different cells of the human spleen," Pathol. Microbiol. 27(1):8 (1964).

98a. P. H. Chin and M. S. Silverman, "Studies on the transfer of myeloma protein synthesis with RNA isolated from the C3H plasma cell tumor (X 5563). II," J. Immunol. 99(3):489 (1967).

99. C. G. Cochrane and F. J. Dixon, "Antibody production by transferred cells," Advan. Immunol. 2:205 (1962).

100. E. P. Cohen and J. J. Parks, "Antibody production by nonimmune spleen cells incubated with RNA from immunized mice," Science 144(3621):1012 (1964).

100a. B. Colombo and C. Baglioni, "Regulation of hemoglobin synthesis at the polysome level," J. Mol. Biol. 16(1):51 (1966).

101. A. H. Coons, "The nature of the secondary response," in: Molecular and Cellular Basis of Antibody Formation, Academic Press, New York (1965), p. 559.

102. A. H. Coons, E. H. Ledue, and J. M. Connoly, "Studies on antibody production. I. A method for the histochemical demonstration of specific antibody and its application of a study of the hyperimmune rabbit," J. Exp. Med. 102(1):49 (1955).

103. P. A. Crabbe, A. O. Carbonara, and J. F. Heremans, "The normal human intestinal mucosa as a major source of plasma cells containing γA-immunoglobulins," Lab. Invest. 14(3):235 (1965).

104. A. Cruchaud and A. H. Coons, "Studies on antibody production. XIII. The effect of chloramphenicol on priming in mice," J. Exp. Med. 120(6):1061 (1964).

105. A. Cruchaud, F. S. Rosen, J. M. Craig, C. A. Janeway, and D. Gitlin, "The site of synthesis of the 19 S γ-globulins in dysgamma-globulinemia," J. Exp. Med. 115(6):1141 (1962).

105a. F. J. Dixon, H. Jacot-Guillarmod, and P. J. McConahey, "The effect of passively administered antibody on antibody synthesis," J. Exp. Med. 125(6):1119 (1967).

106. F. J. Dixon, P. H. Maurer, and M. P. Deichmiller, "Primary and specific anamnestic antibody response of rabbits to heterologous serum protein antigens," J. Immunol. 72(2):179 (1954).

107. F. J. Dixon, P. H. Maurer, M. P. Weigle, and M. P. Deichmiller, "Rates of antibody synthesis during first, second, and hyperimmune responses of rabbits to bovine γ-globulin," J. Exp. Med. 103(4):425 (1956).

108. F. J. Dixon, D. W. Talmage, and P. H. Maurer, "Radiosensitive and radioresistant phases in the antibody response," J. Immunol. 68(6):693 (1952).
109. G. Doria and G. Agarossi, "Effect of thymic action on the precursors of antigen-sensitive cells," Proc. Nat. Acad. Sci. USA 58(4):1366 (1967).
110. R. W. Dutton, A. H. Dutton, M. George, R. Q. Marstron, and J. H. Vaughn, "Phosphate metabolism of spleen cells in antibody formation," J. Immunol. 84(3):268 (1960).
111. R. W. Dutton, A. H. Dutton, and J. H. Vaughn, "The effect of 5-bromuracil deoxyriboside on the synthesis of antibody in vitro," Biochem. J. 75(2):230 (1960).
112. R. W. Dutton and J. D. Pearce, "A Survey of the effect of metabolic antagonists on the synthesis of antibody in an in vitro system," Immunology 5(3):414 (1962).
113. R. H. Egdahl, H. Gress, and J. A. Mannick, "An agar-gel diffusion technique for demonstrating antibody production by lymph node cells in tissue culture," Nature 196(4850):184 (1962).
114. W. E. Ehrich, D. L. Drabkin, and C. Forman, "Nucleic acids and the production of antibody by plasma cells," J. Exp. Med. 90(2):157 (1947).
115. H. N. Eisen, E. S. Simms, E. Helmreich, and M. Kern, "Incorporation of amino acids into a globulin-like protein by cell-free extracts from lymph nodes," Trans. Assoc. Am. Physicians 74:207 (1961).
116. A. Fagraeus, "Antibody production in relation to the development of plasma cells," Acta Med. Scand., Vol. 130, Suppl. 204 (1948).
117. A. Fagraeus, "Inhibition of the effect of cotisone on antibody formation," in: Mechanisms of Antibody Formation, Academic Press, New York (1959), p. 270.
118. J. L. Fahey and S. Sell, "The immunoglobulins of mice. V. The metabolic (catabolic) properties of five immunoglobulin classes," J. Exp. Med. 122(1):41 (1965).
119. J. D. Feldman, "Ultrastructure of immunologic processes," Advan. Immunol. 4:175 (1964).
120. M. Feldman, D. Elson, and A. Goberson, "Antibodies in ribonucleoproteins," Nature 185(4709):317 (1960).
121. M. S. Finkelstein and J. W. Uhr, "Specific inhibition of antibody formation by passively administered 19 S and 7 S antibody," Science 146(3640):67 (1964).
122. M. Fishman, "Antibody formation in tissue culture," Nature 183(4669):1200 (1959).
123. M. Fishman, "Antibody formation initiated in vitro. II. Antibody synthesis in X-irradiated recipients of diffusion chambers containing nucleic acid derived from macrophages incubated with antigen," J. Exp. Med. 117(4):595 (1963).
124. M. Fishman, R. A. Hammerstrom, and V. P. Bond, "In vitro transfer of macrophage RNA to lymph node cells," Nature 198(4880):549 (1963).
125. M. Fishman, J. J. Rood, and F. L. Adler, "The initiation of antibody formation by ribonucleic acid from specifically stimulated macrophages," in: Molecular and Cellular Basis of Antibody Formation, Academic Press, New York (1965), p. 491.

126. J. B. Fleischman, "Synthesis of the γ G heavy chain in rabbit lymph node cells," Biochemistry 6(5):1311 (1967).

127. R. E. Franzl, "Immunogenic subcellular particles obtained from spleens of antigen injected mice," Nature 195(4840):457 (1962).

128. P. C. Frei, B. Benacerraf, and G. J. Thorbecke, "Phagocytosis of the antigen, a crucial step in the induction of the primary responses," Proc. Nat. Acad. Sci. USA 53(1):20 (1963).

129. H. Friedman, "Persistance of antigen in nucleoprotein fractions of mouse spleen cells during antibody formation," Nature 199(4892):502 (1963).

130. H. Friedman, "Acquisition of antibody plaque forming activity by normal mouse spleen cells treated in vitro with RNA extracted from immune donor spleens," Biochem. Biophys. Res. Commun. 17(3):272 (1964).

131. H. Friedman, "Distribution of antibody plaque forming cells in various tissues of several strains of mice injected with sheep erythrocytes," Proc. Soc. Exp. Biol. Med. 117(2):526 (1964).

132 H. P. Friedman, A. B. Stavitsky, and J. M. Solomon, "Induction in vitro of antibodies to phage T2: antigens in the RNA extract employed," Science 149(3688):1106 (1965).

132a. Van R. Furth, "The formation of immune globulins by human tissues in vitro. II. Quantitative studies," Immunology 11(1):13 (1966).

132b. A. E. Gabrielsen and R. A. Good, "Chemical suppression of adaptive immunity," Advan. Immunol. 6:92 (1967).

132c. R. Gallily and M. Feldman, "The role of macrophage in the induction of antibody in x-irradiated animals," Immunology 12(2):197 (1967).

133. M. C. Ganoza, C. A. Williams, and F. Lipman, "Synthesis of serum proteins in a cell-free system from rat liver," Proc. Nat. Acad. Sci. USA 76(1):36 (1965).

134. J. S. Garvey and D. H. Campbell, "Studies on retention and properties of ^{35}S labelled antigen in livers of immunized rabbits," J. Immunol. 76(1):36 (1965).

135. J. Garvey, D. H. Campbell, and M. L. Das, "Urinary excretion of foreign antigens and RNA following primary and secondary injections of antigens," J. Exp. Med. 125(1):111 (1967).

136. B. D. Geller and R. S. Speirs, "Failure of actinomycin D to inhibit antitoxin production to a challenging injection of antigen," Proc. Soc. Exp. Biol. Med. 117(3):782 (1964).

137. D. S. Genthof and J. R. Battisto, "Antibody production in guinea pigs receiving 6-mercaptopurine," Proc. Soc. Exp. Biol. Med. 107(4):933 (1961).

138. R. V. Gilden, and G. L. Rosenquist, "Duration of antibody response to soluble antigen: incorporation of sulphur-35 methionine into normal and antibody globulin," Nature 200(4911):1116 (1963).

138a. A. Globerson and R. Auerbach, "Primary antibody response in organ cultures," J. Exp. Med. 124(5):1001 (1966).

138b. A. L. Goldstein, F. D. Slater, and A. White, "Preparation, assay, and partial purification of a thymic lymphocytopoietic factor (thymosin)," Proc. Nat. Acad. Sci. USA 56(3):1010 (1966).

139. P. Grabar and P. Corvazier, "Formation of antibodies in vitro," in: Ciba

Foundation Symposium Cellular Aspects of Immunity, 1960, Churchill, London (1960), p. 198.

140. H. Green and H. S. Anker, "On the synthesis at antibody protein," Biophys Acta 13(3):365 (1954).

140a. J. P. Gusdon, A. B. Stavitsky, and S. A. Armentraut, "Synthesis of G-antibody and immunoglobulin on polyribosomes in a cell-free system," Proc. Nat. Acad. Sci. USA 58(3):1189 (1967).

140b. S. Han Seong and A. G. Johnson, "Radioautographic and electron-microscopic evidence of rapid uptake of antigen by lymphocytes," Science 153(3732):176 (1966).

141. M. G. Hanna and C. J. Wust, "Actinomycin D effect on the primary immune response in mice. A histologic and serologic study," Lab. Invest. 14(3):272 (1965).

142. T. N. Harris and S. Harris, "Histochemical changes in lymphocytes during the production of antibodies in lymph nodes of rabbits," J. Exp. Med. 90(2):169 (1949).

143. T. N. Harris, K. Hummler, and S. Harris, "Electron microscopic observation on antibody-producing lymph node cells," J. Exp. Med. 123(1):161 (1966).

144. F. Haurowitz, "Immunochemistry," Ann. Rev. Biochem. 29:609 (1960).

145. F. Haurowitz and C. F. Crampton, "The fate in rabbits of intravenously injected I^{131} iodoovalbumin," J. Immunol. 68(1):73 (1952).

146. F. Haurowitz, H. H. Reller, and H. Walter, "The metabolic fate of isotopically labelled proteins, azoproteins and azohaptens," J. Immunol. 75(6):417 (1955).

147. J. D. Hawkins and F. Haurowitz, "The recovery of injected antigens from rat spleens," Biochem. J. 80(1):200 (1964).

148. E. Helmreich, M. Kern, and H. N. Eisen, "The secretion of antibody by isolated lymph node cells," J. Biol. Chem. 236(2):464 (1961).

149. E. Helmreich, M. Kern, and H. N. Eisen, "Observations on the mechanism of secretion of γ-globulins by isolated lymph node cells," J. Biol. Chem. 237(6):1925 (1962).

150. W. J. Holliday and J. S. Garvey, "Hydrocortisone and apparent induction period for antibody formation in vitro," Nature 202(4933):712 (1964).

151. M. Holub, "Morphology of antibody production by different cell systems," Folia Microbiol. 5(3):347 (1960).

152. M. Holub and I. Riha, "Morphological changes in lymphocytes cultivated in diffusion chambers during the primary antibody response to a protein antigen," in: Mechanisms of Antibody Formation, Academic Press, New York (1960), p. 30.

153. M. Hrubesova, "A study of the formation of antibody in newborn animals following transfer of nucleoprotein fractions," Folia Microbiol. 6(1):55 (1961).

154. M. Hrubesova, "The in vitro activity of the microsomal fraction isolated from the spleen and lymph nodes after immunization of the donor," Folia Microbiol. 11(5):347 (1966).

155. M. Hrubesova, B. A. Askonas, and J. H. Humphrey, "Serum antibody and γ-globulin in baby rabbits after transfer of ribonucleoprotein from adult rabbits," Nature 183(4654):97 (1959).

156. M. Hrubesova and Z. Trnka, "Failure to form antibodies following transfer

of nucleoprotein fractions to chick embryos," Folia Microbiol. 6(1):60 (1961).

157. T. Hultin, "Ribosomal functions related to protein synthesis," Intern. Rev. Cytol. 16:1 (1964).

158. T. Hultin and K. A. Abraham, "Amino acid incorporation and protein solubilization by rat liver ribonucleoprotein particles," Biochem. Biophys. Acta 87(2):232 (1964).

159. J. H. Humphrey, "Methods for detecting foreign antigen in cells and their sensitivity," in: Mechanisms of Antibody Formation, Academic Press, New York (1960), p. 44.

160. J. H. Humphrey, "The relationship between antigen and the specific immune response," in: Molecular and Cellular Basis of Antibody Formation, Academic Press, New York (1965), p. 21.

161. J. H. Humphrey and B. D. Sulitzeanu, "The use of ^{14}C-amino acids to study sites and rates of antibody synthesis in living hyperimmune rabbits," Biochem. J. 68(1):146 (1958).

161a. J. H. Humphrey, B. A. Askonas, J. Auzins, J. Schechter, and M. Sela, "The localization of antigen in lymph nodes and its relation to specific antibody-producing cells. II. Comparison of iodine-125 and tritium labels," Immunology 13(1):71 (1967).

162. W. B. Hunt and Q. N. Myrvik, "Demonstration of antibody in rabbit alveolar macrophages with failure to transfer antibody product," J. Immunol. 93(4):677 (1964).

163. T. Ishizaka, D. H. Campbell, and K. Ishizaka, "Internal antigenic determinants in protein molecules," Proc. Soc. Exp. Biol. Med. 103(1):5 (1960).

164. J. S. Ingraham, "Artificial radioactive antigens. I. Preparation and evaluation of S^{35}-sulfanilic acid-azo-bovine-γ-globulin," J. Infect. Diseases 89(2):109 (1951).

165. J. S. Ingraham, "Artificial radioactive antigens. II. The metabolism of S^{35}-sulfanilic acid-azo-bovine-γ-globulin in normal rabbits and mice," J. Infect. Diseases 89(2):117 (1951).

166. J. S. Ingraham, "Artificial radioactive antigens. III. S^{35}-sulfanil azo-sheep red cell stromata: Preparation and gross distribution in normal rabbits and mice," J. Infect. Diseases 96(2):105 (1955).

167. J. S. Ingraham, "Dynamic aspects of the formation of serum antibody in rabbits. Exponential and arithmetic phases in the rise of titer following a reinjection of sulfanil azo-bovine-γ-globulin," J. Immunol. 92(2):208 (1964).

168. J. S. Ingraham and A. Bussard, "Application of a localized hemolysis reaction for specific detection of individual antibody-forming cells," J. Exp. Med. 119(4):667 (1964).

169. B. D. Jankovic, K. Isakovick, and J. Horvat, "Failure of ribonucleoprotein to transfer antibody formation," Nature 183(4672):1400 (1959).

170. N. K. Jerne, "Studies on the primary immune cellular response in mice," in: Molecular and Cellular Basis of Antibody Formation, Academic Press, New York (1965), p. 459.

171. N. K. Jerne and A. A. Nordin, "Plaque formation in agar by single antibody-producing cells," Science 140:405 (1963).

172. A. Juhasz, B. Rose, and M. B. Maude, "Antibody formation. I. Relation be-

tween nucleic acids and antibody synthesis in the primary response in the rat," Can. J. Biochem. Physiol. 41(1):179 (1963).

173. E. A. Kabat, Experimental Immunochemistry, C. C Thomas, Springfield, Ill., (1961).

174. P. A. Kendall, "Labelling of thiolated antibody with mercury for electron microscopy," Biochim. Biophys. Acta 97(1):174 (1965).

175. M. Kern, E. Helmreich, and H. N. Eisen, "A demonstration of antibody activity on microsomes," Proc. Nat. Acad. Sci. USA 45(6):862 (1959).

176. J. Kern, E. Helmreich, and H. N. Eisen, "The solubilization of microsomal antibody activity by the specific interaction between the crystallizable fraction of γ-globulin and lymph node microsomes," Proc. Nat. Acad. Sci. USA 47(6):767 (1961).

177. J. J. Klein, A. L. Goldstein, and A. White, "Enhancement of in vivo incorporation of labelled precursors into DNA and total protein of mouse lymph nodes after administration of thymic extracts," Proc. Nat. Acad. Sci. USA 53(4):812 (1965).

178. P. M. Knopf, R. M. E. Parkhause, and E. S. Lennox, "Biosynthetic units of an immunoglobulin heavy chain," Proc. Nat. Acad. Sci. USA 58(6):2288 (1967).

179. M. Landy, R. P. Sanderson, M. T. Bernstein, and A. L. Jackson, "Antibody production by leukocytes in peripheral blood," Nature 204(4965):1320 (1964).

180. C. Lapresle and T. Webb, "Degradation of a protein antigen by intracellular enzymes," in: Ciba Foundation Symposium Cellular Aspects of Immunity 1960, Churchill, London (1960), p. 44.

181. M. F. La Via, S. A. Urius, and L. A. Ferguson, "Studies on antibody production by spleen explants maintained in vitro," J. Immunol. 84(1):48 (1960).

182. M. F. La Via, "Primary and secondary antibody response in rats to bacterial and sheep erythrocyte antigens," J. Immunol. 92(2):252 (1964).

183. M. F. La Via and S. Urius, "Antibody production by rat spleen explants in vitro," Arch. Pathol. 71(1):28 (1961).

183a. M. F. La Via, A. E. Valter, W. S. Hammond, and P. V. Northrup, "The nature of polysomes isolated from spleen cells of rats stimulated by antigen," Proc. Nat. Acad. Sci. USA 57(1):79 (1967).

184. M. F. La Via, A. Urius, N. D. Barber, and A. E. Warren, "Effect of β-3-thienylalanine on antibody synthesis. I. Preliminary in vitro and in vivo studies," Proc. Soc. Exp. Biol. Med. 104:562 (1960).

184a. V. Lazda and J. L. Starr, "The stability of messenger ribonucleic acid in antibody synthesis," J. Immunol. 95(2):254 (1965).

184b. P. Lonai, E. Kalman, and E. J. Hidvegi, "Demonstration of antibody on polysomes from the spleen of immunized animals," Abstracts III FEBS meeting, Academic Press, New York (1966).

185. E. H. Leduc, A. H. Coons, and J. M. Connolly, "Studies on antibody production. II. The primary and secondary responses on the popliteal lymph node of the rabbit," J. Exp. Med. 102(1):61 (1955).

185a. B. Mach and P. Vassalli, "Biosynthesis of RNA in antibody-producing tissues," Proc. Nat. Acad. Sci. USA 54(3):975 (1965).

185b. B. Mach and P. Vassalli, "Template activity of RNA from antibody-producing tissues," Science 150(3696):622 (1965).

185c. B. Mach, H. Koblet, and D. Gros, "Biosynthesis of immunoglobulin in a cell-free system," Cold Spring Harbor Symp. Quant. Biol., Vol. 32 (1968).

185d. H. O. McDevitt, B. A. Askonas, J. H. Humphrey, J. Schechter, and M. Sela, "The localization of antigen in relation to specific antibody-producing cells. I. Use of a synthetic polypeptide [TG-A-L] labelled with iodine-125," Immunology 11(4):337 (1966).

186. P. D. McMaster and S. S. Hudack, "The formation of agglutinins within lymph nodes," J. Exp. Med. 61(6):783 (1935).

187. P. D. McMaster and J. G. Kidd, "Lymph nodes as a source of neutralizing principle for vaccines," J. Exp. Med. 66(1):73 (1937).

188. P. D. McMaster and H. Kruse, "The persistence in mice of certain foreign proteins and azoprotein tracer-antigens derived from them," J. Exp. Med. 94(4):323 (1951).

189. P. D. McMaster, H. Kruse, E. Sturm, and J. L. Edwards, "The persistence of bovine γ-globulin injected as an antigen into rabbits. A comparison with its previously studied persistence in mice," J. Exp. Med. 100(4):341 (1954).

190. J. M. McKenna and K. M. Stevens, "Studies on antibody formation by peritoneal exudate cells in vitro," J. Exp. Med. 111(4):573 (1960).

191. T. Makinodan and W. J. Peterson, "Growth and senescence of the primary antibody-forming potential of the spleen," J. Immunol. 93(6):886 (1964).

192. G. Manner and B. S. Gould, "Ribosomal aggregates in γ-globulin synthesis in the rat," Nature 205(4972):670 (1965).

193. J. A. Mannick and R. H. Egdahl, "Ribonucleic acid in 'transformation' of lymphoid cells," Science 137(3534):976 (1962).

194. J. Marchalonis and G. M. Edelman, "Phylogenetic origins of antibody structure. I. Multichain structure in immunoglobulins in the smooth dogfish (Mustelus canis)," J. Exp. Med. 122(3):601 (1965).

195. R. C. Mellors and L. Korngold, "The cellular origin of human immunoglobulins (γ_2, γ_1M, γ_1A)," J. Exp. Med. 118(3):387 (1963).

195a. B. Merchant and T. Hraba, "Lymphoid cells producing antibody against single haptens: detection and enumeration," Science 152(3727):1378 (1966).

196. M. C. Michaelides and A. H. Coons, "Studies on antibody production. V. The secondary response in vitro," J. Exp. Med. 117(6):1035 (1963).

197. J. F. A. P. Miller, A. H. E. Marshall, and R. G. White, "The immunological significance of the thymus," Advan. Immunol. 2:111 (1962).

198. J. J. Miller, "An autoradiographic study of the stability of plasma cell ribonucleic acid in rats," J. Immunol. 93(2):250 (1964).

199. J. J. Miller and G. J. V. Nossal, "Antigens in immunity. VI. The phagocytic reticulum of lymph node follicles," J. Exp. Med. 120(6):1075 (1964).

200. J. Mitchell, "Autoradiographic studies of nucleic acid and protein metabolism in lymphoid cells. I. Differences amongst members of the plasma cell sequence. II. The stability and actinomycin sensitivity of rapidly formed RNA and protein," Australian J. Exp. Biol. Med. Sci. 42(3):347 (1964).

201. J. Mitchell and G. J. V. Nossal, "Ribonucleic acid metabolism in the plasma cell sequence," Nature 197(4872):1121 (1963).

202. G. Möller and H. Wigzell, "Antibody synthesis at the cellular level. Antibody

induced suppression of 19 S and 7 S antibody response," J. Exp. Med. 121(6):969 (1965).

203. S. Nakashima, S. Ohi, K. Tsukada, and H. Oura, "Polyribosomes in antibody-forming tissues of hyperimmunized rabbits," Biochim. Biophys. Acta 145(3):671 (1967).

204. H. C. Nathan, S. Bieber, G. B. Elion, and G. H. Hitchings, "Detection of agents which interfere with the immune response," Proc. Soc. Exp. Biol. Med. 107(4):796 (1961).

205. M. Nester, O. Mäkelä, and G. J. V. Nossal, "Studies on the transfer of nucleoprotein fractions from immunized spleens," Transplant. Bull. 28(2):479 (1961).

205a. S. S. Nicolas and A. G. Johnson, "γM-Synthesis during the secondary antibody response in mice," Proc. Exp. Biol. Med. 122(2):355 (1966).

205b. P. V. Northup, W. S. Hammond, and M. F. La Via, "The inhibition of spenic ribonuclease by liver cell extract," Proc. Nat. Acad. Sci. USA 57(2):273 (1967).

206. W. L. Norton, D. Lewis, and M. Ziff, "The effects of progressive immunization on polyribosomal size in lymphoid cells," Proc. Nat. Acad. Sci. USA 54(3):851 (1965).

207. G. J. V. Nossal, "Antibody production in tissue cultures," in: Cells and Tissues in Culture, Academic Press, New York, Vol. 3 (1966).

208. G. J. V. Nossal, G. L. Ada, and C. M. Austin, "Behavior of active bacterial antigens during the induction of the immune response," Nature 199(4900):1257 (1963).

209. G. J. V. Nossal, G. L. Ada, and C. M. Austin, "Antigens in immunity. IX. The antigen content of single antibody forming cells," J. Exp. Med. 121(6):945 (1965).

211. G. J. V. Nossal, J. Mitchell, and W. McDonald, "Autoradiographic studies on the immune response. IV. Single cell studies on the primary response," Australian J. Exp. Biol. Med. Sci. Spec. (Suppl.) 41:423 (1963).

210. G. J. V. Nossal and O. Mäkelä, "Elaboration of antibodies by single cells," Ann. Rev. Microbiol. 16:53 (1962).

212. G. J. V. Nossal, A. Szenberg, G. L. Ada, M. Austin, and M. Caroline, "Single cell studies on 19 S antibody production," J. Exp. Med. 119(3):485 (1964).

213. T. O'Brien and A. H. Coons, "Studies on antibody production. VII. The effect of 5-bromdeoxyuridine on the in vitro anamnestic antibody response," J. Exp. Med. 117(6):1063 (1963).

214. T. O'Brien, M. Michaelides, and A. H. Coons, "Studies on antibody production. VI. The course, sensitivity and histology of the secondary response in vitro," J. Exp. Med. 117(6):1053 (1963).

215. K. Ogata, S. Omori, R. Hirokana, and T. Takahashi, "Biosynthesis of serum albumin in the liver and of serum γ-globulin in the spleen during immunization of animals. Study on cell-free systems," in: Proceedings of the V. International Biochemical Congress Symposium II [in Russian], Academy of Sciences of the USSR, Moscow (1962), p. 194.

216. D. S. Pearlman, "The influence of antibodies in the immunologic response. I. The effect on the response to particulate antigen in the rabbit," J. Exp.

Med. 126(1):127 (1967).
217. B. Pernis and G. Chiappino, "Identification in human lymphoid tissues of cells that produce group 1 or group 2 γ-globulins," Immunology 7(5):500 (1964).
218. S. de Petris, S. Karlsbad, and B. Pernis, "Localization of antibodies in plasma cells by electron microscopy," J. Exp. Med. 117(5):849 (1963).
218a. E. Pick and J. D. Feldman, "Autoradiographic plaques for the detection of antibody formation to soluble proteins by single cells," Science 156(3777):964 (1967).
218b. R. J. Porter, "Secondary macroglobulin antibody responses to plasma proteins in rabbits," Proc. Soc. Exp. Biol. Med. 121(1):107 (1966).
219. M. Pospisil and F. Franek, "Cultivation of lymphatic cells in protein free medium and chemical study of the products," Folia Microbiol. 8(1):9 (1963).
220. Report of the Symposium Committee Concerning the Nomenclature of Cells Responsible for the Immune Reactions, in: Mechanisms of Antibody Formation, Prague (1959), p. 23.
221. A. Rich, J. R. Warner, and H. M. Goodman, "The structure and function of polyribosomes," Cold Spring Harbor Symp. Quant. Biol. 28:269 (1963).
222. M. Richardson and R. W. Dutton, "Antibody synthesizing cells: appearance after secondary antigenic stimulation in vitro," Science 146(3644):655 (1964).
223. M. Richter and F. Haurowitz, "Continuous synthesis of antibody after primary immunization with protein antigens," J. Immunol. 84(4):420 (1960).
224. M. Richter, S. Zimmerman, and F. Haurowitz, "Relation of antibody titer to persistence of antigen," J. Immunol. 94(6):938 (1965).
225. I. Riha and A. Sviculis, "Antibodies against hapten of 7 S and macroglobulin type in chickens," Folia Microbiol. 9(2):45 (1964).
226. A. Roberts, "Quantitative cellular distribution of tritiated antigen in immunized mice," Am. J. Pathol. 44(3):411 (1964).
227. S. Roberts and E. Adams, "Influence of mode of immunization on the relationship between the development of tissue titers and the release of hemolysins in vitro," J. Immunol. 62(2):155 (1949).
228. A. N. Roberts and F. Haurowitz, "Intercellular localization and quantitation of tritiated antigens in reticuloendothelial tissues of mice during secondary and hyperimmune responses," J. Exp. Med. 116(4):407 (1962).
229. R. Robineaux and J. Pinet, "An in vitro study of some mechanisms of antigen uptake by cells," Ciba Foundation Symposium on Cellular Aspects of Immunity, Churchill, London (1960), p. 5.
230. A. Saha, J. S. Garvey, and D. H. Campbell, "The physicochemical characterization of the ribonucleic acid–antigen complex persisting in the liver of immunized rabbits," Arch. Biochem. Biophys. 105(1):179 (1964).
231. K. Sahiar and R. S. Schwartz, "Inhibition of 19 S antibody synthesis by 7 S antibody," Science 145(3630):395 (1964).
232. M. D. Scharff and J. W. Uhr, "Functional ribosomal unit of gamma-globulin synthesis," Science 148(3670):646 (1965).
233. M. D. Schoenberg, V. R. Mumaw, R. D. Moore, and N. S. Weisberger, "Cytoplasmic interaction between macrophages and lymphocytic cells in antibody synthesis," Science 143(3609):964 (1964).

234. M.D. Schoenberg, A.B. Stavitsky, A. D. Moore, and M. J. Freeman, "Cellular sites of synthesis of rabbit immunoglobulins during primary response to diphtheria toxoid Freund's adjuvant," J. Exp. Med. 121(4):577 (1965).

235. J. C. Schooley, "Autoradiographic observations of plasma cell formation," J. Immunol. 86(3):331 (1961).

236. R. Schwartz, A. Eisner, and W. Damashek, "The effect of 6-mercaptopurine on primary and secondary immune responses," J. Clin. Invest. 38(8):1394 (1959).

236a. A. L. Shapiro, M. D. Scharff, J. V. Maizel, and J. W. Uhr, "Synthesis of excess light chains of γ-globulins by rabbit lymph node cells," Nature 211(5046):243 (1966).

236b. A. L. Shapiro, M. D. Scharff, J. V. Maizel, and J. W. Uhr, "Polyribosomal synthesis and assembly of the H and L chains of γ-globulin," Proc. Nat. Acad. Sci. USA 56(1):216 (1966).

237. J. D. Smiley, J. G. Heard, and M. Ziff, "Effect of actinomycin D on RNA synthesis and antibody formation in the anamnestic response in vitro," J. Exp. Med. 119(6):881 (1964).

238. J. D. Smiley and H. E. Jasin, "Incorporation of ^{14}C-hexoses into the carbohydrate of γ_2-globulin in vitro," Immunology 8(1):49 (1965).

239. R. Smith, "Response to active immunization of human infants during the neonatal period," Ciba Foundation Symposium on Cellular Aspects of Immunity, Churchill, London (1960), p. 348.

240. A. Solomon, J. L. Fahey, and R. A. Malmgren, "Immunohistologic localization of gamma-1-macroglobulins, β-2_A-myeloma proteins, 6.6 S gamma-myeloma proteins, and Bence–Jones proteins," Blood 21(4):403 (1963).

241. E. Sorkin and S. V. Boyden, "Studies on the fact of ^{131}I-trace labeled human serum albumin in vitro in the presence of guinea pig monocytes," J. Immunol. 82(4):332 (1959).

242. R. S. Speirs, "Antigenic material: persistence in hypersensitive cells," Science 140(3562):71 (1963).

243. R. S. Speirs, "Examination of the mechanism of antibody formation using nucleic acid and protein inhibitors," Nature 207(4995):371 (1965).

244. R. S. Speirs and E. E. Speirs, "Cellular localization of radioactive antigen in immunized and nonimmunized mice," J. Immunol. 90(4):561 (1963).

245. A. B. Stavitsky, "In vitro production of diphtheria antitoxin by tissues of immunized animals. I. Procedure and evidence for general nature of phenomena," J. Immunol. 75(3):214 (1955).

246. A. B. Stavitsky, "Studies on the in vitro and in vivo initiation of the secondary antibody response," in: Mechanisms of Antibody Formation, Academic Press, New York (1960), p. 295.

247. A. B. Stavitsky, "In vitro studies of the antibody response," Advan. Immunol. 1:211 (1961).

247a. A. B. Stavitsky and J. P. Gudsen, "Role of nucleic acids in the anamnestic antibody response," Bacteriol. Rev. 30(2):418 (1960).

248. A. B. Stavitsky and B. Wolf, "Mechanisms of antibody globulin synthesis by lymphoid tissue in vitro," Biochim. Biophys. Acta 27(1):4 (1958).

249. T. Steahelin, F.O. Wettstein, H. Oura, and H. Noll, "Determination of the

coding ratio based on molecular weight of messenger ribonucleic acid associated with ergosomes of different aggregate size," Nature 201(4916):264 (1964).

250. D. F. Steiner and H. S. Anker, "On the synthesis of antibody protein in vitro," Proc. Nat. Acad. Sci. USA 42(9):580 (1956).

251. K. H. Stenzel, W. D. Phillips, D. D. Thompson, and A. L. Rubin, "Functional ribosomes in antibody-producing cells," Proc. Nat. Acad. Sci. USA 51(4):636 (1964).

251a. K. H. Stenzel and A. L. Rubin, "Biosynthesis of γ-globulin studies in a cell-free system," Science 153(3735):537 (1966).

252. J. Sterzl, "The inductive phase of antibody formation," in: Mechanisms of Antibody Formation, Prague (1959), p. 107.

253. J. Sterzl, "Study of the inductive phase of antibody formation in tissue culture," Folia Microbiol. 4(2):91 (1959).

253a. J. Sterzl, "Inhibition of the inductive phase of antibody formation by 6-mercaptopurine studied by the transfer of isolated cells," Nature 185(4708):256 (1960).

254. J. Sterzl, "Effect of some metabolic inhibitors on antibody formation," Nature 189(4769):1022 (1961).

255. J. Sterzl, "Quantitative and qualitative aspects of the inductive phase of antibody formation," J. Hyg. Epidemiol. Microbiol. Immunol. Prague 7:323 (1963).

256. J. Sterzl and M. Hrubesova, "The transfer of antibody formation by means of nucleoprotein fractions to nonimmunized recipients," Folia Microbiol. 2(1):21 (1956).

257. J. Sterzl, J. Kostka, and A. Lanc, "Development of bacterial properties against gram-negative organisms in the serum of young animals," Folia Microbiol. 7(2):155 (1962).

258. J. Sterzl and L. Mandel, "Estimation of the inductive phase of antibody formation by plaque technique," Folia Microbiol. 9(3):173 (1964).

259. J. Sterzl and Z. Trnka, "Experimental determination of the clonal selection theory of antibody formation," Cs. Epidem. Mikrobiol. Immunol. 10(3):148 (1961).

260. B. Styk and L. Hana, "Formation of 19 S and 7 S types of viral antibodies. The role and nature of 'antibody cofactor,'" in: Molecular and Cellular Basis of Antibody Formation, Academic Press, New York (1965), p. 261.

261. S. E. Svehag, "Antibody formation in vitro by separated spleen cells: Inhibition by actinomycin or chloramphenicol," Science 146(3644):659 (1964).

262. S. E. Svehag and B. Mandel, "The formation and properties of polio virus neutralizing antibody. I. 19 S and 7 S antibody formation: differences in kinetics and antigen dose requirement for induction," J. Exp. Med. 119(1):1 (1964).

262a. R. M. Swenson and M. Kern, "The synthesis and secretion of γ-globulins by lymph node cells. I. The microsomal compartmentalization of γ-globulins," Proc. Nat. Acad. Sci. USA 57(2):417 (1967).

262b. R. M. Swenson and M. Kern, "Synthesis and secretion of γ-globulin by lymph node cells. II. The intracellular segregation of amino acid-labeled and carbohydrate-labeled γ-globulin," J. Biol. Chem. 242(13):3242 (1967).

262c. N. Talal, "Polyribosomes and protein synthesis in the spleen," J. Biol. Chem. 241(9):2067 (1966).

263. W. H. Taliaferro and L. G. Taliaferro, "Further studies on the radiosensitive stages in hemolysin formation," J. Infect. Diseases 95(2):134 (1955).

264. W. H. Taliaferro, L. G. Taliaferro, and E. T. Jansen, "The localization of X-ray injury to the initial phases of antibody response," J. Infect. Diseases 91(2):105 (1952).

265. W. H. Taliaferro, L. G. Taliaferro, and B. N. Jaroslow, Radiation and Immune Mechanisms, Academic Press, New York (1964).

266. W. H. Taliaferro and D. W. Talmage, "Absence of amino acid incorporation into antibody during the induction period," J. Infect. Diseases 97(1):88 (1955).

267. D. W. Talmage, F. J. Dixon, S. C. Bukantz, and G. J. Dammin, "Antigen elimination from the blood as an early manifestation of the immune response," J. Immunol. 67(4):243 (1951).

267a. T. W. Tao and J. W. Uhr, "Primary-type antibody response in vitro," Science 151(3714):1096 (1966).

267b. S. Tawde, M. D. Scharff, and J. W. Uhr, "Mechanisms of γ-globulin synthesis," J. Immunol. 96(1):1 (1966).

268. J. P. Thiery, "Microcinematographic contributions to the study of plasma cells," Ciba Foundation Symposium on Cellular Aspects of Immunity, Churchill, London (1960), p. 59.

269. M. L. Tyan, "Thymus: role in maturation of fetal lymphoid precursors," Science 145(3635):934 (1964).

270. J. W. Uhr, "The heterogeneity of the immune response," Science 145(3631):457 (1964).

271. J. W. Uhr, "Actinomycin D. Its effect on antibody formation in vitro," Science 142(3598):1476 (1963).

272. J. W. Uhr and J. B. Baumann, "Antibody formation. II. The specific anamnestic antibody response," J. Exp. Med. 113(5):959 (1961).

273. J. Uhr and M. Finkelstein, "Antibody formation. IV. Formation of rapidly and slowly sedimenting antibodies and immunological memory to bacteriophage φX174 in guinea pigs," J. Exp. Med. 117(3):457 (1963)

274. J. W. Uhr, M. S. Finkelstein, and J. B. Baumann, "Antibody formation. III. The primary and secondary antibody response to bacteriophage φX174 in guinea pigs," Proc. Soc. Exp. Biol. Med. 111(1):13 (1962).

275. J. W. Uhr, M. S. Finkelstein, and E. C. Franklin, "Antibody response to bacteriophage φX174 in non-mammalian verterbrates," Proc. Soc. Exp. Biol. Med. 111(1):13 (1962).

276. J. H. Vaughn, A. N. Dutton, R. W. Dutton, M. George, and R. Q. Martson, "A study of antibody production in vitro," J. Immunol. 84(3):258 (1960).

277. J. J. Vazquez, "Antibody or γ-globulin forming cells as observed by the fluorescent antibody technic," Lab. Invest. 10(6):1110 (1961).

277a. E. W. Voss, D. S. Bauer, and W. R. Finnerty, "Rapidly sedimenting ribosomes from immunized rabbits," Life Sci. 4(7):731 (1965).

278. J. R. Warner, R. M. Knopf, and A. Rich, "A multiple ribosomal structure in protein synthesis," Proc. Nat. Acad. Sci. USA 49(1):122 (1963).

279. W. O. Weigle, "The elimination of heterologous serum proteins from the blood of animals," in: Mechanisms of Antibody Formation, Prague (1959), p. 53.

279a. E. Weiler, "Differential activity of allelic γ-globulin genes in antibody-producing cells," Proc. Nat. Acad. Sci. USA 54(6):1965 (1965).

280. A. S. Weisberger, T. M. Daniel, and A. Hoffman, "Suppression of antibody synthesis and prolongation of homograft survival by chloramphenicol," J. Exp. Med. 120(2):183 (1964).

280a. I. L. Weissman, "Thymus cell migration," J. Exp. Med. 126(2):291 (1967).

281. H. J. Wellensick and A. H. Coons, "Studies on antibody production. IX. The cellular localization of antigen molecules (ferritin) in the secondary response," J. Exp. Med. 119(4):685 (1964).

282. H. Wigzell, "Antibody synthesis at the cellular level. Antibody-induced suppression of 7 S antibody synthesis," J. Exp. Med. 124(5):953 (1966).

282a. A. R. Williamson and B. A. Askonas, "Biosynthesis of immunoglobulins: the separate classes of polyribosomes synthesizing heavy and light chains," J. Mol. Biol. 23(2):201 (1967).

283. R. W. Wissler, F. W. Fitch, and M. F. La Via, "The reticuloendothelial system in antibody formation," Ann. N. Y. Acad. Sci. 88(1):134 (1960).

284. B. Wolf and A. B. Stavitsky, "In vitro production of diphtheria antitoxin by tissues of immunized animals. II. Development of synthetic medium which promotes antibody synthesis and the incorporation of radioactive amino acids into antibody," J. Immunol. 81(5):404 (1958).

284a. C. J. Wust, "Studies on ribosomes of rat spleen during the immune response," Arch. Biochem. Biophys. 118(3):568 (1967).

285. C. J. Wust and L. Astrachan, "RNA metabolism during antibody synthesis," Federation Proc. 21:380 (1962).

286. C. J. Wust, C. L. Gall, and G. D. Novelli, "Actinomycin D. Effect on immune response," Science 143(3610):1041 (1964).

287. C. J. Wust and M. G. Hanna, Jr., "Time relationship of injection of actinomycin D and antigen to the immune response," Proc. Soc. Exp. Biol. Med. 118(4):1027 (1965).

288. C. J. Wust and G. D. Novelli, "Ribonucleoprotein-bound antibody to the enzyme triose phosphate dehydrogenase," J. Immunol. 90(5):734 (1963).

289. C. J. Wust and G. Novelli, "Cell free amino acid incorporated by rat spleen ribosomes," Arch. Biochem. Biophys. 104(1):185 (1964).

290. D. Zucker-Franklin, E. C. Franklin, and N. S. Cooper, "Production of macroglobulins in vitro and a study of their cellular origin," Blood 20(1):56 (1962).

Chapter VI

Genetic Aspects of Antibody Synthesis

One of the most important achievements of modern biology has been the establishment of the basic stages of protein biosynthesis in cells [17]. It is now generally accepted that the structure of all proteins formed during the development of the organism is determined by the sequence of nucleotides in structural genes of DNA [8]. The activity of these genes can be renewed or curtailed according to necessity at different times during the life of a cell. This regulation is apparently effected by means of special repressors – substances of unknown nature which are able to inhibit synthesis of messenger RNA on DNA [15, 82].

Antigen is the principal regulatory factor in antibody biosynthesis: the inception of antibody formation is determinable only after addition of this alien substance. The intensification of synthesis of an adaptive enzyme after entry of the inducing agent into the organism is one more example of how synthesis of a specific protein is sharply influenced by external factors. In this case, it has been ascertained that the essential mechanism of induction involves reactivation of the relevant sections of the DNA molecule [82]. As a result of a study of this process, a scheme has been newly developed for the regulation of protein synthesis. As will be shown in more detail in Chapter VII, Szilard and many other investigators have suggested that similar processes take place in cells after addition of antigen. However, it is still unclear how far this superficial analogy between synthesis of adaptive enzymes and antibody synthesis holds [103]. Hence, before attempting to discuss current theories of antibody formation, it is necessary to examine any evidence which may provide us with information on the dependence of synthesis of antibodies and γ-globulins on DNA. Several recently published reports have also dealt with this problem [3, 6, 18, 19].

EFFECT OF HEREDITARY FACTORS ON SYNTHESIS OF γ-GLOBULINS AND ANTIBODIES

Hypo- and Agammaglobulinemia

Hypo- and even agammaglobulinemia have been encountered with increasing frequency in recent years by usual methods, such as paper electrophoresis, according to published reports [69, 81, 115]. Patients with such conditions usually suffer from frequent infections and in immunization display a very weak antibody-forming capacity or none at all. As data have accumulated it has become apparent that hypo- and agammaglobulinemia have a variety of causes. One of the classifications of this group of symptoms according to the mechanism of occurrence is given below [81].

Incapable of synthesis

1. Transient in childhood
   Normal, physiological
   Persistent
2. Congenital
   Familial (recessive, sexually determined)
   Familial (autosomic, recessive, or dominant)
   Sporadic
3. Acquired
   Idiopathic
   Nutritionally dependent
   Associated with lymph tissue disease
   a) Drug or radiation damage
   b) Lymphoma
   c) Other diseases

Advanced disturbance or loss

1) Nephrotic syndrome
2) Essential hypercatabolic hypoproteinemia

A committee of experts of the World Health Organization (WHO) proposed the following classification of diseases with primary immunological deficiency [66b]:

1. Infantile sex-linked recessive agammaglobulinemia (Bruton disease)
2. Selective immunoglobulin deficiency
3. Transient hypogammaglobulinemia of infancy
4. Non-sex-linked primary immunoglobulin deficiency with variable onset and expression (primary immunoglobulin aberrations)
5. Thymoma with agammaglobulinemia (Good's syndrome)
6. Thrombopenia, immune deficiency, eczema (Wiskott–Aldrich syndrome)
7. Ataxia telangecstasia (Mrs. Louis Barr's syndrome)
8. Hereditary lymphopenic immunological deficiency (Glauzman and Riniker's lymphocytophthisis, Swiss-type agammaglobulinemia)
9. Autosomal recessive lymphopenia with normal plasma cells and immunoglobulins (Nezelof's syndrome)
10. Autosomal recessive alymphocytic agammaglobulinemia
11. Thymic aplasia (Di George's syndrome)

In most cases, the contents of all types of immunoglobulins in the blood are reduced in such patients; but cases are known in which one or two types of immunoglobulins could not be detected at all [6].

Of the problems dealt with in this chapter, the most interesting are cases of congenital immunoglobulin deficiency. However, it must be kept in mind that even the acquired form may also have as its basis some congenital defects [36, 63]. What, then, are the possible causes of these defects in immunoglobulin synthesis? Fudenberg et al. [63] advanced the following hypothesis. Since the papain-cleaved fragment Fc of every immunoglobulin has its own characteristic features, it is quite reasonable to assume that this chain section (or independent chain) is controlled by a special locus peculiar to each protein. Individual variations in these loci may lead to reduced or completely suppressed synthesis of the corresponding immunoglobulins. The existence of this locus has been demonstrated for γG-globulins (Gm locus), as will be shown below, but for other globulins it has only been postulated. Mutation of the locus Inv, on which depends the synthesis of part of the light chain (see below), leads to deficient synthesis of all immunoglobulins, since the light chains in all immunoglobulins are the same.

Since plasma cells (the chief antibody producers) cannot be detected in cases of immunoglobulin deficiency, it may be asked whether this phenomenon can be explained in another way, namely, by the inability of the transformation of lymphocytes into these cells. It has been found that nonspecific stimulants, such as phytohemagglutinin and streptolysin, cause the same morphological transformations of lymphocytes into large cells, and then, into in both plasma cells and lymphocytes of healthy individuals [64]. However, synthesis of both DNA and RNA in lymphocytes of patients with agammalobulinemia was reduced [58a]. One of the possible causes of inborn immunoglobulin deficiency could be disorders in the activity of the respective genes, which apparently should be classified as hypomorphic genes, i.e., genes whose activity can be partially but not completely suppressed. This can explain both an inborn reduced quantity of immunoglobulins, as well as their total absence.

Inheritable Variations in Antibody Synthesis

As has been shown, immunoglobulin deficiency can be inherited. Its cause is possibly a change in the structural genes of these proteins. It would therefore be valuable to ascertain whether antibody synthesis was dependent on genetic factors.

In this context, the elaboration of antibodies by pure lines of different animal species was studied. It has been found, for example, that marked variations in the degree of immune response to pneumococcal polysaccharides, to ovalbumin, and to staphylococcal toxoid [1] exist among different strains of inbred mice [57]. In rabbits, the capacity to form antibody to protein of the tobacco mosaic virus is also clearly inherited [124]. This point has been studied in detail by Petrov, Man'ko, and Egorov [9, 10] who investigated formation of leptospira agglutinins in ten strains of mice. The two mouse strains C3H and C57BL exhibited sharp differences with respect to their agglutinin-forming capacity. The agglutinin titer was several times higher for the second strain. However, these differences disappeared after a second antigen injection.

Generally similar data were obtained by Dineen [41] who studied the formation of agglutinins to sheep erythrocytes in nine different mouse strains. It was found that the strain contributed 12.8% to the total variability of the titers (i.e., variability from all causes), while variability within the strain contributed only 2.7%. Even more distinct results were obtained by Rokhlin [12] who found that genetic factors contributed 73% to variability. The differences in the titers increased after a secondary immunization. However, the sharpest differences in antibody synthesis were observed after immunization of inbred mice with rat erythrocytes. It was found that CBA and CC57BR mice reacted identically to primary immunization, while after a secondary immunization, CC57BR mice, in contrast to CBA mice, showed an absence of hemolysins in the blood and a sharply reduced number of antibody-producing cells in the spleen [13]. An analysis of three generations of interstrain hybrids showed that the almost total lack of secondary response in CC57BR mice was determined by two dominant reactive genes designated AR and AR' [13a].

Ginzberg-Kalinin [2] has published some very interesting data. In a study of the titers of hemolysins to sheep erythrocytes in identical and nonidentical twins it was found that the titers coincided in 85% of the identical twins, but in only 17% of the nonidentical twins. A determination of natural agglutinins to staphylococcus, dysentery bacteria, and *Salmonella typhosa* gave similar results.

Arquilla and Finn [22] studied the immune response of rabbits and guinea pigs to insulin injection. The experimental procedure was as follows: first insulin linked to cellulose was prepared, and then this immunoadsorbent was treated with rabbit antiserum to insulin. This treatment blocked some antigenic determinants of insulin but not all, since antibodies present in the immune serum from one guinea pig strain were able to combine with insulin. However, animals of another strain were totally incapable of elaborating this type of antibody. Apparently, different guinea pig strains are capable of producing antibodies to different antigenic determinants of insulin.

Inbred mice strains gave different immune response to different variants of influenza virus, as has been shown by Lennox [87a]. Some of them responded poorly to all variants of the virus, although they were immunized well by other antigens.

Sobey et al. [131a] developed a mouse strain of which almost 90% were nonreactive by selective crossbreeding. Nonreactivity was recessive and was probably governed by two or three genes rather than one.

However, in the enumerated experiments a detailed analysis was hindered by the complexity of the antigens used. Hence, investigations in which substances with a known structure were used as antigens are of particular interest. Benacerraf [29, 89-91] et al. used polylysine conjugated with the dinitrophenyl group as antigen. Only about one-fourth of random guinea pigs formed antibody in response to injection of this antigen. In an experiment using pure strains it was found that certain guinea pig strains were totally incapable of an immune response to DNP-polylysine, but that other strains responded with antibody synthesis in 100% of the cases. In genetic studies it was found that formation of antibody to hapten is controlled by a single Mendelian gene. It was also found that these strains did not differ in their capacity to leave DNP-polylysine. However, the results of these experiments do not warrant postulation of the existence of a defect in structural genes of antibody polypeptide chains since the animals which did not form antibodies to the DNP group injected simultaneously with polylysine elaborated anti-DNP antibodies when a DNP group conjugated with proteins was injected. Thus, Benacerraf's experiments rather provided evidence of an unknown genetically controllable metabolic stage which is necessary for antibody synthesis in response to DNP-polylysine or complexes of other haptens with polylysine.

McDevitt and Sela [95a,b] studied the immune response in mice to injection of a branched multichain synthetic polypeptide — poly[tyr, glu]-poly-ala-polylysine. It was found that C57 mice developed a ten times stronger response than CBA mice, although both lines had the same response to injection of bovine albumin. As a result of crossbreeding it was found that the antibody synthesizing capacity is determined chiefly by an autosomic factor. When histidine residues were substituted for tyrosine, CBA mice exhibited a good antibody-synthesizing capacity, while that of C57 mice was weak. Evidently, this genetic control is specific. The capacity to respond to the polymer is transferred by spleen cells [66b]. It is instructive that the capacity for an immune response to both polymers is associated chiefly with the locus of histocompatibility (H_2). The genetic factor controlling synthesis of antibodies of multichain polypeptide antigens, however, differs from the genetic factor controlling synthesis of antibodies to a linear copolymer of glutamyllysine with 5% alaninc, inasmuch as in the latter case no association with the locus of histocompatibility was found [66b]. Besides the above, there are many other observations indicating the presence of an immune response to synthetic antigens in inbred lines of animals [27a, 27b, 115a].

Consequently, the organism does possess certain hereditary characteristics which profoundly influence antibody biosynthesis. However, the concrete mechanisms of this process are unclear [3, 66b]. To ascertain the site of action of the genetic control in the long chain of events occurring after antigen injection and leading ultimately to the synthesis of specific antibodies we may enlist the assistence of the criteria proposed in a report by a group of experts of the World Health Organization on the genetics of the immune response [66b]. The study of the effect of hormones is also of interest – for example, the reduced antibody synthesis in pregnant or lactating rabbits [124].

The rate of synthesis of different types and subtypes of immunoglobulins also varies among different mice lines. Thus, after immunization with hemocyanin and ferritin the macroglobulin content increases in some mice lines; in others the γA-globulin increases, and in still others the 7 S-γ_2-globulin content rises [28a].

Immunoglobulins in Newborn Animals

It would be of great value to ascertain whether animals acquire the ability to synthesize immunoglobulins only after birth (most likely as a result of contact with different antigens) or whether they have this capacity already at birth.

Experiments have shown that many animals have this capacity at birth. For example, in rabbits radioactive methionine was incorporated into γ-globulins within the first 12 hr after birth [143]. A careful study is always able to detect synthesis of γ-globulins in animals raised under sterile conditions [128]. However, the possibility of contact with alien substances during intrauterine life or even during maintenance of animals under sterile conditions cannot be excluded [145]. Hence, experiments on newborn pigs are of special significance. Because of morphological peculiarities of the placenta, these animals cannot come in contact with maternal antigens during the embryonic period. Sterzl conducted a detailed study of these animals [139]. As has been pointed out in Chapters III and IV, the blood of newborn pigs contains only a small quantity of protein with the antigen properties of γ-globulin of adult animals. The molecule of this protein is smaller than the γG-globulin molecule of adults. An analysis of peptide maps shows only a partial similarity between the subunits of this protein and heavy chains of common γG-globulin. At the same time, peptides characteristic of γG-globulin light chains are totally lacking. A careful examination failed to discover antibodies to any kind of tissue, virus, or bacterial antigens. Moreover, newborn pigs, kept under rigidly sterile conditions, began to elaborate antibody macroglobulins after immunization. It appears that at least in young pigs synthesis of the principal types of immunoglobulins begins only after antigenic stimulus. Further research into this problem would, of course, be of great value.

The exact nature of so-called natural antibodies is unknown [32]. They are most probably formed in response to alien antigens entering the organism. Apparently, isoagglutinins are formed in the same way, since group isoantigens are quite prevalent in nature [4, 7, 132].

GENETICALLY CONTROLLABLE ANTIGENIC DETERMINANTS OF IMMUNOGLOBULINS (Allotypes)

As was seen in Chapter III, immunoglobulins also possess antigenic determinants, the properties of which are determined by the genetic peculiarities of the given animal or human. These determinants, designated allotypic by Oudin [111-113] who first described them, are easy to detect by immunization of an animal of the same species but with different genetic characteristics with the given immunoglobulin. However, allotypic determinants can also be detected by heteroimmunization [31]. To detect allotypes, different precipitation procedures are usually used, such as with gels [88, 111], precipitation of ^{131}I- or ^{125}I-labeled immunoglobulin by antiserum to the given allotype, or inhibition of precipitation of a known labeled protein with a known antiserum by addition of the globulin being investigated [45, 79].

The most research has been devoted to allotypes of rabbits, humans, and mice, the properties of which will be described later. Hereditary antigenic variations of immunoglobulins have also been discovered in other animals, such as guinea pigs [30], chickens [130], pigs [117, 141b], etc.

Rabbit Allotypes

At present, seven allotypic determinants controlled by two loci have been studied. Three of them – Aa1, Aa2, and Aa3 – are controlled by the alleles Aa^1, Aa^2, and Aa^3 of one locus a, and the four others (Ab4, Ab5, Ab6, and Ab9) are controlled by the alleles Ab^4, Ab^5, Ab^6, and Ab^9 of the other locus b [43, 47a, 82a]. The A8 specificity is closely associated with the Aa1 specificity [72a, 72b]. Several other allotypic characteristics have been described but as yet have been little studied [82a, 113a]. For example, it is possible that the allele P (A7) is associated with still a third locus.

Genetic studies have shown that the transmission of allotypic properties by heredity strictly follows Mendel's law [46] and that their inheritance is independent of sex. Genotypic properties, controlled by both loci, are independently transmitted; this is evidence that the two loci are not closely interrelated [45, 71].

The research of several groups of investigators has shown that most (but not all) γ-globulin molecules have determinants controlled by the a locus [43]. Allelic determinants, controlled by one locus, i.e., Aa1, Aa2, or Aa3, or by Ab4, Ab5, and Ab6 are not found in the same γ-globulin molecule [43, 44, 66]. It should be noted that not all γ-globulin molecules necessarily contain determinants of both loci; in a small percentage, type a or type b determinants may be completely lacking [43, 56]; these molecules may possibly have other, as yet undiscovered, specificities.

Studies which have established the localization ot allotypic determinants in peptide chain and papain-cleaved fragments of γ-globulin are of prime significance. Determinants of both loci have been found only in Fab fragments [43, 56]. In one molecule, the two univalent fragments are symmetrical with respect to allotypic specificity; i.e., they contain the same determinants [67]. It can now be deemed certain that the determinants of the b locus are situated only in the light peptide chains, while determinants of the a locus are only in heavy chains [138].

The determinant A8 is apparently not located in the same place as other determinants of heavy chains [72a, 72b]. The determinants controlled by the loci a and b are found in all three types of immunoglobulins – γG, γA, and γM [91a, 141]. Macroglobulins have also additional allotypic characteristics [82a, 83]. At present, eight allelic variants of this allotype (Ms I-8) have been found which are apparently associated with at least three loci. It is very interesting that Ms I specificity is always found together in an individual serum with Aa3 specificity, while Ms 3 specificity is always together with Ab4.

Allotypic antigenic variations are apparently the reflection of essential differences in the structure of the corresponding peptide chains. Thus, in a study of the amino acid composition of light chains of the allotypes Ab^4 and Ab^5 differences in 16 amino acid residues involving 7-8 different amino acids were found [118]. The peptide charts of these chains had a whole series of spots which were different [131]. From these experiments it follows that allotypic determinants of the same locus have different primary structures. The pronounced electrophoretic heterogeneity of light chains (cf. Chapter IV) is not associated with allotypic genetic characteristics of the molecule, since for both allotypic variants of light chains the number of bands produced by electrophoresis in a polyacrylamide gel was the same.

Studies on recombination of chains and half-molecules have shown that allotypic antigenic properties are no hindrance to combination of chains and subunits to form a whole molecule. Thus, for example, half-molecules obtained from molecules with Aa3 and Aa4 specificity or from Aa1-Aa5-specific molecules partially combine to form a molecule with the composition Aa1-Aa3-Ab4-Ab5, i.e., a hybrid heterozygotic molecule which is not found in the blood of rabbits [129]. It is not unlikely that the heavy-chain fragments responsible for formation of a dimer are not associated with those fragments which determine allotypic properties. It has also been ascertained that light chains with the determinants Ab4 and Ab5 recombine equally well with the same heavy chains, for example, Aa3 [97]. The light-chain fragments responsible for the bond with heavy chains evidently are also distinct from allotypic determinants.

Little is presently known about the biosynthesis of rabbit allotypes. With the aid of an immunofluorescent tag it has been shown that both allelic determinants can be simultaneously detected in 99% of all stained rabbit lymph node cells with the allotypic characteristics Ab4 and Ab5 [38].

However, careful investigations of Pernis et al. [114a] failed to confirm these data. By the use of immunofluorescence methods it has been ascertained that allotypes determined by allelic genes (Aa1 and Aa2, Ab4 and Ab5, Ab4 and Ab6) are synthesized by different plasma cells in heterozygous rabbits. Both of the two allotypes studied in this experiment could be found in the same cells only in germinal centers of lymphoid follicles. It appears from these experiments that only one allelic gene functions in each cell. From this point of view it is obvious why two different allelic determinants are not encountered in the same γG-globulin molecule. In a certain percentage of cases of homozygous rabbits, cells containing immunoglobulins have been studied having none of the allotypic determinants in either locus. It is probable that the synthesis of peptide chains of such proteins is controlled by some other loci.

It is important to note that in immune rabbits antibody molecules may have less allotypic determinants than specific γ-globulin of the same rabbit [66], or else the proportion of determinant may be different in antibodies and in γG-globulin [119].

Genotypic Factors of Human Immunoglobulins

In 1956, Grubb discovered genetically controlled subclasses of human γ-globulins [72]. Subsequent comprehensive investigations showed that these subclasses differed in their antigenic properties and, consequently, were analogous to rabbit allotypes [20, 23, 65].

The detection of human allotypes was carried out by an indirect procedure, the principles of which can be derived from Fig. 52. First, incomplete γG-antibodies with known allotypic characteristics [Gm(a+) in the given example] and incapable of causing agglutination were added to Rh+ erythrocytes of humans. When a rheumatoid factor, i.e., antibody to γ-globulin with the same allotypic characteristics, was added, agglutination of the treated erythrocytes was observed. This does not take place if an excess of immunoglobulin with the same Gm determinant is added to the system, since it will compete with the rheumatoid factor.

At present, two main systems of allotypic determinants of human immunoglobulins (Gm and Inv) are known. The first type of determinant is found only in heavy chains of γG-globulins, while the second is found in kappa light chains [75, 85, 87]. Both determinant systems are controlled genetically by two dissociated

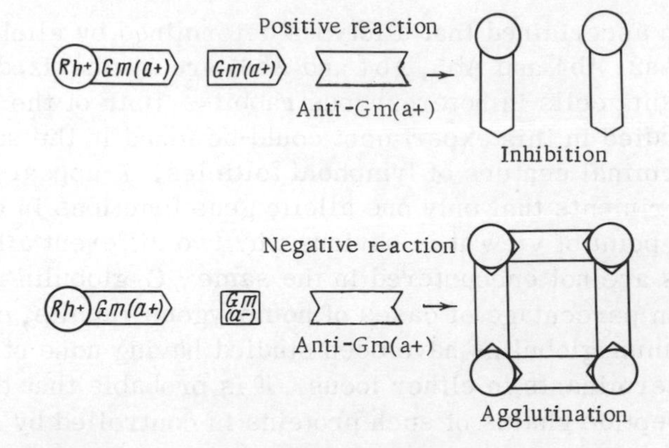

Fig. 52. Reaction scheme used to detect genotypic factors Gm and Inv in human immunoglobulins (after Franklin).

systems of loci. About 23 variants of Gm factors and three variants of Inv factors have been enumerated [68, 73, 74, 84, 98, 116, 121, 122, 133, 135-137, 144]. A special commission, created on the initiative of the Immunology Branch of WHO, proposed the following nomenclature for human allotypes [109a, 103a].

Such a large number of allotypes of the Gm system caused considerable difficulties in the first stages of their investigation

Gm		Gm		Inv	
Original	New	Original	New	Original	New
a	1	b^4	14	1	1
x	2	s	15	a	2
$b^W = b^2$	3	t	16	b	3
f	4	z	17		
$b = b^1$	5	Rouen 2	18		
c	6	Rouen 3	19		
r	7	20	20		
e	8	g	21		
p	9	y	22		
α	10	n	23		
β	11				
γ	12				
b^3	13				

in attempts to elucidate their general hereditary principles. However, as new evidence accumulated, the picture became clearer. It was found that the greater part of the Gm factors is located in the Fc fragment, while the lesser part is found in the Fab fragment (Gm3, Gm4, Gm17) [84, 116, 137]. It was observed that myeloma proteins and antibodies may be distinguished from the bulk of γG-globulins in the same individual by their allotypic characteristics [21, 76, 106]. Later, Kunkel and Norwegian immunologists discovered that the proteins of each of the γG-globulin subgroups contain only certain allotypic determinants [85, 86, 103b]:

Subgroup	γG1(We, γ2b)		γG2(Ne, γ2a)	γG3(Vi, γ2c)	γG4(Ge, γ2d)
Percent	70%		18%	8%	3%
Genetic	a	f	n	b^0g	
Antigen	x	y		b^1	
	z			b^2	
				b^3	
				b^4	
				b^5	

Heavy γ-type chains of different γG-globulin subgroups have common properties on the basis of which they can be regarded as comprising one group. At the same time heavy chains of a given subgroup of γG-globulins differ from heavy chains of other subgroups. In addition to their antigenic properties, they may also differ in the susceptibility of their Fc fragments to proteolysis, in the electrophoretic mobility of these fragments, in the number of SH groups after reduction, in the PCA reaction, and in the complement fixing capacity. Finally, different variants of γ chains differ in their peptide charts and, evidently, in their primary structure. This collected evidence led Kunkel et al. to the conclusion that the synthesis of each γ-chain variant is controlled by its own particular cistron of which there should be no less than four and hardly more than 5-6, since the sum of all described subtypes of γG-globulins is more than 90% of all immunoglobulins of this type. Allotypes of the Gm system are apparently allelic variants of the corresponding γ chains. Actually, allotypes are inherited strictly according to Mendel's law [133]. Each γ-chain variant from the same individual is able to have at most two of the Gm determinants of the chains in a given region. In addition, each given γ-chain has only one determinant in the Fc fragment and one in

TABLE 43. Gene Complexes in Different Racial Groups Arranged in Vertical Fashion According to the Subgroup of Heavy Chain Involved [103b]

	Caucasoid				Negroid				Mongoloid			
γG1		x										
	a	a	y	y	a	a	a	a	a	a	a	y
	z	z	f	f	z	z	z	z	z	z	z	f
γG3	g	g	b^0	b^0	b^0	b^0	b^0	b^0	b^0	g	g	b^0
			b^1	b^1	b^1	b^1	b^1	s	st			b^1
			b^4	b^4	b^4	—	b^4	—	—			b^4
			b^5	b^5	b^5	c^5	b^5	b^5	b^5			b^5
			b^3	b^3	b^3	c^3	c^3	b^3	b^3			b^3
γG2			n									n

the Fd fragment. Thus, an individual γ chain is able to possess one, two (one in the Fc fragment and the other in the Fab fragment), or none of the known Gm determinants.

As peptide chart investigations have shown, the essence of the differences in allotypic variations is to be found in the differences in the primary structures of peptide chains [59, 100]. Thus, for example, γG-globulins with a Gm(a+) characteristic were found to contain a peptide not found in Gm(a–) protein, while Gm(b+) protein contained another peptide not found in Gm(b–) [60]. It is now known that Gm(a+) peptide contains the sequence -Asp-Glu-Leu- in the Fc fragment, while Gm(a–) proteins contain Met-Glu-Glu- in this position [140b]. In light kappa chains with a Inv(b+) characteristic, the 191st residue is valine, while in the same chains in humans with a Inv(a+) characteristic leucine is found in this position (Chapter V).

A study of the frequency of occurrence of the various Gm factors in different peoples has brought the important conclusion that these factors are not inherited randomly but in the form of definite complexes (Table 43).

Evidently, this type of inheritance may take place in the case where the cistons of γ-chain variants are situated close together, which results in their transmission as definite gene complexes.

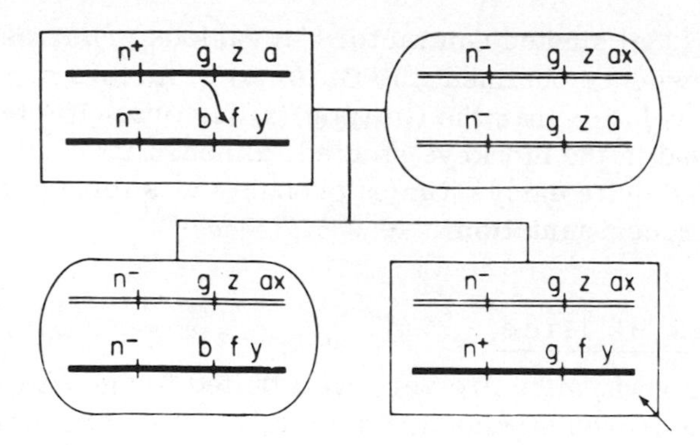

Fig. 53. Gm gene complexes. The unusual gene complex $Gm^{n+}Gm^{g}Gm^{fy}$ in the propositus is indicated along with the postulated cross-over in the father [103b].

Infrequent gene complexes have been found in examination of large groups of people. Thus, for example, the rare complexes Gm(za)Gm(g)Gm(n), Gm(fy)Gm(b−)Gm(n−)Gm(g), etc., have been found in Caucasoids. A study of the allotypic characteristics of γG-globulins in members of families in which such complexes are found led to the conclusion that the appearance of rare complexes is most likely due to cross-overs [103b].

Figure 53, taken from the publication of Natvig et al. [103b], shows the results of an analysis of inheritance of Gm factors in a family of which one of the members had the rare complex, Gm(f,y), Gm(g), Gm(n+). It is evident from the figure that the genesis of such a complex may be explained by recombination, as the arrow shows.

Recombinations have been found between γG1 cistrons and γG2 and γG3 cistrons, and also between γG2 cistrons and γG1 and γG3 cistrons. From this it was concluded that the order of cistrons in gamma-chains is γG2, γG3, and γG1. Their arrangement of genes accords with other data (for example, it has been shown that γG1 and γG3 cistrons are more clearly linked than γG1 and γG2 [103b]). According to Natvig's hypothesis γG4 cistron is most likely situated to the left of γG2 cistron and is followed thereafter by cistrons of other types of heavy chains (α and μ). The total number of cistrons of human immunoglobulin heavy chains is probably 12-15 [103b].

Litwin [93a] studied Gm factors in various primates. According to the evidence obtained, the Gm(z)(a)(b) factors appeared earlier in evolution than the Gm(y)(g)(x) factors. The factor Gm(g) was not found in the monkeys studied, although its allele Gm(b) had appeared quite early. Gm(g) probably was formed from a comparatively recent mutation.

Allotypes of Mice

As is known, mice are very well suited for genetic studies since highly inbred strains are available. They are also suited for the study of allotypes, and all the more so since mice immunoglobulins have been quite well investigated, thanks to Fahey et al. (cf. Chapter III).

At present, four loci are known which control the allotypic determinants found in γG_{2a}- and γG_{2b}-globulins (Ig1 and Ig3), in γA-globulins (locus Ig2), and in γG_1-globulins (locus Ig4) [47, 78, 92, 100a, 144a, 144b, 146]. Determinants of all these types are localized in Fc fragments and consequently in heavy chains [101]. For allotypic determinants of the Ig1 locus eight allelic variations were found; the Ig2 locus had four and the Ig3 and Ig4 had five and two allelic variations, respectively [78, 79, 79a]. All these loci are closely associated with each other, but were not associated with any of the other known markers [79, 79a]. No cross-overs have yet been noted between loci of mouse immunoglobulins [79a, 93]. Thus, it is established that in mice at least one group of loci associated with each other and controlling synthesis of immunoglobulin heavy chains exists.

Data gathered from studies of immunoglobulin allotypy are very important for the understanding of the biosynthesis of these proteins. It has been unequivocally demonstrated that both human and rabbit immunoglobulins possess two independent loci each of which controls the synthesis of at least part of the heavy and part of the light peptide chains. If by the term "locus" is understood a structural gene (analogous to the term cistron), then one section of a chromosome probably contains several of such loci which control the individual allotypic variants of both heavy and light chains [72]. In this way might also, perhaps, be explained the existence of several types of light chains having different electrophoretic mobilities [42].

The problem of allotypes is of prime significance in medical practice. For example, the existence of inherited antigenic variants of immunoglobulins must be taken into account in using γ-globulin preparations for therapeutic purposes. Thus, if a human being receives an administration of a sufficiently large quantity of immunoglobulin with allotypic characteristics alien to him, a severe reaction may set in, with manifestation of headache, stomach ache, tachycardia, etc. [65]. Allotypic antigens of the fetus are capable of inducing antibody formation in the mother. The subsequent passage of these antibodies through the placenta can lead to undesired reactions, for example, hypogammaglobulinemia in the child [62].

GENES CONTROLLING ANTIBODY SYNTHESIS

As was seen in Chapter V, the processes leading to antibody formation consist of several stages in which from all appearances various types of cells participate. Hence, it should be assumed that antibody synthesis is controlled by several types of genes [66]:

1. Structural genes of heavy and light antibody polypeptide chains;
2. Genes responsible for synthesis of enzymes participating in the formation of the carbohydrate component of antibodies and also for the incorporation of this component into the peptide chain;
3. Genes regulating the activity of structural genes;
4. Genes controlling the surface properties of cells which interact during antibody synthesis (for example, macrophages and lymphocytes).

At present, as a result primarily of investigations on allotypes, most is known about structural genes. As has been stated above, two main principles of structural gene localization have been found. First, loci controlling the synthesis of light and heavy chains are not connected and are spatially separated from each other. Secondly, genes controlling the types and subtypes of heavy (and probably, light) chains are connected and are located in close proximity to each other. It may be presumed that the number of structural genes, or in any case the stable parts of these chains, is relatively small, i.e., of the same order as the number of chain variants. However, it is still unclear whether special genes exist for the variable parts of chains and how many of such genes are possible.

Regulation of Gene Expression

Experimental evidence is now available which will provide some conception of the control of the activity of genes responsible for the synthesis of peptide chains of immunoglobulins and antibodies [66b]. Some of the most interesting results have been obtained in experiments indicating the unipotency of lymphoid cells, i.e., their capacity to synthesize basically only one type of antibody and one variant of immunoglobulin polypeptide chain.

The Unipotency of Lymphoid Cells. Several investigators have studied the question to what extent one antibody-synthesizing cell is capable of synthesizing different types of antibodies after immunization of the animal with two or more types of antigens.

Nossal and Mäkelä [96a, 107a] determined antibodies after incubation of cells in a microdrop, after immobilization of bacterial flagellae, or after immune adhesion of bacteria to lymphoid cells. They found that approximately 38% of 7000 cells of animals immunized with two, three, or four antigens formed antibodies. Of these, 98.2% of the cells formed only one type of antibody. In those few cases where a cell synthesized two antibodies (1.8%), one was in very small quantities. Only two cells out of 7000 synthesized both antibodies in identical quantities. Similar data were obtained by Schwartzman et al. [15a]: 97.5% of the cells of animals immunized with two leptospira strains synthesized only one type of antibody. It is interesting that the number of double producers increased with increase in the number of antigens used for immunization [15b]. Similar results were obtained, on the whole, by Mäkelä in experiments with antigens from the tail of phage T6 and with a *Salmonella* polysaccharide antigen [96b, 96c].

Attardi et al. [22a] obtained somewhat different data in their experiments. Although the major portion of cells of rabbit lymph nodes immunized with three phages synthesized antibodies which neutralized only one of the phages, about 20% of the cells formed two antibodies. No cases were found in which one cell synthesized three types of antibodies at once. The differences in the results of the two investigations can be attributed to technical differences in the way the experiments were conducted. In any event, one of the investigators of the latter reference (M. Cohn) maintains that the results of their experiments do not alter the basic thesis that a single cell is capable of synthesizing only one type of antibody in almost all cases [37a]. Experiments of Green et al. [71a] also support this view: in an animal immunized with a complex of DNP-polylysine and albumin, cells synthesized antibody to either hapten or to the carrier protein.

Trentin et al. [141c] investigated how many antibodies the successors of one cell, i.e., one clone, are able to synthesize on the following model. Bone marrow cells from healthy mice were injected into irradiated mice. Discrete colonies appeared in the spleen of the recipients sometime afterwards. The cells from such a colony were transferred to another mouse and the same procedure was repeated again. Ultimately, mice were obtained to which cells from one colony, i.e., the successors of one cell, were transplanted. After immunization of these mice with different *Salmonella* strains, serum albumin, and erythrocytes, these mice synthesized antibodies to all these antigens. These experiments are indirect proof of the multipotency of lymphoid cells. The data of Feldman and Mekori [56a], who studied the immune response of mice into which cells from an individual spleen cell colony had been injected, also provide certain evidence of this. However, it is not thoroughly clear whether the colony is actually the progeny of only one cell or whether cells undergo modifications during proliferation in the recipient. It is obviously very difficult to interpret these experiments. In the experiments of Celada and Vicksell spleen cells from one colony were capable of response to only one antigen [33c].

A quite considerable number of experiments have been conducted on the problem of how many types of immunoglobulin polypeptide chains can be synthesized by one cell. Judging from experiments carried out by Cebra et al. [33a, 33b] with fluorescent methods, with rare exceptions one cell synthesizes only one type of light chain ($\varkappa$ or λ) and only one heavy chain variant (γ, α, or μ). The same principle applies to allotypical variants of chains.

As has already been pointed out, only one allotypical chain type is to be found in one cell of a heterozygous animal [33b, 40a]. It has been mentioned previously that cells containing two variants of heavy or light chains can be found at the same time in germinal centers. However, a special study disclosed that the germinal centers of spleen and lymph node cells contain only one type of heavy chain and one type of light chain [32a]. In some cases, γM- and γG-globulins are synthesized simultaneously in the same cell. In the foregoing chapter we described experiments in which it was shown that such a simultaneous synthesis is observed in cells at that moment when synthesis of macroglobulin antibodies in the organism is superseded by synthesis of γG-antibodies. In a very small percentage of cases cells of one clone were also able to synthesize both of the above-mentioned types of antibodies [33d]. Certain lines of leukemic lymphoid cells are capable of synthesizing two or three types of heavy chains and both types of light chains [56b, 104a]. However, most investigators now consider it the rule that one immunoglobulin-synthesizing cell produces only one light and one heavy chain variant.

The allelic exception, i.e., the capacity of one cell to synthesize only one allelic protein variant is applicable only to those genes which code immunoglobulin polypeptide chains, since the appearance of other genes in the same cells is contrary to this rule. Thus, Celada and Klein [33c] showed that both of the histocompatibility antigens studied can be found in the same lymphocytes in mice which were heterozygous at the H-2 locus. The allelic exception is also known not to apply to hemoglobulin.

It is obvious that there should exist some mechanisms which hinder the activity of the majority of structural genes responsible for the synthesis of immunoglobulin peptide chains. But nothing is yet known of the nature of these mechanisms. In any event, an analytic model is of great value for an understanding of the problem of both allelic and nonallelic gene exceptions.

The Balance of Synthesis of Heavy and Light Peptide Chains. As has been discussed in Chapter V, a slight excess of light chains has been discovered by several investigators, including ourselves and Askonas and Williamson. It is conceivable that inasmuch as loci responsible for the biosynthesis of light and heavy chains are not connected, there is no correlation between the respective amounts of both types of chains formed. However, it is very probable that a certain excess of light chains is formed simply because these chains, being only half as long as heavy chains, are also formed twice as rapidly.

Suppression of Synthesis of Immunoglobulin Allotypes. A study of the heredity of immunoglobulin allotypes showed that it is possible to alter the expression of immunoglobulin genes. This phenomenon, known as allotypic suppression, was first described by Dray et al. [97a]. In these experiments male b^4b^4 (or b^5b^5) rabbits were crossed with b^5b^5 (or b^4b^4) females. Before crossbreeding and during pregnancy the females were immunized with b^4 (or b^5) γG-globulin so that anti-b^4 (or anti-b^5) elaborated by the female rabbit would pass into the blood stream of the fetus. In other experiments, anti-b^4 was injected to newborn rabbits. As a result the activity of the b^4 gene was reduced sharply, and the activity of the b^5 gene increased in compensation. Whereas under normal conditions the b^4:b^5 ratio of molecules was 2.4:1 in rabbits receiving anti-b^5 or in rabbits born of mothers immunized with the paternal allotype b^5, this ratio was 300:1 at an age of five months, 71:1 at the age of one year, 63:1 at the age of two years, and 18.6:1 at the age of 35 months. To suppress the activity of a b^5 gene it was sufficient to inject 0.9 mg precipitating anti-b^5 into newborn rabbits. Immunization had no effect on this process [97b]. Analogous data were obtained with respect to the allele of the a locus which controls synthesis of heavy chains of rabbit immunoglobulins [97c].

Basically similar data were obtained by Herzenberg et al. on mice, with the one difference that the suppression lasted approximately 25 days [78a]. This can be explained by the fact that the Ig^1 locus, the activity of which was determined in the experiments, controlled formation of γG_{2a}-globulin which comprises 50% of mice immunoglobulins, whereas the antigenic determinants controlled by the a and b loci in rabbits comprise over 90% of rabbit immunoglobulins.

Dubiski [46a] injected anti-b^5 antibodies into rabbits which were homozygous with respect to this allotype. This resulted in total suppression of synthesis of b^5 proteins, but on the other hand it stimulated synthesis of any other γG-globulin which did not have a b-locus determinant.

It is presently unclear on which level inhibition of synthesis of allotypic immunoglobulin variants takes place. However, there is no doubt as to the importance of the above-described phenomenon of suppression, since for the first time the investigator has the possibility at will to arrest the synthesis of any immunoglobulin variant. An elucidation of the mechanism of suppression would also be a considerable step forward with regard to an understanding of the activity of other genes. Thus, this problem is of interest to all geneticists.

SYNTHESIS OF DNA BY LYMPHOID CELLS AND ITS STIMULATION

The replication of the genetic material in lymphoid cells, or DNA synthesis, has been extensively studied in recent years. Much interesting data has been accumulated, for example, on the generation time of these cells and the duration of the individual periods of the mitotic cycle. It has been ascertained that antibody-producing cells appear as the result of mitosis of precursor cells, etc. The principal results obtained in such studies are discussed below.

The DNA-synthesizing capacity of various types of cells of lymphoid tissue varies sharply: some may be in a latent state for many days, and even years, whereas others have a very short generation time.

Lymphocytes belong to the first type. From data obtained by various methods [58, 147], the majority of these cells are able to exist for very long periods without dividing or without synthesizing DNA. Only a very small percentage of such cells incorporate a label ^{32}P or ^{3}H-thymidine) into DNA [26, 58, 107]. Radioactive-isotope investigations have established that the lifetime of a small lymphocyte is 100 or more days. Fitzgerald [58] collated the results of several studies on unstable anomalies of lymphocyte chromosomes which can be detected in various individuals over a period of many years and came to the conclusion that a percentage of small lymphocytes exist without mitosis for the same length of time. The notion that lymphocytes are able to store information necessary for a specific immunological response for quite a long time is very well known [70, 147]. After antigenic stimulus they can be transformed into larger cells participating directly in antibody synthesis.

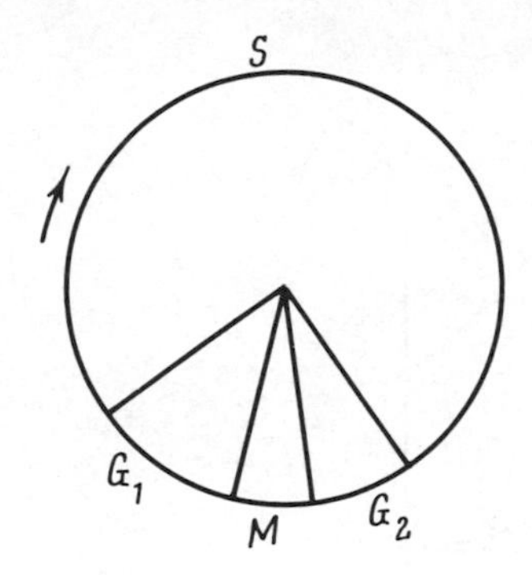

Fig. 54. Cell cycle of blast cells [123]. Periods: G_1, presynthetic (1 hr); S, DNA synthesis (6.8 hr); G_2, postsynthetic (premitotic, 0.7 hr); M, mitosis (0.5 hr). These data were obtained with rabbit spleen cells cultured in diffusion chambers in experiments on irradiated mice. One injection of ^{3}H-thymidine was used.

Blast cells are the complete opposite to these cells. When ^{3}H-thymidine is injected into an animal, all of the rapidly labeled (e.g., within 30 minutes) lymph node cells will be this type of large cell. All blast cells will contain label within a few hours [26, 108]. A more accurate determination of the generation time of cell blasts is quite intricate. It is evidently about nine hours, on the average [123]. Figure 54 is a diagram of the cell cycle of these cells.

Plasma cells, which synthesize the bulk of antibodies, are evidently not capable of mitosis or DNA synthesis; this was first noted by Schooley [126]. Mäkelä and Nossal [96], who made a comprehensive study of this problem, obtained similar data. By injecting the thymidine label one hour before the mice (immunized with *Shigella* antigens) were killed, they were able to show that on the fourth to fifth day after secondary immunization mature plasmocytes in rat lymph nodes were still unlabeled. At the same time, these cells proved to be active antibody producers when cultured in microdrops. On the second to third day after secondary immunization, many antibody-synthesizing cells (mainly plasmoblasts) were capable of incorporating thymidine. Thus, an inverse relationship exists between the DNA-synthesizing and antibody-synthesizing capacities. This relationship is shown in Fig. 55.

Generally similar data on DNA synthesis by cells containing γ-globulin or antibody were obtained by Balfour, Cooper, and Meek [27], who injected labeled thymidine to rats on the second day after immunization with a diphtheria anatoxin and extracted lymph nodes after 45 minutes and five hours. The thymidine label and the total DNA content in nuclei stained by the Feulgen method were determined in cells containing antibody or γ-globulin. Figure 56 presents the results. In the bulk of mature plasma cells

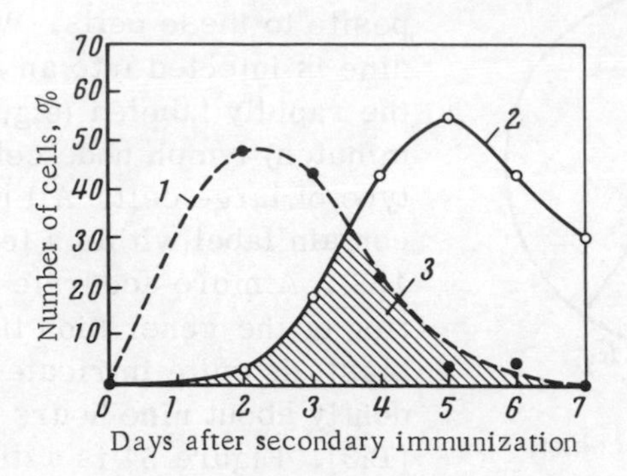

Fig. 55. Capacity of lymphoid cells to synthesize antibodies and DNA [96]. 1) Cells synthesizing DNA; 2) cells producing antibodies; 3) shaded area; cells simultaneously synthesizing antibodies and DNA.

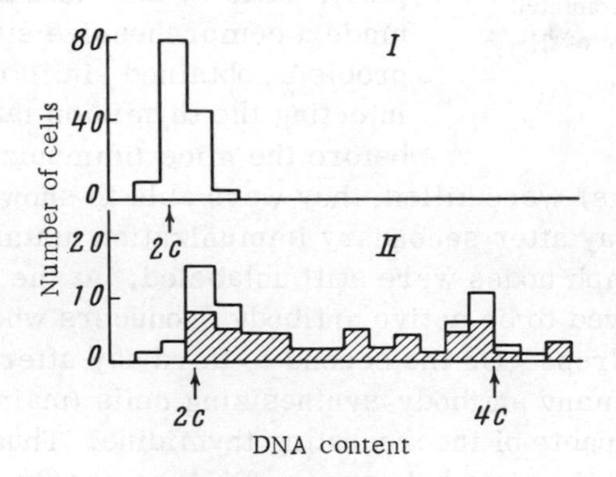

Fig. 56. Distribution of the DNA content in the nuclei of cells containing γ-globulin [27]. The DNA content was determined by the Feulgen reaction with the aid of a microdensitometer. The shaded area represents the number of cells which had incorporated ^{3}H-thymidine (45 minutes after injection). I) Plasma and intermediate cells; II) blast cells. 2c and 4c) Quantities of DNA corresponding to diploid and tetraploid sets of chromosomes.

the quantity of DNA was characteristic for a diploid set (2c). These cells never contained ^{3}H-thymidine. Twenty-four percent of the total number of cells was labeled, and their DNA content was quite evenly distributed between the values 2c and 4c. About 28% of the antibody-containing cells contained label. No correlation was found between the size of nuclei and the DNA content or between the DNA content and the rate of DNA synthesis.

It is important to know which cells are precursors of antibody-synthesizing cells. At present, most investigators agree that antibody-producing cells are formed as a result of division of their precursor cells. However, it is still unclear whether this division begins only after introduction of antigen or whether the cells actively engaged in mitosis even before immunization.

To determine the origin of plasma cells, most investigators have employed autoradiography. Nossal and Mäkelä [108, 109] in experiments discussed in another review [3] injected ^{3}H-thymidine to rats two hours before secondary immunization. In animals which were killed immediately after the second antigen injection, the large and medium-sized lymphocytes and plasmoblasts were labeled, whereas in animals killed on the fourth to fifth day after secondary immunization almost all plasma cells were labeled. It was hence concluded that plasma cell precursors are capable of mitosis even before introduction of antigen. These cells are most probably large lymphocytes and blast cells.

However, Nossal et al. [102] later observed that results of experiments in which the fate of a thymidine label was traced for more than two days can also be interpreted in another way. In setting up such experiments, it is necessary to take into account the possibility of neutralization of labeled DNA. It was found that even on the tenth day after injection of the label, blast cells still contained a significant quantity of ^{3}H-thymidine in spite of the large number of preceding mitoses. During this time, approximately 24 generations ought to have occurred. On the assumption that the generation time was ten hours, the grain count over cells from ^{3}H-thymidine ought to have diminished by 2^{24} or by a factor of more than 16,000,000, if all the cells had undergone mitosis. Actually, the decrease was only by a factor of 600. Apparently, cells can acquire the label not only from the mother cells, but from other sources as well.

Cohen and Talmage studied this problem with a somewhat different experimental setup [37]. The donor mice were killed at different times after secondary immunization, and the spleen cells were incubated for two hours in a medium containing ^{3}H-thymidine. Then the cells were injected into isologous irradiated mice, and after a certain time interval the percentage of labeled antibody-synthesizing cells, presumably the cells of the donor, was determined. It was found that if the donor mice were killed three hours after injection of antigen, only 7% of the antibody-containing cells had incorporated the label. In experiments in which the time between the second antigen injection and cell incubation was six hours, 38% of the cells contained label. The percentage of labeled cells then decreased with increase in this time, so that antibody-containing cells incubated 48 hr after antigen injection contained no label. In other experiments, spleen cells were first incubated with ^{3}H-thymidine five to seven weeks after primary immunization and then injected simultaneously with antigen to the recipient mice. In this case, no antibody-containing cell was labeled.

Since these experiments with labeled thymidine were in vitro, only a very small quantity of label was administered to the recipient mice so that possible reutilization of DNA could be disregarded. From the experimental results discussed above, it can be concluded that the precursors of antibody-synthesizing cells begin to proliferate only after introduction of antigen, this process beginning approximately five hours after immunization. Thus, these results are an indirect confirmation of the hypothesis that cells of the lymphocyte types are transformed into plasmocytes by mitosis after introduction of antigen [70].

All investigators are presently agreed that the majority of plasma cells are not directly formed from a precursor cell, but appear as the result of a series of successive mitoses. Besides the studies discussed, the experiments of Dixon et al. [148], and Makinodan, Gengozian, et al. [33, 104, 142], who studied cell proliferation in diffusion chambers, lend additional support to this position. However, as pointed out in Chapter VI, it cannot be ruled out that some cells may be directly transformed into antibody-producing cells.

Stimulation of DNA Synthesis. A very interesting and important peculiarity of lymphoid cells is their capacity to

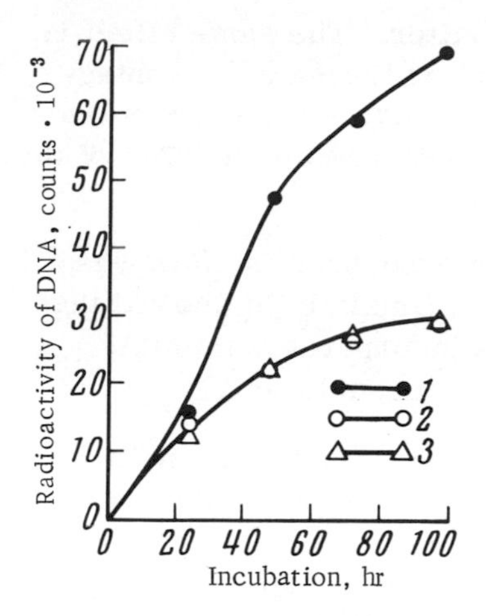

Fig. 57. Stimulation of DNA synthesis by antigen in spleen cells of rabbits primarily immunized with human albumin [53]. To 10^7 cells was added 0.5 mg/ml of (1) human albumin and (2) ovalbumin; (3) without additions.

proliferate and differentiate in a tissue culture after addition of certain substances.

In an extensive series of experiments, Dutton et al. [35, 52] established that 1-2% of lymphoid cells from spleen and lymph nodes of animals which had undergone primary immunization began to synthesize DNA after addition of antigen in vitro. The criterion used in these experiments was incorporation of ^{3}H-thymidine. Figure 57 gives results of one of these experiments. It can be seen that the label incorporation rate began to increase only after introduction of a specific antigen, i.e., the antigen used for primary immunization. Other proteins caused no or only negligible stimulation of DNA synthesis [53]. The stimulatory effect was directly proportional to the amount of antigen added [50]. The time of contact of the antigen with the cells was immaterial.

Among immunologically similar antigens, the greatest stimulation was obtained from those which serologically were more similar to the antigen used in primary immunization [51]. When a conjugated protein, for example DNP-bovine globulin, was used for immunization, DNA synthesis rose sharply upon addition of this antigen. However, the stimulatory effect was less when the protein alone was added without the hapten. Hapten itself did not stimulate synthesis, but inhibited the effect of the conjugated protein. Hapten conjugated with other proteins was also inactive [49].

It is interesting that if cells are first incubated for one hour with ^{3}H-thymidine and then antigen is introduced with included thymidine, the quantity of labeled cells increases by 2-3 times. This indicates that it is precisely those cells which are capable of mitosis which proliferate in the presence of antigen [50]. The addition of cells which have been in contact with antigen to fresh cells stimulates DNA synthesis in the latter. The same effect is obtained if the cells are first irradiated or frozen after contact with antigen. In animals with artificially induced tolerance to any antigen, cells do not synthesize DNA in response to addition of this antigen, but do respond to other antigens [48].

Antigen also induces several other processes in cells when added to a tissue culture. In particular, Drizlikh [5] showed that the quantity of radioactive amino acids incorporated into the protein of antigen-treated cells may rise considerably.

According to certain evidence, addition of antigens, with which the organism had previously been in contact, to lymphocytes of human blood also stimulates mitosis and causes morphological transformations of these lymphocytes into larger blastoid cells (blast-transformation [120]). Such experiments have been conducted with antigens of tuberculosis mycobacteria [95, 114], with diphtheria anatoxin, and with whooping cough vaccine [80]. It should be noted, however, that the same reaction was observed when many other substances, for example, inactivated rabbit serum [127] and human lymphocyte extracts [77], were added to a culture of peripheral leukocytes. It is interesting that when lymphocytes from two different humans or two different rabbits were incubated together, mitosis was also stimulated, but to a much lesser degree than for cells from twins.

Blast transformation may also be induced by addition of specific antisera to all three main immunoglobulin classes to lymphocytes [128c]. Blast transformation is also caused by antiserum to heavy or light chains or to papain-cleaved fragments of γG-globulin [128b]. In this case, 80-90% of all cells gave specific changes. It is evident that practically all lymphocytes contain immunoglobulins. Antiallotypic sera analogously stimulated blast transformation and DNA synthesis in the allotypically corresponding cell population. It was calculated that a large number of antibody molecules (about ten million) is necessary for blast transformation of a lymphocyte.

Addition of a protein isolated from bean seeds (phytohemagglutinin) induced synthesis of DNA and then transformation into blast cells in the overwhelming majority of lymphoid cells [110, 114].

Phytohemagglutinin not only stimulates mitosis and DNA synthesis; after treatment with this protein, an enhanced RNA synthesis, detectable by incorporation of a radioactive label, set in directly after a slight decrease in the total RNA content in the cell.

Sell et al. [128a] have demonstrated the sequence of events taking place after addition of phytohemagglutinin to lymphocytes. First, the rate of incorporation of label into proteins, and then into RNA, increased. In the following stage, a histological picture, characteristic of blast transformation, was apparent; then DNA synthesis began and finally mitosis occurred. According to Mirsky et al. [116a] one of the first stages in the chain of events elicited in cells in response to addition of phytohemagglutinin was acetylation of histones.

Some investigators have observed a stimulatory effect of phytohemagglutinin on γ-globulin synthesis with the aid of Coons' method [55]. Incorporation of radioactive amino acids into protein has also been noted; this process is inhibited by actinomycin [24]. Tao [140] added phytohemagglutinin to lymph node fragments from a primarily immunized rabbit and obtained a secondary immune response in vitro. However, this and other experiments in which stimulation of γ-globulin synthesis by phytohemagglutinin has been observed have been of an extremely approximate nature and require carefully conducted experiments for verification. In any event, Sell et al. [128a] was unable to detect synthesis of γ-globulins in transformed lymphocytes, although these cells did synthesize other cellular proteins.

The mechanism of the blastogenic effect of phytohemagglutinin on lymphoid cells is still somewhat obscure. The suggestion that blast transformation is accompanied by an anamnestic immune response to phytohemagglutinin is not experimentally confirmed. For example, this hypothesis conflicts with data according to which phytohemagglutinin affects cells of embryos and cells of animals which do not feed on beans [94, 125]. A continued investigation of this question would be interesting, since it could shed light on those processes which are initiated after introduction of antigen into the organism and which lead to augmented proliferation and transformation of cells into antibody-producing plasma cells.

Thus, at the present a great deal is known about DNA synthesis by immunocompetent cells and its stimulation. However, it is obvious that research in this area has only just begun, and much more remains to be learned. For example, it is not certain whether lymphoid cells contain metabolically active DNA, i.e., the DNA fraction which is synthesized at a higher rate than other fractions [16]. It is not known whether chromosomes of these cells contain functionally active sections [11] directly involved in antibody synthesis. And finally, the mechanism of the stimulatory effect of specific antigens, as well as nonspecific blastogenic substances, on DNA synthesis remains obscure.

LITERATURE CITED

1. A. K. Akatov, "The use of inbred strains of mice as an experimental model for the study of immunity against staphylococci," J. Hyg. Epid. Microbiol. Immunol. 9:180 (1965).
2. S. J. Ginzburg-Kalinina, "Hereditary influences on the development of immunobiological factors in the organism," Vestn. Akad. Med. Nauk 3:49 (1965).
3. A. E. Gurvich and R. S. Nezlin, "DNA and biosynthesis of antibodies and γ-globulins," Usp. Biol. Khim. 7:150 (1965).
4. Zh. Dosse, in: Immunohematology [in Russian], Medgiz, Moscow (1959), p. 106.
5. G. I. Drizlikh, "Correlation of biosynthesis of nonspecific γ-globulins and antibodies and the effect of antigen on these processes," Biokhimiya 30(4):743 (1965).
6. V. I. Ioffe, "On certain genetic conceptions in immunology," in: Problems of Medical Genetics [in Russian] (1965), p. 29.
7. P. N. Kosyakov, Immunology of Isoantigens and Isoantibodies [in Russian], Meditsina, Moscow (1965).
8. Zh. Medvedev, Biosynthesis of Proteins and Problems of Ontogenesis [in Russian], Medgiz, Moscow (1963).
9. R. V. Petrov, V. M. Man'ko, and I. K. Egorov, "On the different antibody-synthesizing capacities of highly inbred mice strains," Dokl. Akad. Nauk SSSR 153(3):728 (1963).
10. R. V. Petrov, E. I. Panteleev, V. M. Man'ko, and V. S. Egorova, "Interstrain differences in antibody response in inbred mouse strains and their genetic conditioning," Genetika 2(7):78 (1966).
11. A. A. Prokofeva-Belgovskaya, "Replication of DNA in chromosomes," in: Molecular Biology. Problems and Perspectives. (On the 70th Birthday of Academician V. A. Engel'gardt) [in Russian], Nauka, Moscow (1964), p. 97.
12. O. V. Rokhlin, "Role of heredity in antibody synthesis in inbred mouse strains," Byul. Mosk. Obshchestva Ispytatelei Prirody 71(2):131 (1966).
13. O. V. Rokhlin, V. A. Lyashenko, R. J. Vysokodvorova, and R. P. Khromatcheva, "Differences in hemolysin production of inbred strains CBA and CC57BR," Genetika 2(7):71 (1966).

13a. O. V. Rokhlin and R. P. Khromatcheva, "Inheritance of interstrain differences in producing hemolysins against rat erythrocytes," Genetika 2(12):59 (1966).

14. M. T. Tsoneva-Maneva and R. S. Nezlin, "Study of the capacity of phytohemagglutinin from Sax variety bean seeds to stimulate mitosis of leukocytes of the peripheral blood in vitro," Tsitologiya 5(4):458 (1963).

15. V. S. Shapot, "Mechanisms of regulation of protein synthesis in the cell," in: Biosynthesis of Protein and Nucleic Acids [in Russian], Nauka, Moscow (1965), p. 171.

15a. Ja. S. Schwartzman, M. K. Karpov, and A. S. Zuev, "Immune reactions of isolated cells," J. Microbiol. Epidemol. Immunol. Moscow 10:43 (1964).

15b. Ja. S. Schwartzman, "Effect of the number of antigens on the quantity of cells producing several kinds of antibody," Nature 213(5079):925 (1967).

16. Zh. G. Shmerling, "Heterogeneity of DNA and its biological significance," Usp. Sovrem. Biol. 59(1):33 (1965).

17. V. A. Engel'gardt, Certain Problems of Contemporary Biochemistry, Izdat. Akad. Nauk SSSR, Moscow (1959).

18. V. P. Efroimson, "Certain biochemical mechanisms of inherited and acquired immunity," Zh. Vses. Khim. Obshchestva im. D. I. Mendeleeva (3):314 (1961).

19. V. P. Efroimson, Introduction to Medical Genetics [in Russian], Medgiz, Moscow (1964).

20. J. C. Allen and H. G. Kunkel, "Antibodies to genetic types of gamma-globulins after multiple transfusion," Science 139(3553):418 (1963).

21. J. C. Allen, H. G. Kunkel, and E. A. Kabat, "Studies on human antibodies. II. Distribution of genetic factors," J. Exp. Med. 119(3):453 (1964).

22. E. R. Arquilla and J. Finn, "Genetic control of combining sites of insulin antibodies produced by guinea pigs," J. Exp. Med. 122(4):771 (1965).

22a. G. Attardi, M. Cohn, K. Horibata, and E. S. Lennox, "Antibody formation by rabbit lymph node cells. I-IV," J. Immunol. 92(3):335-390 (1964).

23. R. Audran and M. Steinbuch, "Study of a precipitating complex during the Gm/anti-Gm reaction," Compt. Rend. 259(23):4405 (1965).

24. F. Bach and K. Hirschhorn, "Gamma-globulin production by human lymphocytes in vitro," Exp. Cell Res. 32(3):592 (1964).

25. B. Bain and L. Lowenstein, "Genetic studies on the mixed leukocyte reaction," Science 145(3638):1315 (1964).

26. B. M. Balfour, E. H. Cooper, and E. L. Alpen, "Morphological and kinetic studies of antibody-producing cells in rat lymph nodes," Immunology 8(3):230 (1965).

27. B. M. Balfour, E. H. Cooper, and E. S. Meek, "Deoxyribonucleic acid content of antibody-containing cells in the rat lymph node," Nature 206(4985):686 (1965).

27a. S. Ben-Efraim, R. Arnon, and M. Sela, "The immune response of inbred strains of guinea pigs to polylysyl rabbit serum albumin," Immunochemistry 3:491 (1966).

27b. S. Ben-Efraim, S. Fuchs, and M. Sela, "Differences in immune response to synthetic antigens in two inbred strains of guinea pigs," Immunology 12(5):573 (1967).

28. R. N. Baney, J. J. Vazquez, and F. J. Dixon, "Cellular proliferation in relation to antibody synthesis," Proc. Soc. Exp. Biol. Med. 109(1):1 (1962).

28a. N. Barth, C. McLaughlin, and J. L. Fahey, "The immunoglobulins of mice. VI. Response to immunization," J. Immunol. 95(5):781 (1965).

29. B. Benacerraf, "Studies on the nature of antigenicity with artificial antigens," in: Molecular and Cellular Basis of Antibody Formation, Academic Press, New York (1965), p. 57.

30. B. Benacerraf and P. G. H. Gell, "Delayed hypersensitivity to homologous γ-globulin in the guinea pig," Nature 189(4764):586 (1961).

31. P. Bernstein and J. Oudin, "A study of rabbit γ-globulin allotypy by means of heteroimmunizations," J. Exp. Med. 120(4):655 (1964).

32. S. V. Boyden, "Natural antibodies and the immune response," Advan. Immunol. 5:1 (1966).

32a. P. Burtin and D. Buffe, "Synthesis of human immunoglobulins in germinal centers of lymphoid organs," J. Immunol. 98(3):536 (1967).

33. E. E. Capalbo, T. Makinodan, and W. D. Gude, "Fate of ^{3}H-thymidine-labeled spleen cells in in vivo cultures during secondary antibody response," J. Immunol. 89(1):1 (1962).

33a. J. J. Cebra and G. M. Bernier, "Quantitative relationship among lymphoid cells differentiated with respect to class of heavy chain; type of light chain, and allotypic markers," in symposium: Ontology of Immune Response (1966).

33b. J. J. Cebra, J. H. Colberg, and S. Dray, "Rabbit lymphoid cells differentiated with respect to α, γ, and μ-heavy polypeptide chains and to allotypic markers Aa1 and Aa2," J. Exp. Med. 123(3):547 (1966).

33c. F. Celada and G. Klein, "Autonomy of H-2 genes in individual immunocytes," Nature 215(5106):1136 (1967).

33d. F. Celada and H. Wigzell, "Immune response in spleen colonies. II. Clonal assortment of 19 S and 7 S producing cells in mice reacting against two antigens," Immunology 11(5):453 (1966).

34. N. D. Chapman and R. W. Dutton, "The stimulation of DNA synthesis in cultures of rabbit lymph node and spleen cell suspension by homologous cells," J. Exp. Med. 121(1):85 (1965).

35. N. D. Chapman, R. M. E. Parkhouse, and R. W. Dutton, "Antigen stimulated proliferation in lymphoid and myeloid tissues from immunized rabbits," Proc. Soc. Exp. Biol. Med. 117(3):708 (1964).

36. P. Charache, F. S. Rosen, C. A. Janeway, J. N. Craig, and H. A. Rosenberg, "Acquired agammaglobulinemia in siblings," Lancet 1(7379):234 (1963).

37. E. P. Cohen and D. W. Talmage, "Onset and duration of DNA synthesis in antibody-forming cells after antigen," J. Exp. Med. 121(1):125 (1965).

37a. M. Cohn, "Antibody synthesis: The take home lesson," in: Gamma-globulins. Structure and Control of Biosynthesis, Almquist and Wiksell, Stockholm (1967), p. 615.

38. J. E. Colberg and S. Dray, "Localization of immunofluorescence of gamma-globulin allotypes in lymph node cells of homozygous and heterozygous rabbits" Immunology 7(3):273 (1964).

39. D. E. Comings, "A third gamma-globulin chain," Lancet II(7311):786 (1963).

40. H. L. Cooper and A. D. Rubin, "RNA metabolism in lymphocytes stimulated by phytohemagglutinin: Initial responses to phytohemagglutinin," Blood 25(6):1014 (1965).

40a. C. C. Curtain and A. Baumgarten, "Immunocytochemical localization of the immunoglobulin factors Gm(a), Gm(b), Inv(a) in human lymphoid tissue," Immunology 10(6):499 (1966).

41. J. K. Dineen, "Sources of immunological variation," Nature 202(4927):101 (1964).

42. S. Dray, "General discussion," in: Molecular and Cellular Basis of Antibody Formation, Academic Press, New York (1965), p. 650.

43. S. Dray, S. Dubiski, A. Kelus, E. S. Lennox, and J. Oudin, "A notation for allotypy," Nature 195(4843):785 (1962).

44. S. Dray and A. Nisonoff, "Contribution of allelic genes Ab4 and Ab5 to formation of rabbit 7 S γ-globulins," Proc. Soc. Exp. Biol. Med. 113(1):20 (1963).

45. S. Dray and A. Nisonoff, "Relationship of genetic control of allotypic specificities to the structure and biosynthesis of rabbit γ-globulin," in: Molecular and Cellular Basis of Antibody Formation, Academic Press, New York (1965), p. 175.

46. S. Dray, G. O. Young, and A. Nisonoff, "Distribution of allotypic specificities among rabbit γ-globulin molecules genetically defined at two loci," Nature 199(4888):52 (1963).

46a. S. Dubiski, "Suppression of synthesis of allotypically defined immunoglobulins and compensation by another subclass of immunoglubulin," Nature 214(5095):1365 (1967).

47. S. Dubiski and B. Cinader, "A new allotypic specificity in the mouse (MuA2)," Nature 197(4868):705(1963).).

47a. S. Dubiski and P. Muller, "A new allotypic specificity (A9) of rabbit immunoglobulin," Nature 214(5089):696 (1967).

48. R. W. Dutton, "The effect of antigen on the proliferation of spleen cell suspensions from tolerant rabbits," J. Immunol. 93(5):814 (1964).

49. R. W. Dutton and H. N. Bulman, "The significance of the protein carrier in the stimulation of DNA synthesis by hapten–protein conjugates in the secondary response," Immunology 7(1):54 (1964).

50. R. W. Dutton and J. D. Eady, "An in vitro system for the study of the mechanism of antigenic stimulation in the secondary response," Immunology 7(1):40 (1964).

51. R. W. Dutton and G. M. Page, "The response of spleen cells from immunized rabbits to cross-reacting antigens in an in vitro system," Immunology 7(6):665 (1964).

52. R. W. Dutton and R. M. E. Parkhouse, "Studies on the mechanism of antigenic stimulation in the secondary response," in: Molecular and Cellular Basis of Antibody Formation, Academic Press, New York (1965), p. 567.

53. R. W. Dutton and J. D. Pearce, "Antigen-dependent stimulation of synthesis of deoxyribonucleic acid in spleen cells from immunized rabbits," Nature 194(4823):93 (1965).

54. M. W. Elves and M. C. G. Israëls, "Lymphocyte transformation in cultures of mixed leukocytes," Lancet I(7397):1184 (1965).

55. M. W. Elves, S. Roath, G. Taylor, and M. C. Israels, "The in vitro production of antibody lymphocytes," Lancet I(7294):1292 (1963).
56. A. Feinstein, P. G. H. Gell, and A. S. Kelus, "Immunochemical analysis of rabbit γ-globulin allotypes," Nature 200(4907):653 (1963).
56a. M. Feldman and T. Mekori, "Antibody production by 'cloned' cell populations," Immunology 10(2):149 (1966).
56b. J. Finegold, J. L. Fahey, and H. Granger, "Synthesis of immunoglobulins by human cell lines in tissue culture," J. Immunol. 99(5):839 (1967).
57. M. A. Fink and V. A. Quinn, "Antibody production in inbred strains of mice," J. Immunol. 70(1):61 (1953).
58. P. H. Fitzgerald, "The immunological role and long life-span of small lymphocytes," J. Theoret. Biol. 6(1):13 (1964).
58a. D. C. Formey, K. Kamin, and H. H. Fudenberg, "Quantitative studies of phytohemagglutinin-induced DNA and RNA synthesis in normal and agammaglobulinemic leukocytes," J. Exp. Med. 125(5):863 (1967).
59. E. C. Franklin, B. Frangione, and H. Fudenberg, "Differences between Gm(b) and Gm(f) in peptide maps of normal and myeloma γ-globulins," Vox Sanguinis 10(3):368 (1965).
60. E. C. Franklin, H. Fudenberg, M. Meltzer, and D. Stanworth, "The structural basis for genetic variations of normal human γ-globulins," Proc. Nat. Acad. Sci. USA 48(6):914 (1962).
61. H. Fudenberg and E. C. Franklin, "Human γ-globulin. Genetic control and its relations to disease," Ann. Internal. Med. 58(1):171 (1963).
62. H. H. Fudenberg and B. R. Fudenberg, "Antibody to hereditary human γ-globulin (Gm) factor resulting from maternal—fetalincompatibility," Science 195(3628):170 (1964).
63. H. H. Fudenberg, J. F. Heremans, and E. C. Franklin, "A hypothesis for the genetic control of synthesis of γ-globulins," Ann. Inst. Pasteur 104(2):155 (1963).
64. H. H. Fudenberg and K. Hirschhorn, "Agammaglobulinemia. The fundamental defect," Science 145(3632):611 (1964).
65. B. Frangione, "Antigenicity of hereditary human γ-globulin (Gm) factors — biological and biochemical aspects," Cold Spring Harbor Symp. Quant. Biol. 29:463 (1967).
66. P. G. H. Gell and A. Kelus, "Deletions of allotypic γ-globulins in antibodies," Nature 195(4836):44 (1962).
66a. P. G. H. Gell and S. Sell, "Studies on rabbit lymphocytes in vitro. II and III," J. Exp. Med. 122(4):813, 823 (1965).
66b. Genetics of the Immune Response. Report of a World Health Organization Science Group, Geneva (1967).
67. A. M. Gilman, A. Nisonoff, and S. Dray, "Symmetrical distribution of genetic markers in individual rabbit γ-globulin molecules," Immunochemistry 1(2):109 (1964).
68. E. R. Gold, L. Martenson, G. Ropartz, L. Rivat, and P. Y. Rousseau, "Gm(f) — a determinant of human γ-globulin," Vox Sanguinis 10(3):299 (1965).
69. R. A. Good and A. E. Gabrielsen, "Algammaglobulinemia and Hypogammaglobulinemia-relationship to the Mesenchymal diseases. The Streptococcus,

Rheumatic Fever, and Glomerulonephritis," Williams, Wilkins, and Co., Baltimore (1964), p. 368.

70. J. L. Gowans, D. D. McGregor, D. M. Cowen, and C. E. Ford, "Initiation of immune responses by small lymphocytes," Nature 196(4855):651 (1962).

71. U. Grodecka, "Rabbit γ-globulin allotypes," Nature 2204(4958):595 (1964).

71a. J. Green, P. Vassalli, and B. Benacerraf, "Cellular localization of anti-DNP–PLL and anticonveyor albumin antibodies in genetic nonresponder guinea pigs immunized with DNP–PLL albumin complex," J. Exp. Med. 125(3):527 (1967).

72. R. Grubb, "Agglutination of erythrocytes coated with 'incomplete' anti-Rh by certain rheumatoid arithritic sera and some other sera. The existence of human serum groups," Acata Pathol. Microbiol. Scand. 39(3):195 (1956).

72a. R. Hamers and C. Hamers-Casterman, "Molecular localization of A chain allotypic specificities in rabbit IgG (7 S γ-globulin)," J. Mol. Biol. 14(1):228 (1965).

72b. R. Hamers, C. Hamers-Casterman, and S. Lagnaux, "A new allotype in the rabbit linked with As1 which may characterize a new class of IgG," Immunology 10(5):399 (1966).

73. M. Harboe, "A new hemagglutinating substance in the Gm system anti-Gm^b," Acta Pathol. Microbiol. Scand. 47(2):191 (1959).

74. M. Harboe and J. Lundvall, "A new type in the Gm system," Acta Pathol. Microbiol. Scand. 45(4):357 (1959).

75. M. Harboe, C. K. Osterland, and H. G. Kunkel, "Localization of two genetic factors to different areas of γ-globulin molecules," Science 136(3520):979 (1962).

76. M. Harboe, C. K. Osterland, M. Mannick, and H. G. Kunkel, "Genetic characteristics of human γ-globulin in myeloma proteins," J. Exp. Med. 116(5):719 (1962).

76a. G. Harris and R. J. Littleton, "The effects of antigens and of phytohemagglutinin on rabbit spleen cell suspensions," J. Exp. Med. 124(4):621 (1966).

77. M. Hashem and F. S. Rosen, "Mitogenic fractions in human peripheral lymphocyte extracts," Lancet I(7326):201 (1964).

78. L. A. Herzenberg, "A chromosome region for γ_{2A} and β_{2A} globulin H chain isoantigens in the mouse," Cold Spring Harbor Symp. Quant. Biol. 29:455 (1964).

78a. L. A. Herzenberg, L. A. Herzenberg, R. C. Goodlin, and E. C. Rivera, "Immunoglobulin synthesis in mice. Suppression by antiallotype antibody," J. Exp. Med. 126(4):701 (1967).

79. L. A. Herzenberg, N. L. Warner, and L. A. Herzenberg, "Immunoglobulin isoantigens (allotype) in the mouse. I. Genetics and cross reactions of the 7 S γ_{2A} isoantigens controlled by alleles at the Ig-1-locus," J. Exp. Med. 121(3):415 (1965).

79a. L. Herzenberg and N. Warner, "Genetic control of mouse immunoglobunis," in: Regulation of Antibody Response (B. Cinader, ed.), Charles Thomas, Springfield, Illinois (1967).

80. K. Hirschhorn, R. L. Kolodny, M. Hashem, and F. Bach, "Mitogenic action of phytohemagglutinin," Lancet II(7302):305 (1963).

81. C. A. Janeway, "Hypogammaglobulinemia and immunological response," in: Allergology, Pergamon Press, New York (1962), p. 241.

82. F. Jacob and J. Monod, "Biochemical and genetic mechanisms of regulation in a bacterial cell," in: Molecular Biology. Problems and Perspectives. (On the 70th Birthday of Academician V. A. Engel'gardt) [in Russian], Nauka, Moscow (1964), p. 14.

82a. A. S. Kelus, "Rabbit allotypic markers as a model for molecular immunology," in: Gamma-Globulin. Structure and Control of Biosynthesis, Nobel Symposium 3, Almquist and Wiksell, Stockholm (1967), p. 329.

83. A. S. Kelus and P. G. H. Gell, "An allotypic determinant specific to rabbit macroglobulin," Nature 206(4981):313 (1965).

84. G. Kronvall, "Gm(f) activity of human γ-globulin fragments," Vox Sanguinis 10(3):311 (1965).

85. H. G. Kunkel, J. C. Allen, and H. M. Grey, "Genetic characteristic and the polypeptide chains of various types of γ-globulin," Cold Spring Harbor Symp. Quant. Biol. 29:443 (1964).

86. H. H. Kunkel, J. C. Allen, H. M. Grey, L. Martensson, and R. Grubb, "A relationship between the H-chain groups of 7 S γ-globulin and the Gm system," Nature 203(4943):413 (1964).

87. S. D. Lawler and S. Cohen, "Distribution of allotypic specificities of the peptide chains of human γ-globulin," Immunology 8(2):206 (1965).

87a. E. S. Lennox, "The genetics of the immune response," Proc. Roy. Soc. Ser. B. 166(1003):222 (1966).

88. S. Leskowitz, "Immunochemical study of rabbit γ-globulin allotypes," J. Immunol. 90(1):98 (1963).

89. B. B. Levine and B. Benacerraf, "Studies of antigenicity. The relationship between in vivo enzymatic degradability of hapten-polylysine conjugates and their antigenicities in guinea pigs," J. Exp. Med. 120(5):955 (1964).

90. B. B. Levine and B. Benacerraf, "Genetic control in guinea pigs of immune response to conjugates of haptens and poly-L-lysine," Science 147(3657):517 (1965).

91. B. B. Levine, A. Ojeda, and B. Benacerraf, "Studies on artificial antigens. III. The genetic control of the immune response to hapten-poly-L-lysine conjugates in guinea pigs," J. Exp. Med. 118(6):953 (1963).

91a. E. A. Lichter, "Rabbit γA and γM immunoglobulins with allotypic specificities controlled by the a locus," J. Immunol. 98(1):139 (1967).

92. R. Lieberman and S. Dray, "Five allelic genes at the Asa locus which control γ-globulin allotypic specificities in mice," J. Immunol. 93(4):584 (1964).

93. R. Lieberman and M. Potter, "Close linkage in genes controlling γA- and γM-heavy chain structure in BALB/c mice," J. Mol. Biol. 18(3):516 (1966).

93a. S. D. Litwin, "Phylogenetic differences among the Gm factors of nonhuman primates," Nature 216(5112):268 (1967).

93b. S. D. Litwin and H. G. Kunkel, "The genetic control of γ-globulin heavy chains. Studies of the major heavy chain subgroup utilizing multiple genetic markers," J. Exp. Med. 125(5):847 (1967).

94. K. Lindahl-Kiessling and J. A. Böök, "Effects of phytohemagglutinin on leuko-

cytes," Lancet II(7367):1012 (1964).
95. N. R. Ling and E. M. Husband, "Specific and nonspecific stimulation of peri-phenol lymphocytes," Lancet I(7329):363 (1964).
95a. H. O. McDevitt and M. Sela, "Genetic control of the antibody response. I. Demonstration of determinant specific differences in response to synthetic polypeptide antigens in two strains of inbred mice," J. Exp. Med. 122(3):517 (1965).
95b. H. O. McDevitt and M. Sela, "Genetic control of the antibody response. II. Further analysis of the specificity of determinant specific control, and genetic analysis of the response to (H, G)-A-L in CBA and C57 mice," J. Exp. Med. 126(5):969 (1967).
96. O. Mäkelä and G. J. V. Nossal, "Autoradiographic studies on the immune response. II. DNA synthesis amongst single antibody-producing cells," J. Exp. Med. 115(1):231 (1962).
96a. O. Mäkelä and G. J. V. Nossal, "Study of antibody-producing capacity of single cells by bacterial adherence and immobilization," J. Immunol. 87(4):457 (1961).
96b. O. Mäkelä, "Studies on the quality of neutralizing bacteriophage antibodies producedby single cells. I. Evidence indicating that different cells produce different kinds of antibody against the tail of T-6," Immunology 7(1):9 (1964).
96c. O. Mäkelä, "Evidence indicating that different cells of a lymph node produce different kinds of antibody against the polysaccharide antigen 0-9, 12 of Salmonella," Immunology 7(1):17 (1964).
97. M. Mannick and H. Metzger, "Hybrid antibody molecules with allotypically different L-polypeptide chains," Science 148(3668):383 (1965).
97a. R. Mage and S. Dray, "Persistent altered phenotypic expression of allelic γG-immunoglobulin allotypes in heterozygoni rabbits expossed to isoantibodies in fetal and neonatal life," J. Immun. 95(3):525 (1965).
97b. R. G. Mage and S. Dray, "Persistence of altered expression of allelic γG-immunoglobulin allotypes in an "allotype suppressed" rabbit after immunization," Nature 212(5063):699 (1966).
97c. R. Mage, G. O. Young, and S. Dray, "An effect upon the regulation of gene expression: allotype suppression at the "A" locus in heterozygous offspring of immunized rabbits," J. Immun. 98(3):502 (1967).
98. L. Mårtensson, "On the relationship between the γ-globulin genes of the Gm system. A study of the Gm gene products in sera myeloma globulins, and specific antibodies with special reference to the gene Gm," J. Exp. Med. 120(6):1169 (1964).
98a. L. Martensson, "Genes and immunoglobulins," Vox Sanguinis 11(5):521 (1966).
99. L. Mårtensson and H. H. Fudenberg, "Gm genes and γG-globulin synthesis in the human fetus," J. Immunol. 94(4):514 (1965).
100. M. Meltzer, E. C. Franklin, H. Fudenberg, and B. Frangione, "Single peptide differences between γ-globulins of different genetic (Gm) types," Proc. Nat. Acad. Sci. USA 51(6):1007 (1964).
100a. J. D. Minna, G. M. Inverson, and L. A. Herzenberg, "Identification of a gene locus for γG_1-immunoglobulin H chains and its linkage to the H chain chromo-

some region in the mouse," Proc. Nat. Acad. Sci. USA 58(1):188 (1967).

101. R. J. Mishell and J. L. Fahey, "Molecular and submolecular localization of two isoantigens of mouse immunoglobulins," Science 143(3613):1440 (1964).

102. J. Mitchell, W. McDonald, and G. J. V. Nossal, "Autoradiographic studies of the immune response. 3. Differential lymphopoiesis in various organs," Australian J. Exp. Biol. Med. Sci., Suppl. 41(411) (1963).

103. J. Monod, "Antibodies and induced enzymes," in: Cellular and Humoral Aspects of the Hypersensitive States, New York (1959), p. 628.

103a. W. A. Muir and A. G. Steinberg, "On the genetics of the human allotypes, Gm and Inv," Seminars in Hematology 4(2):156 (1967).

103b. J. B. Natvig, H. G. Kunkel, and T. Gedde-Dahl, "Genetic studies of the heavy chain subgroups of γG-globulin. Recombination between the closely linked cistons," in: γ-Globulins. Structure and Control of Biosynthesis, Nobel Symposium 3, Almquist and Wiksell, Stockholm (1967), p. 313.

104. P. Nettesheim and T. Makinodan, "Differentiation of lymphocytes undergoing an immune response in diffusion chambers," J. Immunol. 94(6):878 (1965).

105. J. Newsome, "Synthesis of ribonucleic acid by stimulated human lymphocytes," Nature 206(4988):1013 (1965).

106. U. Nilsson, "Gm characters of 7 S γ-myeloma proteins and corresponding individual normal γ-globulins," Acta Pathol. Microbiol. Scand. 61(2):181 (1964).

107. G. J. V. Nossal, "Cellular genetics of immune responses," Advan. Immunol. 2:163 (1962).

107a. G. J. V. Nossal and O. Mäkelä, "Elaboration of antibodies by single cells," Ann. Rev. Microbiol. 16:53 (1962).

108. G. J. V. Nossal and O. Mäkelä, "Autoradiographic studies on the immune response. I. The kinetics of plasma cell proliferation," J. Exp. Med. 115(1):209 (1962).

109. G. J. V. Nossal, O. Mäkelä, M. L. Engel, and A. Fefer, "Cellular proliferation in immunity," Stanford Med. Bull. 20(1):32 (1962).

109a. "Notation for genetic factors of human immunoglobulins," Bull. World Health Organ. 33:721 (1965).

110. P. C. Nowell, "Phytohemagglutinin: an initiation of mitosis in cultures of normal human leukocytes," Cancer Res. 20(4):462 (1960).

111. J. Oudin, "Reactions of specific precipitation between serums of animals of the same species," Compt. Rend. 242(2489):2606 (1956).

112. J. Oudin, "Allotypy of rabbit serum proteins. I. Immunochemical analysis leading to the individualization of seven main allotypes," J. Exp. Med. 112(1):125 (1960).

113. J. Oudin, "Allotypy of human γ-globulins," Compt. Rend. 225(6):1164 (1962).

113a. J. Oudin, "The genetic control of immunoglobulin synthesis," Proc. Roy. Soc. B, 166(1003):207 (1966).

114. G. Pearmain, R. R. Lycette, and P. H. Fitzgerald, "Tuberculin-induced mitosis in peripheral blood leukocytes," Lancet I(7282):637 (1963).

114a. B. de Permis, G. Chiappino, A. S. Kelus, and P. G. H. Gell, "Cellular localization of immunoglobulins with different allotypic specificities in rabbit lymphoid tissues," J. Exp. Med. 122(5):853 (1965).

115. R. D. A. Peterson, M. D. Cooper, and R. A. Good, "The pathogenesis of immunologic deficiency diseases," Am. J. Med. 38(4):673 (1965).

115a. P. Pinchuk and P. Maurer, "Antigenicity of polypeptides (poly-α-amino acids). XVI. Genetic control of immunogenicity of synthetic polypeptides in mice," J. Exp. Med. 122(4):673 (1965).

116. S. H. Polmar and A. G. Steinberg, "The effect of the interaction of heavy and light chains of IgG on the Gm and Inv antigens," Biochem. Genet. 1(2):117 (1967).

116a. B. Pogo, V. Allfrey, and A. Mirsky, "RNA synthesis and histone acetylation during the course of gene activation in lymphocytes," Proc. Nat. Acad. Sci. USA 55(4):885 (1966).

117. B. A. Rasmussen, "Isoantigens of γ-globulin in pigs," Science 148(3678):1742 (1965).

118. R. A. Reisfeld, S. Dray, and A. Nisonoff, "Differences in amino acid composition of rabbit γG-immunoglobulin light polypeptide chains controlled by allelic genes," Immunochemistry 2(2):155 (1965).

119. R. F. Rieder and J. Oudin, "Studies on the relationship of allotypic specificities to antibody specificities in the rabbit," J. Exp. Med. 118(4):627 (1963).

120. J. H. Robbins, "Tissue culture studies of the human lymphocyte," Science 146(3652):1648 (1964).

121. C. Ropartz, "Hereditary systems of human γ-globulins," Ann. Biol. Clin. 22:445 (1964).

122. C. Ropartz, J. Lenoir, and L. Rivat, "A new inheritable property of human sera: the InV factor," Nature 189(4764):586 (1961).

123. T. Sado and T. Makinodan, "The cell cycle of blast cells involved in secondary antibody response," J. Immunol. 93(4):696 (1964).

124. J. H. Sang and W. R. Sobey, "The genetic control of response to antigenic stimuli," J. Immunol. 72(1):52 (1954).

125. B. S. Sayly, "Phytohemagglutinin and 'sensitized' leukocytes," Lancet II(7362):762 (1964).

126. J. S. Schooley, "Autoradiographic observation of plasma cell formation," J. Immunol. 86(3):331 (1961).

127. R. Schrek and L. M. Elrod, "Lymphoblastoid transformation of rat and human lymphocytes by rabbit sera," Lancet II(7359):595 (1964).

128. S. Sell, "Immunoglobulins of the germ-free guinea pig," J. Immunol. 93(1):122 (1964).

128a. S. Sell, D. S. Rowe, and P. G. H. Gell, "Studies on rabbit lymphocytes in vitro, III," J. Exp. Med. 122(4):823 (1965).

128b. S. Sell, "Studies on rabbit lymphocytes in vitro. V. The induction of blast transformation with sheep antisera to rabbit IgG subunits," J. Exp. Med. 125(2):289 (1967).

128c. S. Sell, "Studies on rabbit lymphocytes in vitro. VI. The induction of blast transformation with sheep antisera to rabbit IgA and IgM," J. Exp. Med. 125(3):393 (1967).

129. S. K. Seth, A. Nisonoff, and S. Dray, "Hybrid molecules of rabbit γ-globulins differing in genotypes at two loci," Immunochemistry 2(1):39 (1965).

130. D. Skalba, "Allotypes of hen serum proteins," Nature 204(4961):894 (1964).
131. P. A. Small, R. A. Reisfeld, and S. Dray, "Peptide differences of rabbit γG-immunoglobulin light chains controlled by allelic genes," J. Mol. Biol. 11(4):713 (1965).
131a. W. R. Sobey, J. M. Magrath, and A. H. Reisner, "Genetically controlled specific immunological unresponsiveness," Immunology 11(5):511 (1966).
132. G. F. Springer, F. E. Horton, and M. Forbes, "Origin of antihuman blood group B agglutinnis in germ-free chickens," Ann. N. Y. Acad. Sci. 78(1):272 (1959).
133. A. G. Steinberg, "Progress in the study of genetically determined human γ-globulin types (the Gm and Inv groups)," Progr. Med. Genet. 2:1 (1962).
134. A. G. Steinberg, "Population, immunogenetic and biochemical studies on the Gm(b) factors of human γ-globulin," Cold Spring Harbor Symp. Quant. Biol. 29:449 (1964).
135. A. G. Steinberg, "Comparison of Gm(f) with Gm(b2), Gm(bw) and a discussion of their genetics," Am. J. Human Genet. 17(4):311 (1965).
135a. A. G. Steinberg, "Genetic variations in human immunoglobulins: The Gm and Inv types," in: Advances in Immunogenetics (ed.: T. J. Greenwalt), Lippincott, Philadelphia (1967), p. 75.
136. A. G. Steinberg and R. A. Goldblum, "A genetic study of the antigens associated with the Gm(b) factor of human γ-globulin," Am. J. Human Genet. 17(2):133 (1965).
137. A. G. Steinberg and S. H. Polmar, "The relation of the S and F fragments and the H and L chains of γ-globulin to the Gm groups," Vox Sanguinis 10(3):369 (1965).
138. G. W. Stemke, "Allotypic specificities of A and B chains of rabbit γ-globulin," Science 145(3630):403 (1964).
138a. G. W. Stemke and R. J. Fischer, "Rabbit 19 S antibodies with allotypic specificities of the a-locus group," Science 150(3701):1298 (1965).
139. J. Sterzl, L. Mandel, I. Miller, and I. Riha, "Development of immune reactions in the absence or presence of an antigenic stimulus," in: Molecular and Cellular Basis of Antibody Formation, Academy Press, New York (1965), p. 351.
140. T. W. Tao, "Phytohemagglutinin elicitation of specific anamnestic immune response in vitro," Science 146(3641):247 (1964).
140a. N. Tanigaki, Ya. Yagi, G. Moore, and D. Pressman, "Immunoglobulin production in human leukemia cell lines," J. Immunol. 97(5):634 (1966).
140b. N. O. Thorpe and H. F. Deutsch, "Studies on papain-produced subunits of human γG-globulins. II. Structures of peptides related to the genetic Gm activity of γG-globulin Fc fragments," Immunochemistry 3(4):329 (1966).
141. C. W. Todd, "Allotypy in rabbit 19 S protein," Biochem. Biophys. Res. Commun. 11(3):1705 (1963).
141a. C. W. Todd and F. P. Inmann, "Comparison of the allotypic combining sites on H-chains of rabbit IgG and IgM," Immunochemistry 4(6):407 (1967).
141b. J. Travnicek, "Allotypic specificities in pigs," Folia Microbiol. 11:13,406 (1966).
141c. J. Trentin, N. Wolf, V. Cheng, W. Fahlberg, D. Weiss, and R. Bonnag, "Antibody

production by mice repopulated with limited numbers of clones of lymphoid cell precursors," J. Immunol. 98(6):1326 (1967).

142. P. Urso and N. Gengozian, "Immunofluorescent direction of proliferating human antibody-forming cells," Nature 203(4952):1391 (1964).

143. A. Winer, J. Robbing, J. Bellanti, D. Eitzman, and R. T. Smith, "Synthesis of γ-globulin in the newborn rabbit," Nature 198(4879):487 (1963).

144. M. Waller, R. D. Hughes, J. J. Townsend, E. C. Franklin, and H. Fudenberg, "New serum group Gm(p)," Science 142(3597):1321 (1963).

144a. N. L. Warner, L. A. Herzenberg, and G. Goldstein, "Immunoglobulin isoantigens (allotypes) in the mouse. II. Allotypic analysis of three γG_2-myeloma proteins from (NZB × BALB/c) F_1 hybrids and of normal γG_2-globulins," J. Exp. Med. 123(4):707 (1966).

144b. N. L. Warner and L. A. Herzenberg, "Immunoglobulin isoantigens (allotypes) in the mouse. IV. Allotypic specificities common to two distinct immunoglobulin classes," J. Immunol. 99(4):675 (1967).

145. B. S. Westman, J. B. Olson, and K. R. Pleasants, "Serum proteins of germ-free rats fed water-soluble diets," Nature 206(4988):1056 (1965).

146. J. Wunderlich and L. A. Herzenberg, "Genetics of a γ-globulin isoantigen (allotype) in the mouse," Proc. Nat. Acad. Sci. USA 49(5):592 (1963).

147. J. M. Yoffey, "The lymphocyte," Ann. Rev. Med. 15:125 (1964).

148. A. Zlotnick, J. J. Vazquez, and F. J. Dixon, "Mitotic activity of immunologically competent lymphoid cells transferred into X-irradiated recipients," Lab. Invest. 11(6):493 (1962).

Chapter VII

Theory of Antibody Synthesis

From the experimental material presented in the foregoing chapters it is evident that many of the chief questions concerning antibody biosynthesis are still unanswered. The undisputably most important of these questions is how the specific configuration of the antibody binding site is formed. Inasmuch as this configuration is, according to current conceptions, associated with the specific structure of the N-terminal parts of peptide chains, it is necessary to elucidate the mechanism responsible for the variability of these chain parts. It should be taken into account that the differences between the variable chain parts are confined to certain chain sections between which are found very stable regions [17, 47-49]. It should also be borne in mind that the differences between the variable parts of the chains are more likely to be due to point mutations rather than to other genetic mechanisms. A second important problem is the way in which antigen exerts its effect. At present, the mechanism of action of antigens is not at all understood.

It is not surprising that many attempts have been made to produce a satisfactory answer to the questions enumerated. The very fact that a large number of theories on antibody synthesis exists indicates that all the presently known facts cannot be explained by one of them alone. Essentially, they are not genuine theories, but merely hypotheses, many important points of which have only been postulated and totally lack experimental verification. Nonetheless, some of them have proven to be very useful in stimulating important experimental research.

Below is presented a brief examination of theories on antibody synthesis proposed at various times. Suggested mechanisms to explain the variety of immunoglobulin peptide chains, as well as the main aspects of immunoglobulin evolution, are also discussed.

TEMPLATE THEORY OF ANTIBODY SYNTHESIS

After Landsteiner demonstrated the possibility of formation of antibodies specific for many synthetic antigens, it became evident that antigen should participate directly in formation of antibody polypeptide chains. Breinl and Haurowitz [10, 34, 35] were the first to present this point of view.

In 1940, Pauling [59] hypothesized that peptide chains of antibodies and γ-globulins had the same primary structure, but that the manner of chain folding differed. Exerting its effect at the moment of chain folding, antigen shapes an antibody binding site secondary structure which is complementary to itself. The first experiments giving positive evidence in support of this hypothesis were conducted by Pauling and others, but subsequent attempts to verify it have often not given convincing results. Recently, another such attempt was made. It was found that the capacity to bind dinitrophenyllysine and the dye Brilliantine Green and Methyl Orange was considerably greater for the heavy chain than for the whole γ-globulin molecule [73]. After the reaction of the heavy chain with these substances and its recombination with the light chain, the newly formed molecule bound the dyes as well as did the heavy chain alone. However, there are probably other explanations for these experiments besides the hypothesis on the formation of a supplementary configuration for haptens by their reaction with a heavy chain.

Another version of this hypothesis was advanced by Karush [30, 31]. Taking into account the high content of disulfide bonds in the immunoglobulin molecule, Karush suggested that the specific structure of the binding site creates a special secondary structure which depends on a definite combination of this type of bond. The appearance of such a combination is determined by antigen at the moment of formation of the antibody molecule from the "standard" peptide chains which are identical for all types of immunoglobulins. The experimental data now available do not support this hypothesis. Thus, for example, experiments on reverse denaturation and reconstitution of completely reduced active antibody fragments (cf. Chapter IV) indicate that the required secondary structure appears even in the absence of antigen.

Recently, Haurowitz, the most consistent defender of the template theory, presented a new conception of the possible participation of antigen in antibody synthesis [36, 39]. There is evidence that different antibodies differ in primary structure (cf. Chapter IV). On this basis, Haurowitz suggested that antigen induces changes not only in the secondary, but also in the primary, structure and that configurational changes due to templates may, in turn, result in changes in the amino acid sequence. Antigen participation concretely involves the following: antigen determinants combine with ribosomes or messenger RNA, as shown in Fig. 58. A determinant forms a complementary configuration of amino acid residues of the growing peptide chain around itself by means of intermolecular forces, and all the unsuitable amino-acyl derivatives of transport RNA are rejected. The remaining section of the peptide chain is determined directly by the sequence of nucleotides in messenger RNA. The principal objection to this hypothesis, which the author himself points out [39], is the absence of notable quantities of ribosomes in the fractions.

Mekler [4] advanced an original hypothesis. He proposed that when an antigen determinant enters a cell, amino acids spontaneously orient themselves around it; "this can ensure a sum bond of maximum firmness, i.e., those which store the most energy."

Further, it is suggested that these amino acids are not in a free state, but are bound with s-RNA (amino-acyl-sRNA). On the basis of these few amino-acyl-sRNA constructed in this manner is formed messenger RNA with the aid of a polymerase enzyme. This m-RNA evidently contains information on the structure of the peptide supplementary to the antigenic determinant. A complex consisting of a determinant, amino-acyl-sRNA, and messenger RNA can form a virus-like particle with a nucleus of messenger RNA. After the entrance of such a "virus" into a lymphoid cell, antibody synthesis begins. A large number of such antibody molecules are formed on the basis of information already present in the cell genome, while the binding site is formed on the basis of the messenger RNA of the "virus." This hypothesis lacks as yet any experimental confirmation.

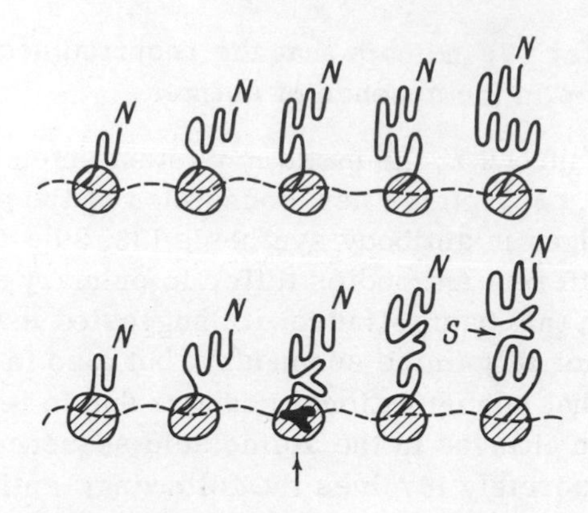

Fig. 58. Diagram illustrating the possible effect of the antigen determinant on formation of the binding site (S) at the moment of formation of an antibody peptide chain on the basis of ribosomes [26]. The antigen determinant is situated on the third ribosome (indicated by arrow).

The following facts are evidence of the possibility of direct participation of antigen in antibody synthesis. Antigen is present in the organism in notable quantities for very long periods commensurable with the duration of antibody synthesis (Chapter V). The last hypothesis of Haurowitz is supported by accumulated evidence that antigen may actually exist in a complex with RNA and that these complexes may play an important role in antibody formation. Finally, it appears probable that the information stored in messenger RNA on the amino acid sequence in a chain will vary under the influence of various external factors. Thus, the influence of alien substances (streptomycin, spermine) and temperature on the composition of peptides in a cell-free system of bacteria in the presence of polynucleotides has been demonstrated to be theoretically possible [20, 27, 60].

As has been shown by Sela (Chapter III) the total charge on antibody molecules stands in a definite proportional relationship to the charge on molecules of the corresponding antigen. These data are evidence in support of the view that antigen participates in some manner in the formation of antibody molecules.

Any theory must explain the absence of synthesis of antibodies to proteins intrinsic to the organism. As Haurowitz correctly suggests [36, 37], the absence of autoantibodies under normal conditions is attributable to the same causes as the impossibility of an immune response in artificially induced tolerance [33]. According to a wide-spread opinion, a large quantity of antigen must be present in the organism for the development of tolerance. The formation of autoantibodies is theoretically possible, however, and they are quite frequently found in various pathological processes [3].

Thus, the template theory, especially in Haurowitz' later version, explains many known facts. Its advantages are its concreteness and the possibility of its experimental verification, for example, in the aspect dealing with the role of RNA-antigen complexes. However, template theories do not explain the variety of N-terminal parts of immunoglobulin peptide chains. It has already been stated that any theory of antibody formation should explain the origin of the variability of these chain segments, since the specific activity of antibodies is governed by their structure.

Most investigators are now of the opinion that the specific structure of antibodies is predetermined and that the information for it, as well as for other proteins in the body, is contained in the determined sequence of DNA nucleotides.

THEORY OF THE PRE-EXISTENCE OF INFORMATION FOR ANTIBODY SPECIFICITY

If it is assumed, as originally done by Ehrlich in 1912, that the information for the antibody binding site exists in the organism before introduction of antigen, this hypothesis can be formulated in two basic versions. According to the first, there is a definite group of cells (clones) which are capable of synthesizing the given antibody, and antigen in some way singles out this group from other groups and stimulates its proliferation (clonal selection theories). In the second version, the cells are assumed to be multipotent, and under the effects of antigen, which inactivates any inhibitory mechanisms, each of them is able to form antibodies to any antigen (theories of derepression).

Jerne was the first after Ehrlich to point out the possible selective function of antigen. He suggested that lymphoid cells synthesize a large variety of γ-globulin molecules with the most varied structures. After immunization, some of these molecules, which have at random acquired a supplementary structure, combine with antigen. These specific complexes are quickly entrapped by phagocytes and then enter antibody-producing cells. Here the immunoglobulin molecule breaks free of the antigen and in some manner stimulates biosynthesis of molecules similar to itself.

In Chapter IV it was recalled that the organisms of nonimmune animals and humans contain antibodies to many antigens. On the other hand, the majority of lymphoid cells contain immunoglobulins, and the addition of specific anti-immunoglobulin antiserums to them provides an impetus to blast transformation.

These facts, however, are not sufficient to confirm the selective role of antigen. The nature of natural antibodies is unclear, but it is possible that they are the result of an immune response to some type of antigens. The presence of immunoglobulins in lymphocytes can be explained as follows: the cells have already come in contact with some type of antigens and synthesize antibodies to them.

Views according to which antigen is selective, not for already synthesized antibody molecules but for specialized cells capable of elaborating the specific antibody, have gained much support.

Burnet [12, 13] has elaborated consistently the hypothesis of antigen selectivity with respect to clones preexisting in the organism. According to Burnet, "the essence of a clonal-selection theory is the assumption that immunity and antibody synthesis are functions of clones of mesenchymal cells" [12]. Cells of a given clone are capable in the presence of antigen of rapid proliferation, conversion to plasmocytes, and response involving antibody formation. In other cases, however, cells may die, and tolerance results.

It is postulated that in the organism there exist "10^8 potential clones, each of which is formed as a result of mitosis of one or a very small number of cambial cells" [13]. There are conceivably two ways for these clones to develop. First, this may take place simply during cell differentiation as the organism develops. Secondly, clones may be formed from cells which have undergone somatic mutation.

The existence of somatic mutations is now considered proven; the number of lymphocytes is so large that the formation of a large number of clones in this way is fully possible [8, 9].

The clonal-selection theory has had a very great influence on the development of our views on antibody synthesis as well as several other immunological phenomena (tolerance, etc.). The introduction of the concept of lymphocyte clones was of theoretical importance. This theory stimulated the undertaking of very important experiments. However, not all experimental data can be explained by this theory. Thus, it has been shown that the action of an antigen is not restricted solely to stimulation of proliferation of precursor cells capable of an immune response. As has been discussed in Chapter V, the latent period of antibody elaboration in newborn animals cannot be shortened by increasing the number of transplanted cells.

It is also difficult to imagine that such an important and well-developed mechanism as the synthesis of specific antibodies is developed during the lifetime of an individual and then only on the basis of random processes, i.e., somatic mutations. However, attempts are now being made to form a conception of the concrete mechanisms of these processes.

Hypothesis of the Somatic Origins of Variability

In recent years, a number of hypotheses have been advanced on the possible origins of the variability of peptide chains of immunoglobulins during the development of lymphoid cells.

Smithies [68] proposed that the special variability of genes controlling immunoglobulin synthesis was due to intrachromatid rearrangements taking place during mitosis of lymphoid cells. For example, as Fig. 59A shows, intrachromatid inversion can result in a change in the sequence in a section of the DNA chain, since a part of the chain is replaced by its complementary part. However, current data on the primary structure of immunoglobulin peptide chains indicate that inversion is not the cause of variability [18].

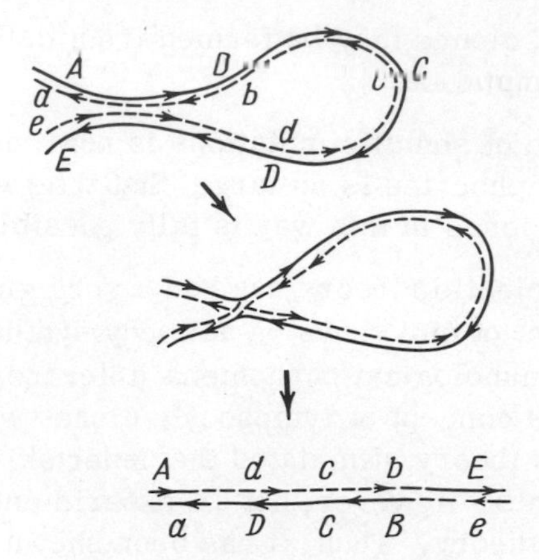

Fig. 59A. Chart illustrating disruption of the sequence of information with DNA due to inversion [42].

Later, Smithies proposed that antibody peptide chains are controlled by a gene pair comprising a "master gene" and a "scrambler" gene [69]. The first codes the entire peptide chain (VC) while the second controls only the variable region (V'). V and V' differ in certain residues. It is hypothesized that somatic recombination of these two nonallelic genes can result in a quite large group of new genes (Fig. 59B). Thus, if V and V' differ in 20 residues, 2^{20} different recombinations are possible. However, in this simplest form the hypothesis does not explain the facts – for instance the existence of hypervariable positions at certain points in the chain. Two genes are apparently not sufficient to explain the formation of all observed variations by means of somatic recombination [18].

Edelman and Gally [24] suggest that the diversity of the variable parts of immunoglobulin chains is attributable to somatic recombination of tandem duplicated genes which appeared during the course of evolution. The differences between these genes are due to point mutations. It is further postulated that a light chain gene (VC) coding the variable and constant parts is able to recombine at its V-section at some stage during lymphocyte maturation with a group of tandem duplicated V genes which differ from each other. The number of such V genes need not be very large. Crossing over between the V region of the VC gene and a V gene

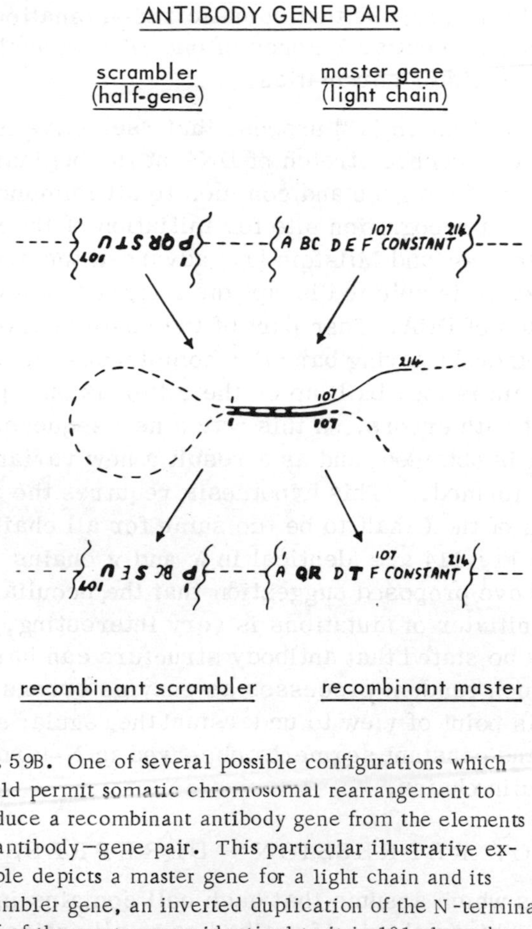

ANTIBODY GENE PAIR

scrambler (half-gene) master gene (light chain)

recombinant scrambler recombinant master

Fig. 59B. One of several possible configurations which would permit somatic chromosomal rearrangement to produce a recombinant antibody gene from the elements of an antibody–gene pair. This particular illustrative example depicts a master gene for a light chain and its scrambler gene, an inverted duplication of the N-terminal half of the master gene identical to it in 101 places but differing from it in 6: ABCDEF versus PQRSTU. The original chromosome is shown at the top of the figure, the synaptic configuration in the middle, and the rearranged chromosome below. Note the recombinant antibody gene, AQRDTF CONSTANT differing from the nonrecombinant only in the "variable" half of the molecule [69].

produces a new sequence in DNA, and, as a result, a new peptide chain variant. According to the authors' calculations, four V genes, each containing 10 unique codons, can produce 10^{12} individual sequences under conditions of unlimited crossing over.

Some authors propose that somatic recombinations may take place between consecutive V genes of one DNA stretch [40, 81] but not between sister chromatids.

Cohen and Milstein [18] propose that "selective mutation could be achieved if a specific stretch of DNA at the beginning of the invariable part of the gene and common to all immunoglobulin genes acted as a recognition site for initiation of the mutation process." Brenner and Milstein [11] advanced the hypothesis that a special enzyme is able to disrupt the integrity of a chain at a certain stretch of DNA. Then part of the chain is broken down by exonuclease thereby laying bare the homologous chain. The decomposed chain is then built up on the latter as on a polymerase template, but with errors. In this way a new sequence of nucleotides in DNA is obtained, and as a result a new variant of the peptide chain is formed. This hypothesis requires the sequence at the beginning of the C half to be the same for all chains. Indeed, the residues 110-114 are identical in λ and κ chains [18]. Although the above proposed suggestion that the peculiar sequence in DNA is the initiator of mutations is very interesting, it should nevertheless be stated that antibody structure can hardly be dependent on such random processes as enzyme errors. It is difficult from this point of view to understant the regular succession of variable and constant segments observed in N-terminal parts of immunoglobulin chains.

THEORY OF REPRESSION – DEREPRESSION

Theories which assume that each cell contains information necessary for biosynthesis of antibodies to all antigens merit special attention. The role of antigen is reduced to the specific function of relieving derepression of mechanisms responsible for antibody synthesis. This notion was first clearly presented by Szilard [76] and has been supported by Gurvich and Efroimson [8, 9].

Szilard assumed that antigen consists of a "nonantigen" component and several groups of metabolites linked to it. These latter are identical to one of the metabolites whose transformation in the body under normal conditions is catalyzed by a definite enzyme. The gene controlling synthesis of antibody to an antigen is formed as a result of mutations of the gene controlling synthesis of this enzyme. The activity of the gene responsible for antibody synthesis is inhibited by means of a special repressor. After introduction into the organism, antigen combines with the contact site of the special enzyme responsible for synthesis of the repressor, thereby blocking its activity. Synthesis of the repressor is thereby interrupted, and formation of antibodies specific for the introduced antigen begins [76].

This hypothesis cannot now be completely accepted since it is difficult to imagine that the synthesis of such different proteins as antibodies and enzymes can be regulated by similar genes. However, the main idea of the release of repression of specific genes by an introduced antigen is of considerable interest. Several difficulties can be avoided if it is assumed that the bulk of antibodies is synthesized under the control of the same genes which are responsible for the synthesis of the contact areas of enzymes [2]. From this point of view, it is not completely clear whether there exist antibodies to D-amino acids, i.e., to substances for which special enzymes are practically nonexistent in the organism [66, 74]. Furthermore, there is some doubt whether the contact area of enzymes is constituted on one section of a peptide chain, or whether it, like the enzyme binding site, is formed from a special tertiary structure of the protein [6].

At first glance, there seems little likelihood that every cell could contain information on antibodies to all possible antigens, whether natural or prepared in chemical laboratories; this is indeed an enormous number of substances. However, it should be taken into account that a certain number of antigens may have similar immune properties, which should manifest themselves as cross reactions. It is quite probable that substances having different chemical structures contain sections which are similar in configuration and which, consequently, are able to cross-react with antigen determinants. Haurowitz proposes that there exists "not more than 10^4-10^5 different antigenic determinants, and that 10^5 different immunoglobulin types would be sufficient to combine with all of the known antigenic determinants [38].

If it is assumed that the information on the structure of all types of antibodies is genetically predetermined in each cell, the question arises as to whether too much information is not needed for this DNA. However, simple calculations show that this possibility is very real [4]. The nucleus of one spleen cell contains $6.5 \cdot 10^{-12}$ g DNA [78]. Since information can be read from only one of two DNA chains [31, 54], this figure must be divided by two ($3.25 \cdot 10^{-12}$ g). This amount of DNA contains about $7 \cdot 10^9$ nucleotides. If it is assumed that only 10% DNA nucleotides contain data on protein structure, the cell DNA will contain information for formation of peptide chains with a total length of $2 \cdot 10^8$ amino acid residues (assuming that one residue is coded by three nucleotides), or for formation of 300,000 proteins with a molecular weight of about 70,000. Thus, a cell can theoretically contain information for the biosynthesis of a very large quantity of antibodies even if each antibody has its own characteristic light and heavy chains. This possibility becomes even more probable if it is assumed that the synthesis of only a specific small section, i.e., the antibody binding site, is controlled by a special gene (V), while the main part of all antibodies is synthesized under the control of general genes (C) [2, 22].

Hood, Dreyer et al. [22, 23, 43, 44] assumed that the chemical evolution of immunoglobulins is in principle similar to evolutionary processes known for such well-studied proteins as cytochrome C or hemoglobin. Genes which have appeared during evolution and which code the different peptide chains of hemoglobin are fixed in a genotype. These investigators advanced the hypothesis that genes formed as a result of duplication and subsequent point mutations for all the variable parts of immunoglobulin peptide chains are also contained in cell genotypes. This proposition does not now seem unlikely since it has been shown that the same genes may be duplicated several times [81]. Thus, for example, cistons, which code both types of ribosomal RNA, are duplicated several times in a cell genome [72].

It may now be regarded as certain that peptide chains of immunoglobulins are formed as integral wholes. If each chain is coded by two different genes, combination can occur either during RNA synthesis or in the already synthesized RNA molecules which contain information on the amino acid sequence of immunoglobulin peptide chains. It is possible that a mechanism exists which makes possible the formation of a single messenger RNA chain on a template consisting of DNA fragments removed from each other [2]. There is evidence that the variable sections of

$\varkappa$-chains have two variants which are controlled by two different genes. However, the stable section of chains of this type, judging from the majority of the evidence, is identical and is consequently controlled by a single gene [44]. This confirms the above-mentioned hypothesis that light chain synthesis is actually controlled by two types of genes "which are expressed as a single continuous polypeptide chain" [44]. The hypothesis of two genes coding one immunoglobulin chain explains the paradoxical inheritance of allotypical determinants of heavy chains of rabbit immunoglobulins. As was pointed out in Chapter VI, these determinants, inherited as allels, are found in all the chief types of immunoglobulins present in every animal. If these determinants are localized in the variable part of the chains controlled by specific genes common to all types of chains, the paradox is resolved.

There are several objections to the hypothesis of the existence of a large number of genes formed during the course of evolution [19, 53]. Thus, in this case it appears highly unlikely that allotypes are strictly inherited since when many genes are situated in sequence the crossing-over probability is increased [69]. Hence, several investigators are of the opinion that some type of special mechanisms are required to prevent crossing-over.

Some objections to the hypothetical existence of a large number of V genes are based on evolutionary conceptions [18]. Many facts indicate that the V segments in different animal species have common precursors. It is conceivable that the "conservative" invariable parts of the N-terminal halves of chains persist throughout the course of evolution and that, consequently, they are the same for all animal species. However, this is not what is observed; these parts are different in the λ and $\varkappa$ chains in the same animal species.

It is known that cells which form immunoglobulins are monopotent, i.e., they are able to synthesize only one variant of a light or heavy chain. It may be asked how this fact can be made to fit with the potential multipotency of cells. It is probable that after entering a lymphoid cell antigen behaves in such a way that the cell becomes incapable of responding to other antigens [76] and, in consequence, only one chain variant is selected.

The majority of so-called nonspecific serum immunoglobulins

are probably antibodies to some type of antigens. It is reasonable to assume that cells synthesizing immunoglobulins in determinable quantities have already come into contact with antigens and have thereby become monopotent. The blocking effect of antigen on a cell with respect to other antigens explains the phenomenon of antigen competition, in particular, as shown by Sela and Schechter in experiments with polypeptide antigenic determinants [64-66].

In what way does an antigen exert its depressing effect? It is conceivable that it acts on one of the stages of the process which ultimately leads to formation of antibody peptide chains: namely, on transcription, translation, or on another as yet unknown process [2, 36, 37]. It is possible that a small quantity of antibodies is formed in cells which have not had contact with antigen [77]. For example, it has been shown that in principle repressed genes may be effectively derepressed by DNA reduplication, and that as bacterial cells proliferate, considerable quantities of the repressed enzyme β-galactidase [32] are formed. On contacting a lymphoid cell antigen is able to form a complex with the precursor antibody, and in precisely this form exerts a specific derepressing and stimulatory effect.

What could be the nature of the specific repressors inhibiting the activity of antibody genes? This problem has not yet been elucidated for the synthesis of other proteins. However, most data indicate that the regulation of synthesis of messenger RNA on the basis of DNA is controlled by protein substances [7, 29]. These substances are synthesized in the same way as all other proteins on a template of special messenger RNA which is formed, probably, on the basis of regulator genes [7]. It is not impossible that histones serve as these regulators, but as yet their specificity has not been proven [15].

Finch advanced the hypothesis that the repressor of synthesis of a given antibody may be its heavy peptide chain [25]. Figure 60 shows a chart of antibody biosynthesis according to Finch's views. Antigen enters the cell and blocks the repressor. Messenger RNA synthesis begins on the basis of the liberated operator, and this leads to formation of antibody peptide chains. The heavy chains combine immediately with light chains so that they have no opportunity to block the operator.

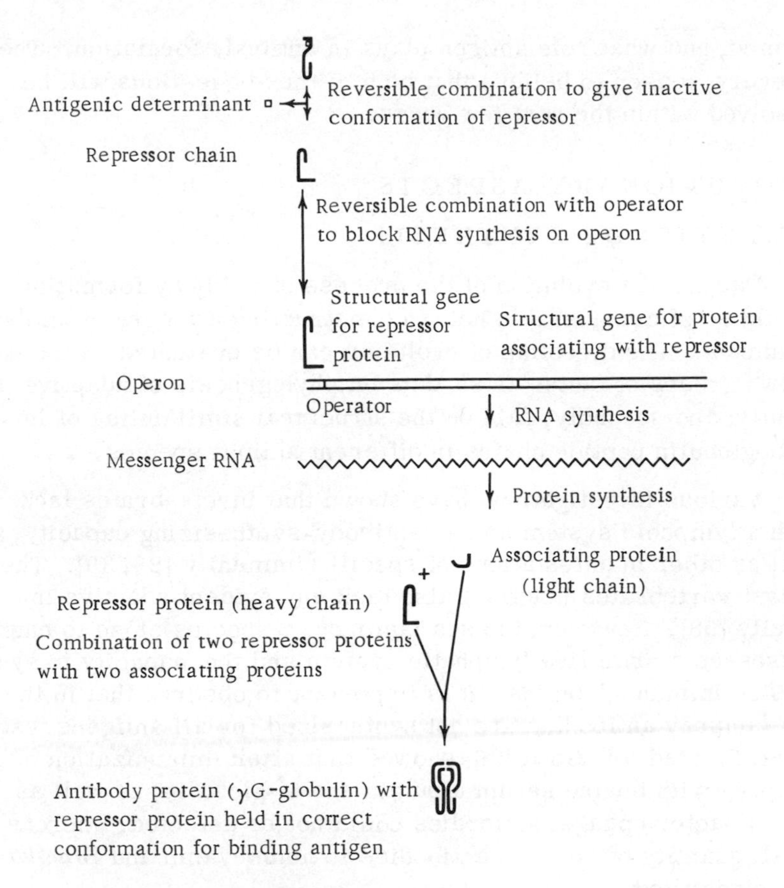

Fig. 60. Scheme of antibody biosynthesis and its regulation according to Finch [19].

This simple and consistent schema does not completely accord with the known facts. Thus, it is known that the loci of heavy and light chains are not closely bound (Chapter VI).

In conclusion, it should be noted that there is no decisive evidence in support of any of the theories discussed above. Perhaps their most important common feature is the acknowledgment that on the whole antibody synthesis takes place according to the same principles as synthesis of all other proteins. There still remains some dispute as to how that segment of immunoglobulin genes ultimately responsible for the structure of antibody binding sites is formed, and what role antigen plays in antibody formation. There is every reason to believe that both of these questions will be resolved within the next few years.

EVOLUTIONARY ASPECTS OF ANTIBODY SYNTHESIS

Data on the evolution of the process of antibody formation should be of considerable help in constructing a theory of antibody synthesis. The direction of evolution can be evaluated on the basis of two groups of facts: first, data on phylogenesis of adaptive immunity; and secondly, data on the structural similarities of immunoglobulin peptide chains in different animal species.

Various investigations have shown that invertebrates lack both a lymphoid system and an antibody-synthesizing capacity, as well as other manifestations of specific immunity [28, 30]. The lowest vertebrates (hagfish) also lack any evident adaptive immunity [58]. However, the sea lamprey, a species close to hagfish, possesses a primitive lymphatic system and the capacity to synthesize immunoglobulins. It is important to observe that in the sea lamprey antibodies are not synthesized for all antigens tested. Thus, Finstad and Good [26] showed that after immunization of sea lamprey with bovine serum albumin and γ-globulin, as well as with a bacteriophage, antibodies could not be detected, whereas a small quantity of specific antibodies to hemocyanin and *Brucella* was elaborated.

Further along the evolutionary scale, in sharks and rays, which are separated from future mammals by more than 200 million years, the lymphoid system is better developed, and they are able to respond to a greater percentage of antigens. In certain types of shark, antibody properties have been studied in detail. It was found that with respect to molecular weight antibodies are classified as 19 S and 7 S types. The molecules of both types of antibodies consist of heavy and light chains. However, the heavy chains of both types of antibodies are identical in both molecular weight (about 70,000) and fingerprints, electrophoretic properties, antigenic properties, and carbohydrate content (about 8%). Thus, the heavy chains of shark immunoglobulins resemble the μ chains of mammals [16, 55, 56, 63].

TABLE 44. Amino Acid Sequences of N-Terminal Pentapeptides for Paddlefish, Shark, Mouse, and Man [61]

Paddlefish H-chains	Residue number and mole % yield					
	1	2	3	4	5	6
From pooled 19 S immunoglobulins	Asp 67%	Ile	Val 71%	Ile (Leu)	Thr	
Paddlefish L-chains	Asp 9%	Ile 14%	Val 10%	Ile (Leu)	Thr	
Shark H chains from 7 S and 19 S immunoglobulins	Glu	Ile	Val	Leu	Thr	Gln
Shark L chains	Asp Glu	Ile	Val	Leu Val	Thr	
Mouse kappa chains and human kappa chains	Asp	Ile	Val	Leu Met Val	Thr	Gln
Light chains of human 7 S myeloma protein	Glu	Ile	Val	Leu	Thr	Gln

* Major residues identified. Yields calculated as moles PTH amino acid per mole of light chain (22,000 g), of heavy chains (70,000 g) (uncorrected for procedural losses).

A study of the primary structure of the N-terminal part of light and heavy chains of immunoglobulins of the leopard shark [75] disclosed that light and heavy chains of 19 S and 7 S immunoglobulins have much in common both among themselves as well as with light $\varkappa$ chains of mammals (Table 44), which is an indication of the possible existence of a common precursor to them. The paddlefish shows an analogous picture [61].

Two types of immunoglobulins – 19 S and 7 S – also exist in amphibians; however, heavy chains in these proteins are no longer identical [57].

In recent years, much knowledge has been acquired on the primary structure of immunoglobulins. A comparison of the sequence of amino acid residues of different types, of light chains, and of light and heavy chains revealed a considerable degree of homology; from this fact very important conclusions can be drawn concerning the evolution of immunoglobulins. Thus, a comparison

VARIABLE REGION OF LIGHT CHAINS

	CYS	CYS		TOTAL
Human K	22	65	19	107
λ	20-21	64-67	22	108-112
Mouse K	22	65-69	20	107-111
Rabbit F_c Heavy chain	(up to 36)	(50-60)	20	(111)

CONSTANT REGION OF LIGHT CHAINS

	CYS	CYS		
Human K	26	60	20	107
λ	26	58	19	105
Mouse K	26	60	20	107
Rabbit F_c Heavy chain	25	57	21	(105)

Fig. 61. The homologous intrachain disulfide bridge structures of the variable and constant regions of human and mouse light chains compared to corresponding half-cystine positions in rabbit Fc heavy chain fragment [62].

of the sequence of the stable half (C end) of human light λ chains with human and mouse light κ chains disclosed 40 positions in which identical residues were found (39%). These residues were distributed rather evenly over the chain length. Even more positions with identical residues (60 or 64%) were found among the stable parts of human and mouse light κ chains. These data indicate a quite early evolutionary differentiation of the two types of light chains [62].

A certain structural similarity was also found between heavy and light chains. This was especially pronounced among intrachain disulfide bridges (Fig. 61). A similar number of amino acid residues were also found in the stable and variable parts of human and mouse light chains and in analogous sections of the Fc fragment within the disulfide bridge, as well as on both sides of the bridge [62]. This stability of disulfide bridges is fully applicable since their position has a great influence on the overall chain configuration. Hill et al. [41, 42] discovered a homology in the position of amino acid residues between the Fc fragment of rabbit γG-globulin and the variable and stable parts of λ and κ human light chains. Of the 118 residues of an Fc fragment, 56 (47%) were identical with light chain fragments, while another 34 (29%) had a similar chemical structure.

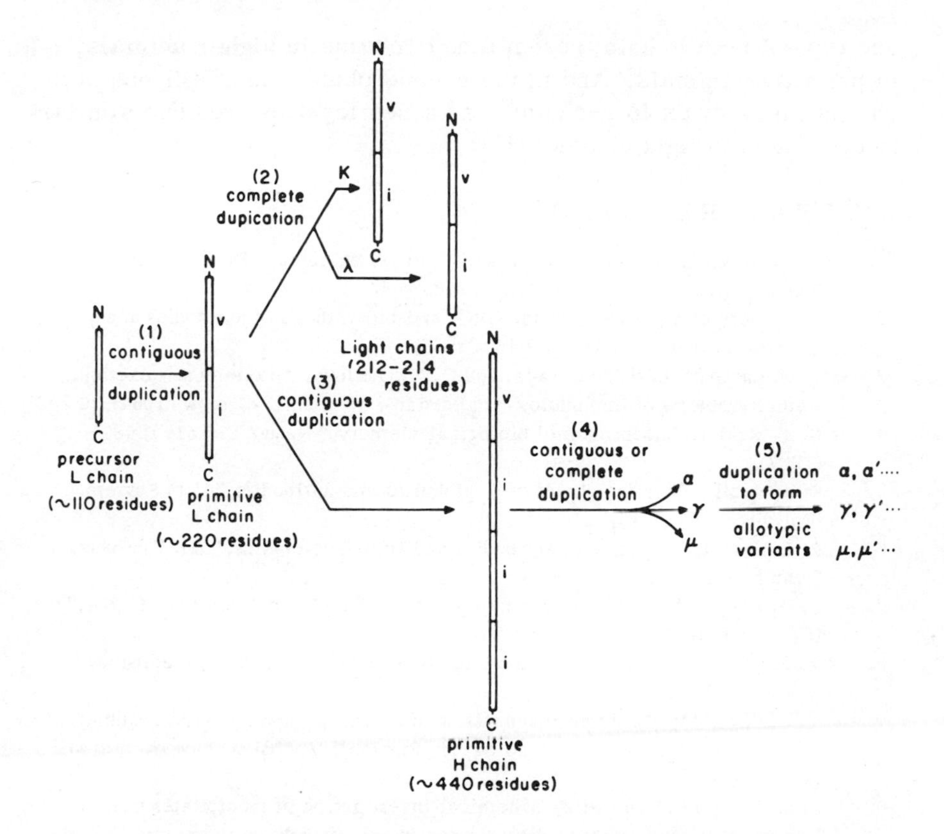

Fig. 62. A proposed mechanism for the evolution of the immunoglobulins. The details are described in the text [42].

If the structural homology of heavy and light chains is taken into account, the hypothesis of their descendance from a common precursor appears probable. Figure 62 shows an evolutionary chart of immunoglobulin peptide chains [42]. It is hypothesized that the precursor chain had a size approximately equal to one half of the present light chain. Then by gene duplication a primitive light chain was formed which then was differentiated into a λ chain and a κ chain; in addition, a primitive heavy chain appeared as a result of duplication. This hypothesis is confirmed by the data in Table 44, which shows the structural similarity of light and heavy chains in primitive vertebrates.

Evidence indicates that type μ chains were the first heavy chains to appear. In the first place, we have the existence of only one type of heavy chain, resembling μ chains in higher animals, in primitive animals. And in the second place, the COOH end of μ chains, in contrast to γ-chains, has a semicystinyl residue similar to that found in light chains [21].

LITERATURE CITED

1. A. E. Gurvich, "On antibody synthesis," in: Virology and Immunology [in Russian], Nauka, Moscow (1964), p. 243.
2. A. E. Gurvich and R. S. Nezlin, "DNA and biosynthesis of antibodies and γ-globulins," Usp. Biol. Khim. 7:150 (1966).
3. I. M. Lampert, L. V. Belezkaya, and O. N. Grislova, Autoimmune Reaction. Actual Problems of Immunology [in Russian], Meditsina, Moscow (1964), p. 202.
4. L. B. Mekler, "Mechanism of biological memory," Nature 215(5100):481 (1967).
5. R. S. Nezlin, "Peptide chains of γ-globulins and antibodies," Usp. Sovrem. Biol. 58(5):201 (1964).
6. R. S. Nezlin, "Structure of antibodies and immunoglobulins," Zh. Vsesoyuz. Obshchestva Khim., No. 4 (1968).
7. R. V. Khesin, "Role of proteins in regulation of biological activity of DNA," Usp. Sovrem. Biol. 59(1):12 (1965).
8. V. P. Efroimson, Introduction to Medical Genetics [in Russian], Meditsina, Moscow (1963).
9. V. P. Efroimson, "Genetic mechanisms of hereditary and adaptive immunity," in: Actual Problems in Immunology [in Russian], Meditsina, Moscow (1964), p. 31.
10. F. Breinl and F. Haurowitz, "Chemical investigation of precipitates from hemoglobulin and antihemoglobulin serum and remarks on the nature of antibodies," Z. Physiol. Chem. 192:45 (1930).
11. S. Brenner and C. Milstein, "Origin of antibody variation," Nature 211(5046):242 (1966).
12. F. Burnet, The Integrity of the Body, Harvard University Press, Cambridge Mass. (1962).
13. F. A. Burnet, "A Darwinian approach to immunity," Nature 203(4944):451 (1964).
14. F. M. Burnet, "Evolution of the immune process in vertebrates," Nature 218(5140):426 (1968).
15. H. Busch, Histones and Other Nuclear Proteins, Academic Press, New York (1965).
16. L. W. Clem and P. A. Small, "Phylogeny of immunoglobulin structure and function. I. Immunoglobulins of the lemon shark," J. Exp. Med. 125(5):893 (1967).
17. S. Cohen and C. Milstein, "Structure of antibody molecules," Nature 214(5087):449 (1967).
18. S. Cohen and C. Milstein, "Structure and biological properties of immunoglobulins," Advan. Immunol. 7:1 (1967).

19. M. Cohn, "Antibody synthesis: The take home lesson," in: Gamma-Globulin. Structure and Control of Biosynthesis, Almquist and Wiksell, Stockholm (1967), p. 615.
20. J. Davies, W. Gilbert, and L. Gorini, "Streptomycin, suppression and the code," Proc. Nat. Acad. Sci. USA 51:883 (1964).
21. R. F. Doolittle, S. J. Singer, and J. Metzger, "Evolution of immunoglobulin polypeptide chains: carboxy-terminal of an igM heavy chain," Science 154(3756):1561 (1966).
22. W. J. Dreyer and J. Bennett, "The molecular basis of antibody formation: a paradox," Proc. Nat. Sci. USA 54(3):864 (1965).
23. W. J. Dreyer, W. R. Gray, and L. Hood, "The genetic, molecular, and cellular basis of antibody formation: some facts and a unifying hypothesis," Cold Spring Harbor Symp. 32:353 (1967).
24. G. M. Edelman and J. A. Gally, "Somatic recombination of duplicated genes: an hypothesis on the origin of antibody diversity," Proc. Nat. Acad. Sci. USA 57(2):353 (1967).
25. L. R. Finch, "Gamma-globulin operon — a hypothesis for the mechanism of the specific response in antibody synthesis," Nature 201(4926):1288 (1964).
26. J. Finstad and R. A. Good, "The evolution of the immune response. III. Immunologic responses in the lamprey," J. Exp. Med. 120(6):1151 (1964).
27. S. M. Friedman and J. B. Weinstein, "Lack of fidelity in the translation of synthetic polyribonucleotides," Proc. Nat. Acad. Sci. USA 52(4):588 (1964).
28. "Genetics of the immune response," Report of a World Health Organization Scientific Group, WHO, Geneva (1967).
29. W. Gilbert and B. Müller-Hill, "Isolation of the Lac-repressor," Proc. Nat. Acad. Sci. USA 56:6 (1891).
30. R. A. Good and B. W. Papermaster, "Ontogeny and phylogeny of adaptive immunity," Advan. Immunol. 4:1 (1964).
31. W. K. Guild and M. Robinson, "Evidence for message reading from a unique strand of pneumococcal DNA," Proc. Nat. Acad. Sci. USA 50(1):106 (1963).
32. P. Hanawalt and R. Wax, "Transcription of a repressed gene: evidence that it requires DNA replication," Science 145(3636):1061 (1964).
33. M. Hasek, A. Lengerova, and T. T. Hraba, "Transplantation immunity and tolerance," Advan. Immunol. 1(1) (1961).
34. F. Haurowitz, Chemistry and Biology of Proteins, Academic Press, New York (1950).
35. F. Haurowitz, "The template theory of antibody formation," in: Conceptual Advances in Immunology and Oncology, Harper and Row, New York (1963), p. 32.
36. F. Haurowitz, "Antibody formation and the coding problem," Nature 205(4974):847 (1965).
37. F. Haurowitz, "Antibody formation," Physiol. Rev. 45(1):1 (1965).
38. F. Haurowitz, "The evolution of selective and instructive theories of antibody formation," Cold Spring Harbor Symp. 32:559 (1967).
39. F. Haurowitz, Immunochemistry and the Biosynthesis of Antibodies, Wiley, New York (1968).

40. N. Hilschmann, "The structure of the immunoglobulins (L-chains) and the antibody problem," in: Gamma-Globulins. Structure and Control of Biosynthesis, Nobel Symposium, 3, Almquist and Wiksell, Stockholm (1967), p. 33.

41. R. L. Hill, R. Delaney, R. E. Fellows, and H. E. Lebovitz, "The evolutionary origins of the immunoglobulins," Proc. Nat. Acad. Sci. USA 56(6):1762 (1966).

42. R. L. Hill, H. E. Lebovitz, R. E. Fellows, and R. Delaney, "The evolution of immunoglobulins as reflected by the amino acid sequence studies of rabbit Fc fragment," in: Gamma-Globulins. Structure and Control of Biosynthesis, Nobel Symposium, 3, Almquist and Wiksell, Stockholm (1967), p. 109.

43. L. E. Hood, W. R. Gray, and W. J. Dreyer, "On the mechanism of antibody synthesis: a species comparison of L-chains," Proc. Nat. Acad. Sci. USA 55(4):826 (1966).

44. L. Hood, W. R. Gray, B. G. Sanders, and W. J. Dreyer, "Light chain evolution," Cold Spring Harbor Symp. 32:133 (1967).

45. N. K. Jerne, "The natural selection theory of antibody formation," Proc. Nat. Acad. Sci. USA 41:849 (1955).

46. N. K. Jerne, "Summary: Waiting for the end," Cold Spring Harbor Symp. 32:591 (1967).

47. E. A. Kabat, "A comparison of invariant residues in the variable and constant regions of human K, human L, and mouse K Bence–Jones proteins," Proc. Nat. Acad. Sci. USA 58(1):229 (1967).

48. E. A. Kabat, "The paucity of species-specific amino acid residues in the variable regions of human and mouse Bence–Jones proteins and its evolutionary and genetic implications," Proc. Nat. Acad. Sci. USA 57(5):1345 (1967).

49. E. A. Kabat, Structural Concepts in Immunology and Immunochemistry, Holt, Reinhart, and Winston, New York (1968).

50. F. Karush, "The role of disulfide bonds in antibody specificity," J. Am. Chem. Soc. 79(19):5323 (1957).

51. F. Karush, "Disulfide pairing and the biosynthesis of antibody," in: Immunochemical Approaches to Problems in Microbiology, Rutgers University Press, New Brunswick (1961), p. 368.

52. J. Lederberg, "Genes and antibodies," Science 129(3364):1649 (1959).

53. E. S. Lennox and M. Cohn, "Immunoglobulins," Ann. Rev. Biochem. 36:365 (1967).

54. J. Marmur and C. M. Greenspan, "Transcription in vivo of DNA from bacteriophage SP8," Science 142(3590):387 (1963).

55. J. Marchalonis and G. M. Edelman, "Phylogenetic origins of antibody structure. I. Multichain structure of immunoglobulins in the smooth dogfish (Mustelus canis)," J. Exp. Med. 122(3):601 (1965).

56. J. Marchalonis and G. M. Edelman, "Polypeptide chains of immunoglobulins from the smooth dogfish (Mustelus canis)," Science 154(3756):1567 (1966).

57. J. Marchalonis and G. M. Edelman, "Phylogenetic origins of antibody structure. II. Immunoglobulins in the primary response of the bullfrog, Rana catesbiana," J. Exp. Med. 124(5):901 (1966).

58. B. W. Papermaster, R. M. Condie, and R. A. Good, "Immune response in the California hagfish," J. Am. Chem. Soc. 62(10):2643 (1940).

59. L. Pauling, "A theory of the structure and process of formation of antibodies," J. Am. Chem. Soc. 62(10):2643 (1940).
60. S. Pestka, R. Marshall, and M. Nirenberg, "RNA codewords and protein synthesis. V. Effect of streptomycin on the formation of ribosome–sRNA complexes," Proc. Nat. Acad. Sci. USA 53(3):639 (1965).
61. B. Pollara, A. Suran, J. Finstad, and R. A. Good, "N-terminal amino acid sequences of immunoglobulin chains in polyodon spathula," Proc. Nat. Acad. Sci. USA 59(4):1307 (1968).
62. F. W. Putnam, "Structure and evolution of kappa and lambda light chains," in: Gamma-Globulins. Structure and Control of Biosynthesis, Nobel Symposium 3, Almquist and Wiksell, Stockholm (1967), p. 45.
63. V. M. Sarich and G. M. Edelman, "Polypeptide chains of immunoglobulins from the smooth dogfish (Mustelus canis)," Science 154(3756):1567 (1966).
64. I. Schechter, "Competition of antigenic determinants," Biochem. Biophys. Acta 104(1):303 (1965).
65. I. Schechter and M. Sela, "Preferential formation of antibodies specific toward d-amino acid residues upon immunization with poly-dl-peptidyl proteins," Biochemistry 6:897 (1967).
66. M. Sela, "Use of synthetic polypeptides in studies on the control of antibody formation," in: Gamma-Globulins. Structure and Control of Biosynthesis, Nobel Symposium 3, Almquist and Wiksell, Stockholm (1967), p. 455.
67. S. J. Singer and R. F. Doolittle, "Antibody active sites and immunoglobulin molecules," Science 153(3731):13 (1966).
68. O. Smithies, "Gamma-globulin variability: a genetic hypothesis," Nature 199(4900):1231 (1963).
69. O. Smithies, "Antibody variability," Science 157(3786):267 (1967).
70. J. Sterzl, "Actual problems of immunology," Usp. Sovrem. Biol. 51(3):337 (1961).
71. J. Sterzl and A. M. Silverstein, "Developmental aspects of immunity," Advan. Immunol. 6:337 (1967).
72. J. Stevenin, J. Samec, M. Jacob, and P. Mandel, "Determination of the fraction of the coding genome for ribosome RNA and messenger RNA in the brain of the adult rat," J. Mol. Biol. 33(3):777 (1968).
73. G. T. Stevenson, "Binding properties of A chains of plasma γ-globulins," Nature 206(4980):163 (1965).
74. Y. Stupp and M. Sela, "Further studies on the immunogenicity of d-amino acid polymers," Biochim. Biophys. Acta 140(2):349 (1967).
75. A. A. Suran and B. W. Papermaster, "N-terminal sequences of heavy and light chains of leopard shark immunoglobulins: evolutionary implications," Proc. Nat. Acad. Sci. USA 58(4):1619 (1967).
76. L. Szilard, "The molecular basis of antibody formation," Proc. Nat. Acad. Sci. USA 46(3):293 (1960).
77. D. W. Talmage and D. S. Pearlman, "The antibody response: a model based on antagonistic actions of antigen," J. Theoret. Biol. 5(2):321 (1963).
78. R. W. Thomson, F.C. Heagy, W. C. Hutchinson, and J. N. Davidson, "The deoxyribonucleic acid content of the rat cell nucleus and its use in expressing

the results of tissue analysis, with particular reference to the composition of liver tissue," Biochem. J. 53(3):460 (1953).

79. K. Titani, M. Wikler, and F. W. Putnam, "Evolution of immunoglobulins: structural homology of kappa and lambda Bence–Jones proteins," Science 155(3764):828 (1967).

80. Z. Trnka, "Experimental data on the clonal theory of immunity," J. Epidemiol. Microbiol., Immunol., Prague 8:79 (1961).

81. H. L. K. Whitehouse, "Crossover model of antibody variability," Nature 215(5099):371 (1967).

Index

Acetic acid 48
N-acetyl-D-glucosamine 52
N-acetyl-D, L-homocysteinethiolactone 39
Actinomycin D 233-235, 238
Adsorbent 2
Adsorption 10
Affinity labeling 165, 177
Agammaglobulinemia 306
Agar 30
Agglutination 1, 16
Agglutinins 308
Alanine 127, 183, 310
Albumin 2, 4, 52
Alcohol dehydrogenase 283
Aldolase 264
Alizarin 42
Alkali solution 33
Alkyl halide 46-48
Allergens 43
Allotypes
  human 317-319
  mouse 320-321
  primate 320
  rabbit 313
Alpha helix 82-83
Aluminum oxides 42
Ambrosia 53
Amines 43, 52
Amino acids
  dicarboxylic 85
  hydroxy 85
Amino cellulose 47, 48
Amino groups 43
p-Amino benzoic acid 39, 107
p-Amino benzoyl group 38
p-Amino benzyl cellulose 45
p-Amino hippurate 37
p-Amino phenylalanine 130
p-Amino phenyl-β-lactoside 163
p-Amino phenylarsonic acid 39
p-Amino phenyl-azo-p-phenylarsonic acid 43-44
Amylase 52
Anaphylaxis 3, 81, 215, 272
Anatoxin 51, 216, 222, 235
Animals, newborn 311-312
Anisotropy 19
Antibody
  amino acid composition 85-88
  amino acid terminal residues 88-89, 127, 135
  -antigen complex 4, 31
    bond 32
    dissociation 32-41
    interaction 27
  antidiphtheria 12
  antiinfluenza 36, 48-50
  antihapten 3, 41, 43
  antiovalbumin 36, 40
  antipneumococcal 31-34
  antipolysaccharide 28, 33, 188
  antitissue 31-32
  binding site 65, 184, 347
  biosynthesis 11, 209-286
  bound 9
  chemical properties 85-91
  clonal selection theory 353
  conjugated 19, 32
  coupled 14, 49

Antibody (Continued)
crystalline 37
donkey 267
dynamics of synthesis 240-249
electrophoresis 77-81
evolution 362-366
fluorescence 84-85
formation by single cells 258-259
formation by organs 220
genes 321-326
guinea pig 148
heterogeneity 96-111
horse 54, 89, 92
human 89
incomplete 92
inhibitors 280-286
irradiated 18
media for 259
molecular hybrids 156-166
molecular weight 75-77
mouse 148
natural 312
nitrogen 3
nonprecipitating 8, 12, 20, 92
optical properties 81-85
peptide maps 187-191
primary structure 184-186
priming 282
precipitating 4
pre-existing information theory 351-356
rabbit 17, 33, 54, 75, 84, 85, 87, 89, 96, 108, 129, 134
radioactive 10-13, 20, 32, 46, 252, 262
repression-derepression theory 356-362
7 S 249-254
19 S 249-254
secondary structure 191-194
sensitizing cells 220-228
size 71-75
shape 71-75
structure 184-186, 191-194
synthesis 308-311

Antibody (Continued)
template theory 348-351
theories 347-366
titration 2
valence 91-92
virus 12-13
virus neutralizing 241
Wassermann 32

Antigen
adsorbed 1, 42-43
allotypic 312-314
antibody complex, see Antibody
azo- 38
bacterial 226, 228
binding site 175
bound 47
cellulose linked 9
coupled 2, 9, 13
denatured 13, 40
determinants 312
dysentery 228
fixed 10-15, 266
fragments 213
homologous 1
insoluble 11, 40
labeled 3, 14, 214
linked (to insoluble support) 43-54
neutralization 3
pathways 210
precipitating 104
properties 93-95
purified 13, 27-63
Shigella 210
soluble 210
synthetic 348
transformation 210
typhoid 228

Antiinfluenza serum 14

Antitoxin 37, 51, 78, 95
diphtheria 51, 78, 95, 144, 157, 158, 163, 222, 261, 265
tetanus 78, 95, 158, 216

Arginine 43, 86

Arsanilazo protein 103

p-Arsanilic acid 17, 39, 43, 45

Aspartic acid 86, 127, 150
Attardi method 255
Autoradiography 329
Autosome 310
8-Azaguanine 282
6-Azauridine 284
p-Azobenzoate 129, 137, 159-160
4-Azonaphthalene-I-sulfonate group 166
p-Azophenylarsenate 19, 39, 86, 87, 132, 165, 169, 177
p-Azophenyl-β-lactoside 35, 129
p-Azophenyltrimethylammonium 86-87
Azoprotein 38, 43

Bacteria 12, 13, 32, 54, 210, 221, 222, 226-228, 236, 240, 250, 253, 259, 309, 322, 323
Bacteriophages 3, 159, 215, 221, 234-237, 241, 250, 322
Bence-Jones protein 65, 70, 74, 93, 155-156
Bentonite 2
Benzene 48
Benzidene
  bis-diazotized 2, 40, 41, 45
  protein complex 2
  stain 260
  tetra-diazotized 45
Bio-gel polyacrylamide 30
  R-200 30
  R-300 30
3, 6-Bis-(acetoxymercurimethyl)-dioxane 40
Bismuth gallate 42
Blast cells 327
Blood groups
  anti-A 52, 102
  anti-B 52
Bonds
  chemical 43-54
  covalent 19, 165
Bonds (Continued)
  diazo 38, 43, 44, 52
  disulfide 127, 130, 131, 137, 139-143, 150, 166-168, 171, 174, 191, 193,
  noncovalent 159
  peptide 52, 86
Botulinum toxin 18
Bovine serum albumin 13, 33-35, 39, 41, 44, 45, 52, 53, 108
  gamma globulin, see Globulin
Boyden method 1
Breinl-Haurowitz hypothesis (of antibody synthesis) 348
Bromacetyl cellulose 51
5-Bromdeoxyuridine 285
Bromelain 138
Bromsulfalein 6
Brownian movement 19
Brucella 236, 245, 362
Burnet hypothesis (of antibody synthesis) 352

Carbodiimides 44
Carbon 2
Carbon dioxide 35
Carbon disulfide 47
Carboxypeptidase 52
Casein 43
Cellophane dialysis 16
Cellulose 9, 11, 44-46
  amino- 47-48
  p-aminophenylbutyryl-aminoethyl-51
  bromacetyl- 51
  carboxymethyl- 30, 35, 44, 45, 128-133, 143, 189, 260
  chitosan-treated 42
  DEAE- 29, 35, 66, 67, 96, 133, 138
  diazotized- 48
  ether- 51
  particles 9, 11
  powder 9, 43, 45, 47
  TEAE 30
Ceruloplasmin 30

Charcoal 42
Chicken
  embryo 336
  immune 336
Chloramphenicol 282-283
Chloride group 43
2-Chloroethanol 83
Cholesterol 42
Chromatography
  descending 9
  ion-exchange 28
Chymotrypsin 83, 187, 188, 266
Chymotrypsinogen 266, 269
Cohen & Milstein hypothesis (of antibody synthesis) 356
Colchicine 284
Cold hemagglutinins 32
Collagen 141
Collagenase 37
Collodion 2, 42
Colorimeter
  photoelectric 6
  Russian FEK-M 6
Complement fixation 3
Coon immunofluorescence method 220-223, 233
Copper ammonium solution 47-48
Coprecipitation 8
Cortisone 243
Cotton cloth 42
Counter
  end-window 11
  scintillation 11
Coxsackie virus 241
Cryoglobulinemia 156
Cyanogen bromide 168-169
Cyclohexane carboxylic acid 107
Cyclophosphamide 282
Cysteine 43, 128, 130, 132, 136, 150
Cytidine 231-232

DEAE
  cellulose 29, 35, 66, 67, 96, 133, 138
  Sephadex 29, 30, 96, 276
Decomplementation 31
Deoxyribonucleic acid (DNA) 193, 224, 228, 305, 326, 329, 331-334
Detergents 35, 142, 145, 149
Determinants
  allotypic 312
  antigenic 312
Dextran 38, 181
  gel 30, 102
Dextranase 38
Dialysis 3, 132, 192, 193
  equilibrium analysis 3, 15-17, 129, 158
Diaminodiphenylamine 2
Diazoacetamide 176
Diazo groups 45, 47
Dicyclohexylcarbodiimide 44
1, 3-Difluoro-4, 6-dinitrobenzene 2
2, 4-Dinitrophenyl groups 35, 97, 158, 159, 178, 192, 272, 310
2, 4-Dinitrophenyl-lysine 19, 44
Diphtheria
  anatoxin 51, 327
  antitoxin 51, 78, 95, 144, 157, 158, 163, 222, 261, 265
  toxin 2, 261, 284
Disulfide
  bonds 127, 130, 131, 137-143, 150, 166, 192, 193
  bridges 41, 151, 167, 364
DNA, see Deoxyribonucleic acid
Dodecylsulfate 35
Drude equation 82
Dysgammaglobulinemia 226

Edelman & Galby hypothesis (of antibody synthesis) 354-355
Egg albumin, see Ovalbumin
Ehrlich hypothesis (of antibody synthesis) 351
Electric field 2
Electrophoresis 2, 28, 34, 133, 144-145, 148
  antibody 78-81
  immuno- 2
  paper 77

Electrophoresis (Continued)
starch 77, 144-145
Tiselius 77-79
Endoplasmic reticulum 222
Enzymes
adaptive 305
alcohol dehydrogenase 283
aldolase 264
amylase 52
carboxypeptidase 52
collagenase 37
chymotrypsin 83, 187, 188, 266
dextranase 38
lysozyme 36, 37, 179, 183, 189, 191, 266, 269
nagarase 190
papain 35, 94, 128-139, 151-154, 161, 174, 183, 188, 189
pepsin 52, 127, 134-139, 151, 155, 157, 167, 169, 189
proteases 37, 105, 132, 138, 189
ribonuclease 40, 52, 83, 96, 107, 191, 229, 231, 236-238, 279
trypsin 37, 96, 138, 149, 150, 187, 189
urease 38, 52, 127
Equilibrium dialysis 3, 15-17, 129, 158
Erythrocytes 1, 32, 169, 210, 219, 226, 238, 239, 250, 260, 275, 283, 308, 309
conjugated 2
coupled 2
stroma 43, 45, 52
tanned 261
Ethanol 28-29
Ethylchloroformate 41
Ethylmercaptan 169
Euglobulin 28
Exanthematous fever 12

F-groups 152
Fats 29
Fatty acids 249
Ferric oxides 42
Ferritin 210-212, 222, 224, 311
Feulgen stain 327
Fever, exanthematous 12
Fibrin 40
Fibrinogen 40, 43
Finch hypothesis (of antibody synthesis) 360-361
Fingerprint method 187
Flagella, bacterial 221, 322
Flagellin 211
Fluorescein isothiocyanate 19, 37
Fluorescence
intrinsic 19
methods 3, 18-20
polarization 19, 20, 84
quenching 18, 19
spectrum 18
Formaldehyde 41
Formic acid 143, 144, 158, 169, 266
Freund adjuvant 217
Fucose 91, 137, 152

Galactosamine 151
D-Galactose 52
Gaussian function 104, 111
Gel
diffusion 2
filtration 29, 30, 36, 266, 268, 269
methods 2
see also Sephadex
Gelatin 37
Gelatin polytyrosyl 37
Genes
expression 322-326
hypomorphic 308
Mendelian 310
Glass beads 42
Globulins
alpha 30
bovine 73, 80, 81, 89, 133-135, 158, 243
carbohydrate content 91
fractions 28
gamma 15, 28, 33, 43, 44, 53, 67, 70, 71, 83, 87, 96, 97, 109-111, 128, 129, 134, 138, 139, 148

Globulins (Continued)
gamma-A 30, 65-68, 72-79, 169, 227, 248
gamma-D 69
gamma-E 69
gamma-G 30, 65-81, 88-98, 108, 111, 127, 131-132, 136, 164, 226, 227, 267
gamma-M 30, 65-73, 77, 79, 90-92, 95, 108, 111, 132, 169, 226, 227, 248, 267
guinea pig 74, 79, 94, 133, 165
horse 73, 80-83, 89, 94, 95, 133-135, 146
human 72, 80, 81, 89, 90, 94, 96, 127, 128, 132-136, 148, 150, 152, 154, 315-320
immune 33, 41, 65-71
macro 30, 67, 92, 167, 170
micro 69-71, 266-271
monkey 74
mouse 74, 94, 133, 136
pig 30, 70, 73, 80, 89, 127, 133-135, 170
pseudo 38
rabbit 72, 80, 83, 89-91, 94, 95, 127-136, 146, 148, 150-153, 167-168
serum 69, 71
Glucagon 77
Glutamic acid 150
Glutamyl lysine 310
Glycine 30, 43, 109, 150, 212, 242
Glycinin 43
Glycoprotein 85
Gonadotropin 54
Grabar method 78
Granuloma 217, 220
Guanidine
chloride 191-193
groups 43
Guinea pig 78, 81, 110, 250, 309
Gurvich method 9
^{3}H-toxin 212
Hagfish 362
Haptens 3, 15, 35, 42, 103, 104, 106
-antibody reaction 15-17
bound 18-19
conjugated 132, 137, 189
divalent 75
DNP-lysine 18
dye 16, 104, 180
fixed 43
free 19
Haptens
labeled 36
univalent 16
Haurowitz theory (for antibody synthesis) 348-351
Heat 32
Heidelberger precipitation method 2, 3-8, 12, 260
α-Helix 82-83, 139
Hemagglutination 2, 3, 261
inhibition 13, 36, 48, 50
method 2
Hemagglutinins, cold 32
Hemocyanin 44, 107, 211, 213, 215, 236, 237, 311, 362
Hemoglobin 82
Hemolysins 217, 219, 226, 238, 242, 260, 309
Hemolysis
plaques 219, 223, 238, 239, 260
zones 239
Heteroimmunization 312
Hexoses 137, 152
Hexosamine 137, 152
Histidine 43, 187, 212, 310
Histocompatibility 310
Histones 333
Homoserine 168
Homotransplant 236
Hormone 243, 311
Horse
antitoxin 147, 157
serum 28
Human, serum 29, 32
Hydrocortisone 263
Hydroxylamine 176

4-β-Hydroxyethylsulfanyl-2-aminoanisole 51
Hydroxylysine 85
p-Hydroxyphenyl-azo-phenylarsonic acid 17
Hypoglobulinemia 306
Hypomorphic genes 308

Imido esters 178
Immunization 7, 107, 108, 230, 233, 242, 247, 330
Immune lysis 3-4
Immunoadsorbent 3, 8-15, 41
Immunoelectrophoresis 2
Immunoglobulins 65-71
  deficiency diseases 306-308
  in diseases 66, 155-156
  types 65-71
Immunosuppressant 282-283
Indole 43
Influenza virus 13, 35, 36, 49, 52, 96, 228, 309
Insulin 43, 169, 309
Iodine 13, 40, 132, 211
Iodoacetate 189
Iodoacetamide 140, 168
Iodobenzoate 129
p-Iodophenylsulfonyl group 148
Ion-exchange resin 45, 51-54
Iprite nitrite 286
Irradiation 243
Isoagglutinins 311
Isocyanate groups 43
Isoleucine 86, 131-133
Isomaltohexose 102, 181-182
Isomaltose 102
Isomaltopentose 181-182
Isomaltotetrose 181-182
Isomaltotriose 102, 181-182

Jerne hypothesis (of antibody synthesis) 352
Jerne method 219, 222, 226, 238, 262
Jerne-Nordin method 260

Kaolin 42
Karush hypothesis (of antibody synthesis) 348
Kholchev method 29

D-Lactose 52
β-Lactoside 43, 76
Latent period 240
Latex 2
Leptospira, agglutinins 308, 322
Leucine 86, 130-133, 233
Liquid scintillation method 266
Liver 32, 212-215
Losozyme 96
Lowry method 5, 6, 8, 10, 11, 275
Luminescence 19
Lungs 217, 218, 255, 263, 265
Lymph nodes 65, 216-218, 220, 222-224, 227, 231, 237, 253, 261, 262, 266, 322, 327
Lymphocytes 210, 211, 221-225, 236, 237, 307, 326, 330, 332
Lymphoid cells 110, 211, 212, 225, 228-235, 322-324, 326-334
Lymphoid tissues 220
Lymphopoiesis 224
Lysine 43, 178, 273
Lysis 1
Lysozyme 36, 37, 179, 183, 189, 191, 266, 269

Macroglobulinemia 156, 226
Macrophages 2, 11, 215, 225, 236, 237
Meckler theory (of antibody synthesis) 349
Membranes 211, 257
Mendel's law 313, 317
β-Mercaptoethanol 140, 141, 145, 161, 169, 189, 193, 226, 253
Mercaptoethanolamine 136, 142, 167
2-Mercaptoethylamine 166, 168
Mercaptophenol 132
6-Mercaptopurine 243-245, 254, 280, 282

Mercapto-succinyl groups 41
Mercuripapain 132
Merthiolate 11
Methionine 168, 249, 311
Mouse 212, 215, 216, 219, 224, 237-239, 308, 320, 321, 330, 362
Microglobulins 266-271
Microsomes 211, 229, 230, 249, 272-279
Mitochondria 211, 249
Mitomycin C 286
Mitosis 226, 243, 261, 264, 327, 329 332, 333
Moffit-Yang equation 82, 83
Myeloma 65, 70, 95, 133, 134, 154-156, 185, 226, 238, 279
Mustelus canis 255
Myelomatosis, multiple 156
Myleran 280
Myosin 82

Nagarase 190
β-Naphthol 47
Neuraminic acid 91
p-Nitrobenzylchloride 46
N-(m-Nitrobenzyloxymethyl)-pyridine chloride 9, 13, 46
Nitrosoalkylurea 283
N-Nitrosoethylurea 283
N-Nitrosomethylurea 283
N-Nitrosopropylurea 283
Nucleic acids see Deoxyribonucleic acid and Ribonucleic acid

Oligosaccharides 38
Olovnikov method 2
Optical rotatory dispersion 82, 83
Optical rotation 192
Ornithosis 13
Ovalbumin 2, 4, 33-35, 40, 42, 43, 54, 80, 100, 128, 136-138, 212, 217, 218, 264, 266, 269, 284, 308

Paddle fish 363
Papain 35, 94, 128-139, 151-154, 161, 174, 183, 189
Paper
  discs 9, 13, 47
  strip 42
Pasteurella pestis see *Yersinia pestis*
Pauling theory (of antibody synthesis) 348
Pepsin 52, 127, 134-139, 151, 155, 157, 167, 169, 189
Peptide map 187
Perchloric acid 215
Performic acid 187
Perfusion
  apparatus 257
  experiments 265
Phagocytosis 210, 225
Phenol 43, 215, 229, 238
Phenyl-(p-azobenzoylamine)-acetate 35, 43
β-Phenyllactoside 86, 87
Phytohemagglutinin 308, 332, 333
Pig 30, 70, 73, 148, 240, 311
Pinocytosis 210, 211
Plasma 65, 69
Plasma cell 211, 220-223, 226, 231, 233, 307, 327, 329
Plasma tumor cell 238, 278, 279
Plasmin 138
Plasmoblast 211, 217, 226, 327
Plasmocyte 327, 330
Pneumococci 28, 32, 82, 96, 101, 187, 217, 250, 255, 265, 272, 308
Polarography 17
Poliomyelitis 13, 235, 241, 251, 284
Polyacrylamide 30, 54
Polyaminostyrene 53
Polyelectrolyte 35
Polyhapten 39
Polylysine 310, 322
Polymer, synthetic 51
Polymethacrylate 35
Polynitropolystyrene 53
Polyoma virus 74
Polypeptide
  alpha-chain 154
  amino acid analysis 151, 185
  carbohydrates 152

Polypetide (Continued)
heavy chain 94, 143-171, 186, 270, 271, 279, 324
kappa-chain 186
light chain 94, 143-171, 185, 270, 271, 279, 324, 364
N-terminal amino acid 143, 185
synthetic 96, 211
Polyribosome 272-279
Polysaccharide 28, 33, 35
pneumococcal 28, 82, 96, 101, 187, 250, 308
Polystyrene 13, 35, 42, 45, 52, 53, 192
Poly-1-tyrosyl gelatin 37, 189, 190
Porter method 131-133
Porter model 172
Potassium chloride 34
Potassium iodide 33
Potassium periodate 52
Potassium phthalate 28
Precipitation 1, 2, 15, 28, 100, 265
Precipitin 31, 38, 48, 96, 100, 129, 242
Prednisolone 243
Proline 85
Propionic acid 143, 167
Proteases 37, 105, 132, 138, 139
Protein
Bence-Jones 65, 70, 74, 93, 155, 156
determination 4, 5, 6, 8, 10, 11
guinea pig 79, 93
Lowry method 5-8, 10-11
mouse 93
myeloma 65
rabbit 93, 155
rat 155
serum 27
unbound 9
virus 13
Proteolysis 132
Puromycin 264
Pyrrolidone carboxylic acid 150

Q-fever 12

Rabbit 4, 17, 38, 39, 43, 74, 86, 132, 160, 188, 212, 213, 216, 222, 236, 237, 240, 244, 245-247, 251, 265, 308, 309, 314
Radioautography 2
Rat 40, 77, 237, 253
Resorcinol 43
Ribonuclease 40, 52, 96, 107, 191, 193, 229, 231, 236, 237, 238, 279
Ribonucleic acid (RNA) 209, 215, 221, 225, 228-235, 332
Ribosome 221, 225, 228, 271, 272-279
Rickettsia prowazekii 12, 13
Rivanol 30
Ropheocytosis 210

Salmonella 222, 227, 236, 240, 250, 253, 259, 309, 322, 323
Schweizer reagent 48
Sea lamprey 362
Sedimentation analysis 146
Sendai parainfluenza virus 51
Sephadex
A-50 29
C-200 30
DEAE 29, 30, 96, 276
G-25 102, 267
G-75 36, 143, 156
G-100 36, 38, 143-145, 157, 168, 170
G-200 36, 67, 75, 145, 251, 252, 266, 268, 269, 272
isothiocyanate phenoxyhydroxypropyl 54
S-200 30
Serine 127
Serum 1, 5
bovine 13, 33-35, 39, 53
guinea pig 79, 81
horse 28, 39, 43, 49, 51, 69, 78, 79, 127, 128
human 29, 33, 34, 43, 50-53, 66, 79, 96, 132
hyperimmune 28
immune 27

Serum (Continued)
mouse 79
placenta 30
rabbit 43, 51, 53, 74, 78, 96, 97, 100, 130
syphilitic 32
Shark 363
Shigella 210, 327
Sialic acid 91, 152
Silk fibroin 183
Smithies hypothesis (of antibody synthesis) 353-354
Sodium acetate 47
Sodium arsanilate 39
Sodium dodecyl sulfate 142, 145, 149
Sodium hydrosulfite 48
Spectrofluorometry 178
Spleen 212, 213, 217, 220, 226, 236, 237, 242, 245, 262, 264-266, 310, 323
Staphylococcus 308, 309
Streptolysin 308
Streptomycin 236
Styrene 53
Subfilisin 187, 188
S-groups 152
Sulfanilic acid 212
^{35}S-Sulfanilic-azoalbumin 215
S-carboxymethylcysteine 167, 174
Sulfitolysis (S-Sulfonation) 140, 142, 149, 157, 158
Sulfobromophthalein 6
Szilard hypothesis (of antibody synthesis) 356-358

Tannic acid 2
Tetanus antitoxin 78, 95, 158, 216
Tetradiazobenzidine 40
Threonine 151
Thymidine 327, 329, 330, 331
Thymosine 224
Thymus 217, 223-224
Thyroid gland 220
Tiselius apparatus 77-79
Tissue
culture 255-258, 261, 271, 331
Tissue (Continued)
protein 43
tumor 40
Tobacco mosaic virus 308
Toluene-2, 4-diisocyanate 2
Toxin 37
diphtheria 37
bacterial 43
Transplantation, immunity 236
Trichloracetic acid 266
6-Trichloromethylpurine 45
Triethylthiophosphamide 282
2, 4, 6-Trinitrophenyl groups 35
Triplet code 278
Tris-(1(2-methyl)-aziridinyl)-phosphinoxide 41
Trypsin 37, 96, 138, 149, 150, 187, 189
Tryptophan 18, 19, 43, 149, 178, 187
Tyramine 45
Tyrosine 37, 43, 86, 165, 176, 187, 212, 310

Ultracentrifuge 33, 36
Ultraviolet light 18, 82
Urea 34, 35, 82, 131, 132, 140-144, 146, 149, 158, 283
Urease 38, 52, 127
Uridine 231, 232, 234, 236, 237
Urine 65, 69-71, 215

Vaccination 209
Versenate 282
Vicilin 43
Viruses
see Bacteriophage
Coxsackie 241
fixed 48
influenza 13, 35, 36, 49, 52, 96, 228, 309
neutralization 20
ornithosis 13
poliomyelitis 13, 235, 241, 251, 284
polyoma 74
protein 13, 239
Sendai parainfluenza 51

Viruses (Continued)
  tobacco mosaic 308
  wart 74

Wart virus 74

X-rays 242-243

Yeast protein 242
Yersinia pestis 54

Z-type molecule 133
Zinc acetate 30